Social Learning

Contents

Social Learning

Chapter 1

Introduction

The study of social learning sits at the interface of a truly astonishing number of academic disciplines. How many other fields could boast being central to both social anthropology and human evolution; core material for both experimental psychologists and theoretically minded economists; or emerging influences in the fields of both cognitive neuroscience and artificial intelligence?

The observation that many animals, including humans, acquire valuable life skills and knowledge through copying others has been the focus of attention of animal behaviorists dating back to Darwin. Likewise, social learning, the diffusion of innovations, conformity, and social influences on child development have been key concepts within the social sciences for over a century. However, in recent decades, the field of social learning has received such unprecedented attention spread across the sciences, social sciences, and humanities, and experienced such growth, that researchers within the field have referred to an "explosion of interest" in the topic (Galef and Giraldeau 2001; Shettleworth 2001). Long gone are the days when research on imitation could (ungenerously) be characterized as the esoteric province of an obscure branch of comparative psychology. Now, social learning is a rapidly growing subfield of animal cognition research; while biological anthropologists and archaeologists are constructing models of cultural evolution, economists are frequently talking norms and herding behavior, neuroscientists are mapping circuitry associated with social influences on decision making, and engineers are building imitating robots.

The diverse backgrounds of the researchers studying social learning contribute to the field's controversies. East African but not West African chimpanzees use stalks to fish for termites, while western but not eastern chimps crack open nuts with stone hammers. Capuchin monkeys in Costa Rica exhibit extraordinary social conventions, such as sniffing each other's hands and placing fingers in each other's mouths. Humpback whales and chaffinches sing different songs than their fellow conspecifics living in different regions. If these, and other, animal traditions are acquired through social learning, are behavioral scientists justified in speaking of animals possessing "culture"? Or are anthropologists correct to assert that human cultures are so imbued with meaning, so permeated with symbolism, and so reliant on uniquely human aspects of cognition, that to liken them to the behavioral traditions of animals is, frankly, ridiculous? Is human social learning shaped by evolved structure in the mind then biased to acquire content—from choosing sugar rich foods to admiring specific body shapes—that proved adaptive among our Pleistocene ancestors, as suggested by many evolutionary psychologists? Or is human learning dominated by general rules (e.g., copying the highest payoff behavior or conforming to the local norm) that are for the most part acquired independently of their content, as claimed by cultural evolutionists? Is imitation critically dependent on the ability to take another individual's perspective, to understand their goals, or on complex cognition? Are socially learned traditions constrained in order to be adaptive? Can cultural processes support a viable form of group selection? And so on, and so forth. The controversies are multiple, ripe, and engaging, enriched by the varied standpoints that characterize adjacent disciplines struggling to understand a common topic.

The disparate backgrounds of those drawn into the field have also contributed to the newly emerging methods that are appearing to address these challenges. Until recently, experimental studies of social learning were restricted to behavioral investigations, typically conducted in the laboratories of comparative or developmental psychologists, and focused on very specific questions, such as: "Can animals imitate?" or "Do children acquire violent dispositions from others?" Similarly, with the exception of some early experiments on birdsong learning, biologists' interest in social learning was pursued almost entirely through observations and recordings of the natural behavior of animals, largely by ethologists and primatologists.

In recent years, however, new methods have become available, considerably expanding the social learning researcher's toolbox. These include (*i*) experimental and statistical methods that allow researchers to categorize cases of social learning according to their underlying psychological processes and learning mechanisms; (*ii*) neuroscientific methods for identifying the brain structures, neural circuitry, and physiological processes underlying both social learning and social influences on decision making; (*iii*) mathematical and statistical methods for identifying social learning when it occurs in natural populations (or in naturalistic contexts in captivity); (*iv*) experimental and statistical methods to predict and explain when humans and other animals copy, from whom they learn ("social learning strategies"), and to also detect the strategies deployed; and (*v*) mathematical methods for predicting the pattern of diffusion of novel learned innovations, and for modeling cultural evolution and gene-culture coevolution. Finally, numerous and diverse tools and procedures are available for applying social learning methods outside of academia. These range from commercial hatcheries training hatchery-reared fish to recognize predators and thereby enhance re-stocking efforts, to predicting the pattern of spread of technological innovations and identifying likely targets for uptake.

This book is designed to be a complete and accessible practical guide for the social learning researcher and their students, as well as for others whose interest in social learning is less central. As it currently stands, there is no single source that reviews the aforementioned conceptual and methodological developments, and the field's new theory and tools are to be found in a diverse collection of articles in academic journals. This book is first and foremost a monograph on social learning concepts and methods; it seeks to summarize and extend new developments in the field, rendering the new tools available to a broader constitu-ency and heightening awareness of this emerging research topic. However, such is the growth of interest in social learning that we envisage the material contained in this book to be of interest to many individuals, academic and nonacademic, who are not directly connected to the immediate field. Social learning and imita-tion, including their mechanisms, methods, and models, have become central to a broad range of disciplines, and significant progress in their scientific under-standing is potentially of widespread interest, both within and outside academia.

In order to discuss the conceptual foundations of social learning some defini-tions will be necessary from the outset. Certain terms, including "social learn-ing," "imitation," "innovation," "tradition," and "culture," will appear repeatedly throughout this book. In almost all cases, there exists no universally accepted definition, and in many instances the terms are the focus of considerable debate. Accordingly, at the outset we provide initial definitions for these labels, though later in the book we will suggest refinements. A summary of these definitions is given in box 1.1, with a justification presented below.

1.1 What Is Social Learning?

The most commonly used definition of social learning is "learning that is influ-enced by observation of, or interaction with, another animal (typically a conspe-cific) or its products" (Heyes 1994); this draws on a similar earlier generic definition by Box (1984). This definition is extremely broad, and indeed might be regarded as problematically so. The concern here is that the phrase "influenced by" lacks specificity. To take matters to the extreme, the presence of another individual might even impede learning, yet would still meet the above definition, because the other's activities constitute an influence on learning. It is then perhaps not surprising that others have used the term "social learning" to refer to a more specific concept. For example, Lonsdorf and Bonnie (2010) restrict social learning to cases in which see-ing another individual performing a behavior pattern causes the observer to learn the same pattern; this approximates our use of the term "social transmission."

This ambiguity in the meaning of the term "social learning" has the poten-tial to lead researchers into the logical error of equivocation, which occurs when the meaning of a term is changed during a line of reasoning. For example, those

> **Box 1.1**
> **Preliminary definitions**
>
> **Social learning** is learning that is facilitated by observation of, or interaction with, another individual or its products.
>
> **Social transmission** occurs when the prior acquisition of a behavioral trait T by one individual A, when expressed either directly in the performance of T or in some other behavior associated with T, exerts a lasting positive causal influence on the rate at which another individual, B, acquires and/or performs T.
>
> An **innovation** (sensu **product**) is a new or modified learned behavior not previously found in the population.
>
> **Innovation** (sensu **process**) is a process that introduces novel behavioral variants into a population's repertoire and results in new or modified learned behavior.
>
> A **tradition** is a distinctive behavior pattern shared by two or more individuals in a social unit that persists over time and that new practitioners acquire in part through socially aided learning.
>
> **Cultures** are those group-typical behavior patterns shared by members of a community that rely on socially learned and transmitted information.

definitions of "culture" (see below) that emphasize geographic differences in behavior caused by social learning (e.g., Laland and Hoppitt 2003) typically have in mind a narrower conception of social learning, more akin to the Lonsdorf and Bonnie definition than the broader Heyes definition. There could be a danger that researchers might prematurely make the claim of culture if variant forms of social learning in the broader sense were identified for a given species. This is particularly problematic if different researchers interpret the term "social learning" with variant levels of breadth, unaware of discrepancies in usage.

On the face of it, a solution might be to adopt a narrower definition of social learning, with another term for the broader category; for example, "socially biased learning" could be used (e.g., Humle et al. 2009). However, there are two concerns here. First, the broad definition is widely used and accepted, and hence it may lead to further confusion if we to advocate a change in usage. Second and more importantly, narrower definitions could miss out on some phenomena of interest to those who would consider themselves social learning researchers. For example, a definition that requires matching behavior would not include cases where an observer learns what not to do, perhaps after watching another individual make a mistake. Consequently, on balance, we maintain that a broad and generic term still has some currency and propose the following definition, modified from Heyes (1994):

> Social learning is learning that is facilitated by observation of, or interaction with, another individual (or its products).

Our use of the term "facilitated" restricts forms of social interaction that generate social learning to those that have a positive influence on the observer learning. Instances where social interaction impedes learning are better captured by a different term, such as "social inhibition" (Brown and Laland 2002). Even with this refinement, the definition of social learning remains broad, meaning a researcher's primary task is to investigate the manner in which learning is social, rather than to answer whether learning is social learning or not. This kind of reasoning justifies researchers' tendency to focus on general social processes thought to be widely important in promoting social learning, potentially across different contexts and modalities, rather than on isolated or specialized instances.

Throughout this book we will refer to the individual who learns socially as the "observer" and the individual that they learn from as the "demonstrator." While we acknowledge drawbacks to both of these terms, there are no obviously superior alternatives.[1]

1.2 Social Transmission

Notwithstanding the utility of a generic term, researchers clearly require more specific terms to describe different types of social learning. In recent years there has been a concerted effort to develop methods to "detect social learning" in natural populations (see chapters 5–7). Such methods are not really aimed at detecting social learning in the broad sense, but rather to detect what we term "social transmission." This term was initially coined by Galef (1976), and subsequently described as "cases of social learning that result in increased homogeneity of behavior of interactants that extends beyond the period of their interaction" (Galef 1988, 13). Familiar examples include the spread of sweet potato washing in Japanese monkeys (Kawai 1965) and hybrid-corn use by American farmers (Ryan and Gross 1943). Our definition of social transmission is very much in the same spirit as Galef's, but is designed to confer greater precision:

> Social transmission occurs when the prior acquisition of a behavioral trait
> T by one individual A, when expressed either directly in the performance
> of T or in some other behavior associated with T, exerts a lasting positive
> causal influence on the rate at which another individual B acquires and/or
> performs T.[2]

---

[1] The term "observer" would mislead if it implied that social learning is solely reliant on vision; however, our usage is not constrained to any sensory modality. Likewise, "demonstrator" might imply the deliberate production of the target behavior, an implication that we do not wish to be drawn. In the majority of instances of animal social learning the "demonstrator" makes no attempt to transmit knowledge. In this regard, the term "'model" is arguably preferable, being more neutral, but it brings with it the potential for confusion in discussions of mathematical and statistical models of social learning. A host of other terms ("actor," "transmitter," "receiver," "source," "sink," "learner") are used in the social learning literature but all suffer from drawbacks.

[2] This definition is adapted from Hoppitt, Boogert, and Laland (2010), in which we provided a definition of social transmission that also allowed "information" to be transmitted instead of a behavioral trait. In principle, the definition could be used for a piece of information, but in practice it might be difficult to detect unless the information is associated with an observable behavioral trait.

Here a behavioral trait is a target behavior pattern, which might be specific to a particular context (e.g., pigeons pecking at a lever in response to a green light; Dorrance and Zentall 2002). Inclusion of the phrase, "expressed either directly in the performance of T or in some other behavior associated with T" is recognition of the fact that acquired information and behavior can be transmitted with or without performance of the trait. Examples of the latter are not uncommon. In the case of humans, for example, we can provide written or spoken instructions on how to perform the trait, instead of demonstrating it directly. For example, social transmission of a recipe between chefs might occur when one chef passes a recipe to another. The number of times a chef cooks the recipe might have no causal influence on the acquisition of the recipe by others, but we would still consider this social transmission (see fig. 1.1c). In the case of other animals, signals such as food calls (e.g., the vocalizations of callitrichid monkeys that recruit infants to a desirable or novel food source [Rapaport and Brown 2008], or mother hens that warn their chicks from consuming a toxic foods [Nicol and Pope 1996]) or the opportunity teaching exhibited by several carnivores (e.g., when meerkat helpers provision pups with disabled prey; Thornton and McAuliffe 2006) can facilitate social transmission without direct performance of the trait by a demonstrator.

We also recognize two types of social transmission, which we call social transmission of trait acquisition (where the effect of social transmission is on the rate at which B acquires T) and social transmission of trait performance (where the effect is on the rate at which B performs T). In practice, many cases are likely to involve both the social transmission of trait acquisition and of trait performance—where individuals who observe a demonstrator both acquire the trait sooner and perform it more frequently once they acquire it, than individuals who do not observe a demonstrator. Nonetheless, we see the two as logically distinct. Observers might acquire T sooner, but may not perform T any more frequently than individuals who have acquired it asocially.[3]

It might seem strange to some readers to discuss a rate at which acquisition of a trait occurs, when each individual can only acquire a trait once (except when an individual has forgotten a trait and then does learn it again). Here our use of the term "rate" is in the sense of a "hazard rate" in time to event or in "survival" models—comparable to the rate of death or the rate at which a machine component fails—which therefore refers to the pattern of learning spread across a sample of individuals.[4] For example, if a large sample of individuals all had a learning rate of 0.5 per hour, we would expect the mean time to acquisition to be $1/0.5 = 2$ hours.

---

[3] When we use the term "asocial" in contrast to "social transmission," we mean that it has not been socially transmitted; we do not mean that it has not been affected by any social cues.

[4] We suggest the rate at which individuals acquire a trait as the most general criterion. Note, were we instead to utilize the probability that another acquires the trait, this would not allow for the fact that individuals who are not influenced by informed individuals might also reliably acquire the trait, but tend to do so later. Nonetheless, rates can easily be converted into a probability of acquisition for a specified time period, so the probability of acquisition within an individual's lifetime or an experimental trial can still be used to detect social transmission.

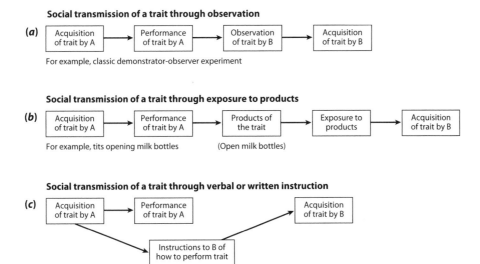

Figure 1.1. Three causal pathways through which information can be socially transmitted from individual A to individual B. An arrow represents a direct causal influence between two types of event (i.e., one that is not mediated through another type of event present in the graph). For example, in (a) if B observes A performing the trait, then A's action must be a cause of B observing the trait. In this case, observation of the trait then causes B to acquire the trait. In (a) the causal influence of the performance of the trait by A on the acquisition of the trait by B is mediated entirely through observation, whereas in (b) it is mediated entirely through exposure to a product. In reality, both causal pathways might operate. See Shipley 2000 for an accessible introduction to causal graphs (see also chapter 6).

Our definition also covers the case in which there are two or more traits that are alternative variants for the same function, and (1) the variant possessed by A causes B to be more likely to acquire the same variant,[5] and/or (2) the preferences of A for one variant causes B to develop a similar preference. Definitions in terms of rates also have the advantage that they allow researchers to model trait acquisition or performance as a stochastic process, and thereby infer social transmission from relatively complex patterns of acquisition or performance (see chapter 5). As indicated above, we interpret the phrase "performance" broadly to encompass transmission mediated by an animal's products, including, in the case of humans, signs, signals, and linguistic information.

There are clearly a number of causal routes by which a trait T could be socially transmitted between individuals A and B. Figure 1.1 illustrates three such cases. Perhaps the most obvious is social transmission through observation: B observing trait T being performed by A, causes B to acquire T (fig. 1.1a). It is this causal pathway that most traditional observer-demonstrator experiments aim to detect.

---

[5] For example. if each individual sticks with whichever variant it acquires first, the probability that it will acquire variant a rather than another variant, is $p_a = \lambda_a / \sum_x \lambda_x$, where λ_x is the rate of acquisition of variant x. Therefore, a positive causal influence on λ_a is also a positive causal influence on p_a.

There are also a number of documented cases where the physical products of the performance of T (excluding visual and auditory cues) can play a role in social transmission (fig. 1.1b). For example, Heyes and Dawson (1990) found that in rats the trait of pushing a joystick in a particular direction for a food reward was socially transmitted from trained demonstrators to observers, and that olfactory cues left on the joystick were sufficient for this effect to occur (Mitchell et al. 1999; see also Galef and Beck 1985; Laland and Plotkin 1991). Some researchers argue that social transmission mediated through physical traces left by animals in their environment is likely to be highly important in natural populations (Fragaszy, 2012b). We discuss several such examples in chapter 3.

It is also possible that both types of social transmission could operate indirectly through the behavior of an intermediate individual. A possible example of such transmission is provided by brown-headed cowbirds (*Molothrus ater*). Freeberg (1996, 1998) found that female cowbirds prefer to mate with males from the same cultural background, and that this preference seems to be based largely on the males' vocalizations. It has also been shown that females respond to preferred song types with the wing stroke display, which causes males to perform those song types more frequently (West and King 1988). This suggests the songs heard by a female during development cause her to respond to those types later by performing the wing stroke display. This in turn causes the males with which she interacts to perform those song types more. This would mean that song type performance is transmitted between males through the mating preferences of females, and mating preferences are transmitted between females through the singing of males.

Note also, we place emphasis on social transmission as a whole, rather than on a more specific subset of social learning, such as social transmission through direct observation. We do this because the term identifies with precision the subset of social learning that can result in the spread of a trait through a population of animals, in terms of either acquisition or frequency of performance. There is clearly some utility for a generic term for processes that can underpin such a diffusion.

An obvious experimental procedure for detecting social transmission in a laboratory setting is the traditional demonstrator-observer experiment, which we discuss in detail in chapters 3 and 4. Briefly, the experimenter exposes naïve observers to demonstrators performing the trait in question (for example, the solution to a foraging task), and compares the observers' rate of acquisition or performance with that of control subjects. The exact way in which the experiment is set up determines which types of social transmission can be detected—for example, if observers are not allowed access to the products of a demonstrator's behavior the experimenter will not detect the social transmission pathway depicted in figure 1.1b.

Historically, demonstrator-observer experiments have frequently sought to determine whether subjects have a capacity for a specific mechanism of social learning, such as imitation. In chapter 4 we discuss such mechanisms and describe how they can be distinguished. However, laboratory experiments do not tell us how important social transmission is in natural settings where animals can freely interact; experiments have little to say about behavioral traits that arise naturally. In some cases it might be possible to set up comparable field experiments to elucidate the role of social transmission in development (see section 3.2 and chapter 8). In other cases, this might not be possible, and the researcher may be limited to inferring social transmission from the distribution of traits in individual or group repertoires, or from the pattern of spread of a behavioral trait. As always, the absence of experimental manipulation makes inferring causal pathways more difficult (Shipley 2000). In such cases the researcher will be limited to detecting statistical patterns that are consistent with social transmission, and that may also rule out any plausible alternative causal hypotheses. In recent years a number of candidate methods have arisen that attempt to accomplish this, which we discuss in chapters 5–7.

1.3 Imitation

In chapter 4 we discuss twelve processes through which a demonstrator can influence the behavior of an observer in a manner that increases the probability that the observer learns; these are listed with definitions in table 4.1. Two of these processes possess the word "imitation" in their name: "contextual imitation" (learning to use an established action in a novel context) and "production imitation" (the learning of a new motor pattern). Each constitutes a highly specific means through which the activities of one individual may trigger learning in another individual (see chapter 4 for more precise definitions). We are, of course, aware that the word "imitation" is used in a more generic sense in some academic fields, notably economics, where its use as an umbrella term to refer to all forms of social learning could be considered synonymous to our use of "social transmission." Throughout this book, to avoid confusion, we restrict our use of the term "imitation" to the narrower meanings defined in chapter 4.

1.4 Innovation

The appearance of sweet potato washing in Japanese monkeys (Kawai 1965) and of hybrid-corn use by American farmers (Ryan and Gross 1943) are both also examples of "innovation"—the invention of new behavior patterns. Humans, of course, have devised literally millions of novel innovations, from computers to soufflé, to sonatas (Basalla 1988; E. Rogers 1995; Sternberg 1999; Simonton 1999), in many cases building on earlier inventions. There are also hundreds of innovations reported in animals, ranging from the incorporation of new items into foraging repertoires to novel elements in the songs of birds, to novel courtship displays, feeding behaviors, and tool use in primates (Byrne 1999; Casanova 2008; Lefebvre et al. 2004; Reader and Laland 2003a). Each of the many thousands of reports of social transmission in the animal behavior and human diffusion of innovation literatures typically involve initiation by a single individual's invention. While at the mechanistic level, the consanguinity of animal and human innovation is a matter of debate, from a functional perspective, these phenomena exhibit clear similarities in humans and nonhumans (Reader and Laland 2003).

Reader and Laland (2003b) distinguished between *innovation sensu product*, which refers to the new idea or invention that is learned, and *innovation sensu process*, which is the inceptive process that led to the new product. Following these authors, throughout this book *an innovation (sensu product) is a new or modified learned behavior not previously found in the population*, while *innovation (sensu process) is a process that introduces novel behavioral variants into a population's repertoire and results in new or modified learned behavior*. This contrasts with some definitions of human innovation, which refer to acquisition of a novel act by any route as innovation, and the initial inception as "invention" (E. Rogers 1995).

Fragaszy and Perry (2003, xiii) defined a "tradition" as *a distinctive behavior pattern shared by two or more individuals in a social unit, which persists over time and that new practitioners acquire in part through socially aided learning*. We interpret the phrase *persists over time* as meaning that to qualify as tradition the behavior must be maintained through repeated bouts of social transmission, such that each learner is then potentially capable of passing on the acquired information to others, thereby allowing the behavior to become characteristic of the population or of a subgroup within it. Whiten et al. (2011) note that "this makes sense insofar as a continuum is possible, from mere fads and fashions (perhaps lasting only weeks or even much less) to those that pass down very many generations"; they add that "particularly robust evidence of traditions comes from those that are of long duration, or rely on multiple transmission events, whether between generations or within them." We will be utilizing this term to refer to human traditions and the natural traditions of animals (e.g., the use of stone hammers to open nuts among Tai forest chimpanzees; Whiten et al. 1999). We will also use the term to refer to laboratory traditions, such as the consistent pathways through a maze taken by laboratory populations of guppies (Laland and Williams 1997), and humans (Reader et al. 2008), in spite of turnover in the populations' composition (Laland and Williams 1997).

The concept of "culture" is particularly tricky, because it has proven extremely difficult for social scientists to operationalize or derive a consensual definition (Kroeber and Kluckholm 1952; Durham 1991), and many social anthropologists seem to have given up on the notion of culture altogether (Kuper 2000; Bloch 2000). Conversely, biologists and students of animal behavior are giving unprecedented attention to the "culture" of other animals, and like it or not, the term is ever present in various literatures. Among behavioral scientists, some authors essentially equate "culture" with "tradition," and references to population-specific vertebrate traditions for singing particular songs, exhibiting specific feeding behavior, and the like as "cultural transmission" are common (e.g., Slater 1986). However, other authors apply more stringent criteria to the use of the term "culture," arguing that animal cases such as these might otherwise be too readily assumed to be reliant on mechanisms homologous (that is, sharing evolutionary ancestry) with human culture, when they might really be merely analogous (dependent on different forms of social learning, for example) (Galef 1992). Perhaps not surprisingly, there is a trend for those with a more functionalist perspective to emphasize the functional continuity of animal tradition and human culture, and for those from a background of studying mechanisms to emphasize the mechanistic discontinuity.

Nonetheless, two important aspects of culture are relatively noncontentious. First, culture is built upon socially learned and transmitted information; "culture" does not apply to genetic information or to knowledge and skills that individuals acquire alone. Second, socially transmitted information can underpin group-typical behavior patterns, which may vary from one population to the next. Thus, culture helps to explain both continuity within groups and diversity among groups. These considerations led us to the following definition (Laland and Hoppitt 2003, 151): *Cultures are those group-typical behavior patterns shared by members of a community that rely on socially learned and transmitted information.* It will be apparent that we make no real distinction between tradition and culture, and we will use these terms interchangeably throughout the book. This stance is entirely pragmatic—being designed to foster exploration of the evolutionary roots of human culture (Laland and Hoppitt 2003)—and is obviously not intended to imply that the "culture" of chimpanzees or chaffinches is identical to that of humans. We envisage that all "cultural" species will have their own distinctive species-typical modes of communication, learning, and social interaction that render their culture unique. Nonetheless, these concepts of culture have properties in common, particularly at the functional level, which are of broad interest.

1.5 Why Study Social Learning?

Current interest in social learning draws from a number of pressing research challenges and controversies. First, attention partly derives from a more general interest among behaviorists and neuroscientists in cognitive neuroscience and cognition: that general interest sets out to understand the biological and psychological bases of cognitive processing, often using animal models. Imitation has long been regarded of interest because of the enduring challenge of explaining what is known as "the correspondence problem"—in essence, how the brain converts the perception of an observed act into an enacted body movement (Heyes 2009). The nature of the challenge is most apparent for perceptually opaque actions, such as imitating facial expressions. The discovery of mirror neurons—neural networks that are active during observation and execution of the same actions—has renewed this interest, in part by exciting a further debate as to whether or not mirror neurons can be viewed as a solution to the correspondence problem, and whether mirror neurons are best viewed as preexisting enablers of imitation or as a byproduct of social learning (Rizolatti and Craighero 2004; Heyes et al. 2009).

Second, interest in social learning is the product of vigorous debates among psychologists, including whether social and asocial learning are reliant on the same underlying processes, whether evidence for some forms of social learning (notably "imitation") might be indicative of animals possessing complex psychological capabilities (Heyes 1994; Tomasello and Call 1997; Hurley and Chater 2005), and whether imitation is best accounted for by transformational or associative theories (Heyes and Ray 2000; Heyes 2009; chapter 4). Also of interest here is the debate over neonatal imitation and, more generally, the role that imitation plays in child development (Carpenter 2006; Meltzoff and Prinz 2002), both of which we introduce in the following chapter. Recent experiments also suggest that imitation may be linked to prosocial attitudes and behavior (Van Baaren et al. 2009), a finding that has stimulated some excitement.

Third, social learning is of interest to ethologists, behavioral ecologists, primatologists, and cetologists, amongst others, because it seems to allow animals both to make adaptive decisions and to learn about their environments efficiently (Lefebvre and Giraldeau 1996). However, it can also sometimes propagate arbitrary and even maladaptive variants, and generate spatial variation partially disconnected from ecological distributions (Laland et al. 2009). For example, bluehead wrasse mating sites cannot be predicted from knowledge of environmental resource distributions (Warner 1988, 1990); rather, removal and replacement experiments demonstrate that mating sites are maintained as traditions. Cultural processes, like gene-frequency clines, can generate geographical patterns in behavioral phenotypes. Examples include the languages of Micronesia, which exhibit a correlation between geographic and linguistic distance (Cavalli-Sforza and Wang 1986); and bird and whale vocalizations, which exhibit similar patterns (Catchpole and Slater 1995; van Schaik et al. 2003; Janik and Slater 1997). Another challenging feature of cultural transmission is that, although it typically propagates adaptive behavior, both theory and empirical data suggest that under restricted circumstances arbitrary and even maladaptive information can spread. Once again, this is well documented in humans (Richerson and Boyd 2005). One case is informational cascades, where individuals base behavioral decisions on prior decisions of others (Bikhchandani et al. 1992; Giraldeau et al. 2002).

Fourth, evolutionary biologists have recognized that social learning can affect evolutionary dynamics and equilibria, and are increasingly exploring cultural evolution mathematically (Cavalli-Sforza and Feldman 1981; Boyd and Richerson 1985; Richerson and Boyd 2005) and through laboratory experiments (McElreath et al. 2005; Mesoudi and O'Brien 2008). This is most obvious in humans, and a great deal of mathematical theory has investigated gene-culture coevolution (Boyd and Richerson 1985; Laland et al. 2010; Richerson et al. 2010). By homogenizing behavior across a population, and by allowing rapid changes in behavior, cultural processes typically increase rates of evolutionary change, although a reduction in rates of change is also possible (Feldman and Laland 1996; Laland et al. 2010). In other animals, theoretical models of mate-choice copying reveal that learned preferences could plausibly coevolve with gene-based traits (Kirkpatrick and Dugatkin 1994; Laland 1994); models of birdsong suggest that song learning affects the selection of alleles influencing song acquisition and preference (Lachlan and Slater 1999); and other analyses have found that song learning could lead to the evolution of brood parasitism and facilitate speciation (Beltman et al. 2003, 2004).

Fifth, as described in chapter 2 and discussed again in subsequent chapters, biologists, psychologists, and anthropologists are engaged in a lively debate over the parallels and differences between animal traditions and human culture. These have been fueled by high-profile reports (*Nature, Science, PNAS*) of inter- and intrapopulation variation in the behavioral repertoires of animal populations, and claims of "culture" in apes (Whiten et al. 2005; van Schaik et al. 2003), monkeys (Perry et al. 2003), and cetaceans (Rendell and Whitehead 2001; Krützen et al. 2005).

Sixth, researchers studying robotics and artificial intelligence are paying attention to animal social learning as part of endeavors to develop "imitating robots" and related technology (Dautenhahn and Nehaniv 2002; Acosta-Calderon and Hu 2004). Part of the technological challenge in delivering imitating robots relates to the specification of what should and what should not be imitated by machines, without detracting from the flexibility that such machines are designed to confer. These sorts of insights are emerging from social learning studies in animals and humans.

Seventh, economists are increasingly interested in what they term "imitation," specifically in whom one imitates, because it allows individuals to economize on computational costs, cash in on superior information, and increase learning efficiency (Schlag 1998; Apesteguia et al. 2003, 2005; Benhabib et al. 2011a, 2011b). There is agreement that social learning can influence economic decision making, and that a better understanding of such decision making is central to economics, business, and commerce. Moreover, there is widespread current interest in the processes that underlie cooperative behavior in humans, and one of the most prevalent theories emphasizes the role of strong reciprocity, which includes a conformist social learning element (Fehr and Fischbacher 2003). Moreover, the recent global financial crisis has stimulated interest in "herding" in financial institutions (Guarino and Cipriani 2008; Raafat et al. 2009), and in how social influences on individual decision making can sometimes trigger information cascades (Bikhchandani et al. 1992; Cao et al. 2011).

Finally, an eighth reason for the interest in social learning is that researchers from diverse backgrounds have been intrigued by the idea that it might play an important role in driving brain evolution and intelligence (Wyles et al. 1983; Wilson 1985; Boyd and Richerson 1985; Reader and Laland 2002; Whiten and van Schaik 2007; Reader et al. 2011; van Schaik and Burkart 2011; van Schaik et al. 2012). Allan Wilson (1985, 1991) suggested that through social learning individuals expose themselves to novel environmental conditions, which increases the rate of genetic change. He argued that the ability to invent new behavior and to copy the good ideas of others would give individuals an advantage in the struggle to survive and reproduce, and assuming that these abilities had some basis in neural substrate, this would generate selection for brain expansion. Boyd and Richerson (1985) view human cognition as having evolved to be specifically

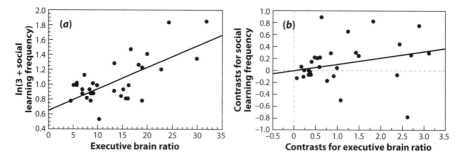

Figure 1.2. Social learning covaries with brain size across primates. Based on figure 1 in Reader and Laland (2002).

adapted to the acquisition of cultural knowledge, a view now termed the "cultural intelligence" hypothesis and supported by a comparative analysis (Herrmann et al. 2007). A related hypothesis argues that evolved changes in the reliance on social learning favored the evolution of enhancements in other cognitive capacities, such as individual learning (Whiten and van Schaik 2007; van Schaik and Burkart 2011).

Consistent with this cluster of related ideas, it has been found that across primates, a species' degree of reliance on social learning covaries strongly with measures of brain size (Reader and Laland 2002; see fig. 1.2). However, several other factors, ranging from social group size to diet (Clutton-Brock and Harvey 1980; Dunbar 1995; Deaner et al. 2000; Barton 2006), also correlate positively with primate brain volumes; this implies that the observed relationship between social learning and brain evolution may not be the result of a direct causal pathway between the two. Reader et al. (2011) provide a simple resolution to this paradox by showing that a large number of cognitive measures covary strongly across primates, suggesting that social, cognitive, and ecological capabilities may have coevolved. Social learning, tool use, and behavioral innovation form part of a correlated composite of cognitive traits evocative of general intelligence. This composite of traits covaries with species' performance in laboratory tests of learning and cognition, as well as with brain volume measures. The analysis suggests that high levels of general intelligence evolved independently at least four times in primates, with independent responses to selection in capuchin, baboon, macaque, and ape lineages. It is striking that these are precisely the groups renowned for their social learning and traditions.

1.6 Summary of the Book

This book is designed to provide a practical guide for the social learning researcher, as well as a variety of useful tools, methods, concepts, and analyses for individuals in adjacent fields. In chapter 1, we have introduced the field of social learning research and presented the key ideas and concepts. Chapter 2 will provide a précis of the history of interest in this topic. Chapter 3 describes research methods for investigating social learning in the laboratory, presenting traditional social learning experimental designs, transmission chain approaches, and controlled diffusion studies. It also draws attention to some recent neuroscientific analyses of social learning, which extend the study of social learning beyond the behavioral level. Chapter 4 presents a classification of social learning mechanisms, building on our earlier framework (Hoppitt and Laland 2008a), but extending the scheme to incorporate teaching. The chapter also describes how these mechanisms can be distinguished empirically. Chapters 5 to 7 depict methods for detecting and characterizing social learning in natural contexts, such as in the wild, or in natural social groups in captivity. Chapter 5 concentrates on diffusion methods, which endeavor to isolate social learning through characteristic signatures of socially mediated spread, such as network-based diffusion analysis. Chapter 6 describes repertoire-based methods for inferring social learning. These rely on the distribution of traits among individuals to ascertain whether social learning might account for behavioral differences within and among groups. Chapter 7 describes field experiments and developmental studies that provide additional means for identifying where social learning has occurred. Chapter 8 discusses social learning strategies—functional rules specifying what, when, and who to copy. As well as presenting the evidence for some of the better-studied learning heuristics, the chapter describes statistical procedures for identifying which social learning strategies are being deployed in a data set. Chapter 9 introduces a variety of diverse approaches to modeling social learning, cultural evolution, and gene-culture coevolution, including population-genetic, game-theoretical, agent-based, and spatially explicit models. Finally, in chapter 10 we end with a brief summary of which methods to apply in what circumstances.

Chapter 2

A Brief History of Social Learning Research

The history of research into social learning and imitation dates back to the Greek philosopher Aristotle, who was perhaps the first person recorded as explicitly making the claim that animals acquire behavior through imitation and other forms of social learning. Aristotle was particularly impressed with the human imitative tendency. He wrote that humans are "the most imitative of living creatures"; Aristotle added that through imitation, a human "learns his earliest lessons; and no less universal is the pleasure felt in things imitated" (Butcher 1922, 15). These three insights made in the fourth century BC—that humans are uncharacteristically reliant on imitative learning compared to other animals, that young children in particular acquire important aspects of their behavioral repertoire through copying, and that imitation appears intrinsically rewarding to children—are strikingly relevant to contemporary social learning research (Carpenter 2006; Herrmann et al. 2007). Aristotle also provided the first evidence of the social learning of song in birds.

Social learning and imitation have likewise been a focus of interest for many eminent comparative psychologists, developmental biologists, sociologists, anthropologists, and evolutionists, over many centuries, and for many quite distinct reasons. Miller and Dollard (1941) noted that accounts of the processes underlying social learning followed the interpretive traditions of the science at the time they were propounded, and that remains true today. A historical perspective is of value to the extent that it provides context to contemporary debates, or to current challenges, and helps prevent the repetition of past mistakes. Here we will try to provide a précis of the major themes in these histories, picking out the conceptual lineages that we regard to be of greatest relevance. Below, we describe how investigations of social learning have been central to research into the evolution of mind, the mechanisms of learning, the origins of culture, the diffusion of innovations, child development, and cultural evolution.

2.1 The Evolution of Mind

The greatest difficulty which presents itself, when we are driven to the above conclusion on the [animal] origin of man, is the high standard of intellectual power and of moral disposition which he has attained.

Darwin 1871

In making a case for human evolution, Darwin and his followers needed to undermine the widespread belief, frequently attributed to the French philosopher and mathematician Rene Descartes, that animals were merely machines driven by "instincts," while human beings alone were capable of reason and advanced mental processing. Overcoming this perceived gulf was, as the above quote acknowledges, a considerable challenge. Accordingly, early evolutionists sought to demonstrate that the differences in mental ability between humans and other animals were not as great as widely believed. Darwin and his followers set out to make the case that both animal and human behavior resulted from a combination of unlearned "instincts" and experience, thereby attacking the Cartesian dichotomy from both sides. Imitation (broadly conceived) was to play two important roles in the evolutionists' argument, as a result of which its study took on a new kind of relevance and became the immediate focus of attention following the publication of Darwin's *The Descent of Man* (Boakes 1984).

The first role that imitation played was as an intellectual "stepping-stone," or bridge, between "animal instinct" and "human reason." Charles Darwin reported in *The Descent of Man* (1871, 161): "Apes are much given to imitation . . . and the simple fact previously referred to, that after a time no animal can be caught in the same place by the same sort of trap, shews [sic] that animals learn by experience, and imitate each other's caution." Similarly, early evolutionists, including Alfred Russel Wallace, George Romanes, Conwy Lloyd Morgan, and James Baldwin, placed great emphasis on social learning as a source of adaptive behavior.

Wallace (1864) argued that the human brain evolved to the point where it became so large and powerful, and conferred such a potent ability to control and regulate the environment (e.g., through our ancestors' manufacture of fire, clothes, tools, and shelter) that natural selection no longer acted on our species. Wallace attributed human success to our ability to acquire knowledge and skills from others through social learning. For Wallace, then, understanding human evolution meant comprehending how knowledge and skills accumulated in a society. In studying the human mind the important challenge was not to understand the mechanisms of biological heredity, but to understand social transmission. However, Wallace (1870) did not regard the passing on of skills and habits from one generation to the next to be restricted to humans, and saw a great deal of similarity between the processes underlying the construction of nests by birds (deemed to be learned, in part, through imitation) and the building of shelters by humans. Wallace also provided early evidence that the means through which young birds develop their songs was critically dependent on social learning.

For Romanes (1884), demonstrations of social learning in animals provided evidence for the evolutionary origins of human cognition, with the capacity for imitation viewed as ancestral to the human capability for cultural learning. He presented numerous examples of what he claimed was imitative behavior in animals, from birdsong learning, to animals learning appropriate responses to the alarm calls of heterospecifics, or puppies reared by cats and acquiring habits like face washing and hunting mice. He interpreted such examples as reflecting an understanding of the utility of the acquired behavior, and therefore indicative of some degree of intelligence on the part of the animal concerned. Romanes claimed that many mammals, including cats, dogs, and monkeys, were capable of rational (i.e., conscious) imitation, and since these species were somewhat closely related to humans, this argument was central to the case for mental continuity among humans and other animals.

Romanes also regarded imitation as the critical means by which animals, particularly mammals, and especially young children, refine their instincts. Referring to several reports that cats had learned to open doors by turning the handles, Romanes (1882) attributed this capability to the "rational imitation" of their human owners, which in this instance he believed involved an understanding of the mechanical properties of the door. Conwy Lloyd Morgan too stressed birdsong dialects and traditional food preferences in animals as suggesting continuity of mental abilities among humans and other animals (C. L. Morgan 1896a). Further, C. L. Morgan (1896b), Baldwin (1896), Spalding (1873), and Osborne (1896) independently suggested that organisms could survive ecological challenges through their acquired knowledge and skills, frequently learned from others, and that this would then result in natural selection favoring unlearned versions of the same adaptive behavior (see Bateson and Gluckman 2011 for a modern treatment of this idea, commonly known as the "Baldwin effect").

Late in the nineteenth century, animals were studied primarily because of the aforementioned interest in the evolution of the human mind, but also as a means to evaluate Lamarckian inheritance. The latter occurred by addressing the question of whether any behavioral tendency acquired by an individual animal could be passed on (genetically) to its offspring (Boakes 1984); this constitutes the second key role that imitation research played in the study of the evolution of mind—as a mechanism of inheritance. Romanes (1884) interpreted observations of behavioral similarity across generations, such as kittens that adopt a begging habit from their mothers, as evidence for Lamarckian inheritance. In contrast, C. L. Morgan (1896a) forcefully argued that such examples—to the extent that they could be trusted—should be interpreted as evidence for social transmission. Morgan paid particular attention to birdsong. He concluded that in many birds the song characteristic of that particular species develops only if it is heard during a sensitive period of infancy, an argument confirmed by recent research (Catchpole and Slater 2008).

Romanes' claims of rational imitation in animals were widely accepted at the end of the nineteenth century (Boakes 1984), but were subsequently challenged by early learning theorists, notably Edward Thorndike and John Watson. Thorndike's landmark experiments on imitation were conducted as a graduate student at Harvard University. The procedure he adopted was to place an inexperienced "observer" animal (usually a cat or dog) in a compartment adjacent to a puzzle box containing an experienced conspecific "demonstrator." The demonstrator animal had been trained to pull a chord or lift a latch to escape from the box, and Thorndike expected to show that watching the solution to the task would accelerate the rate of learning in the observer. To his surprise, Thorndike found no effect of the observational experience on the rate of learning, and in his published report he suggested that cats and dogs, and indeed animals in general, were incapable of imitation (Thorndike 1898). Thorndike later noted (1911, 76) the general imitative faculty observed in humans who "from an act witnessed learn to do the act," which he claimed was "not possible for animals"; he distinguished this from "certain pseudo-imitative or semi-imitative phenomena" (77), such as the rapid loss of fear of trains by birds, or the tendency of a sheep to jump when it sees another do so. According to Thorndike, these were quite different, largely instinctive, processes.

Thorndike's conclusion was received with considerable dismay (Boakes 1984), and initiated a debate that continues to this day. Many behavioral scientists, dating back to Hobhouse (1901), have found it difficult to believe, as Thorndike implied, that there exists only one kind of animal intelligence based on associative learning. Critics of the "Thorndikean" stance have sought to demonstrate that some animals can imitate; these critics, with many claims of positive findings, were frequently motivated by an ideology that assumed mental evolution implies variation in mental ability across animals (Hobhouse 1901; Haggerty 1909; Warden and Jackson 1935). In contrast, Thorndike's negative conclusions on imitation were published in the context of a much-admired and novel treatise on animal learning through the accumulation of stimulus-response associations (*S-R* theory) (Thorndike 1898). Other researchers were inspired by Thorndike's experimental approach, and the power of his *S-R* theory framework. Subsequently, Thorndike's conclusion that animals don't imitate was reinforced by the further negative findings of experimentalists Lawrence Cole on raccoons (1907) and John Watson on monkeys (1908). Watson tested four monkeys on several tasks, trying both human and monkey demonstrators, and was forced to conclude: "There was never the slightest evidence of inferential imitation manifested in the actions of any of these animals."

This line of research has continued to the present day, with some prominent researchers (e.g., Tomasello 1996) remaining skeptical of the ability of nonhuman animals, including monkeys and apes, to imitate. Others claim to have found strong evidence for imitation in animals (e.g., Whiten et al. 1996; Whiten 1998; Akins and Zentall 1996). The rigorous, experimental associative-learning tradition initiated by Thorndike continues to produce champions of the one-intelligence-for-animals position (McPhail 1982), and the more naturalistic, evolutionist stance yields contemporary advocates of diverse animal intelligences (Byrne 1995). Nonetheless, social learning remains central, as it has from the outset, to attempts to understand the evolution of mind and cognition, and humanity's place in nature.

2.2 Social Learning Mechanisms

The history of the study of social learning mechanisms can be viewed to a large degree as reflecting repeated attempts by individuals to demonstrate "true" imitation in an animal, only for another researcher to come along subsequently and claim that the data can be explained by an alternative social learning process. Conwy Lloyd Morgan was also the first person to try to classify imitative phenomena. He initially distinguished between two types of imitation, "instinctive" and "intentional" (C. L. Morgan 1896b). Instinctive imitation was a class of instinct in which the sight of an action elicited the same action; for example, he claimed the observation of conspecifics pecking automatically triggers pecking behavior in chicks. In contrast, C. L. Morgan thought intentional imitation referred to the conscious learning of a novel behavior pattern through observation of another individual performing that behavior. He later (1900) subdivided the category of intentional imitation into "intelligent imitation," which was reliant on learning but did not require conscious reflection, and "reflective imitation," which referred to the deliberate and conscious acquisition of a motor pattern through observation. In making these distinctions, C. L. Morgan was to launch a tradition for the continual splitting of imitative phenomena into more and more categories with repeated classifications, which has continued to the present (McDougall 1945; Thorpe 1956; Galef 1988; Whiten and Ham 1992; Heyes 1994; Hoppitt and Laland 2008a).

For example, Spence (1937) suggested that preceding experimental claims of imitation in primates could be accounted for by a simpler mechanism that he called "stimulus enhancement." He argued that successful demonstrator animals acted in such a way as to enhance the stimulus value of parts of the apparatus that they manipulated, thereby making the observer more likely to respond to it.

Thorndike, while skeptical as to whether other animals were capable of imitation, was convinced that human imitation was learned and critically dependent on relevant prior experience. He wrote (1913, 175): "The idea of an act may by previous training lead to the act, and then the sight of others doing the act may cause the thought of the act which then leads to the act."

In a classification with an impressively contemporary quality, psychologist William McDougall (1945, 88–91) categorized imitation into five types. He noted the unlearned tendency of babies to reproduce tongue protrusions, which he referred to as a highly "specific motor tendency" and an "innate disposition." This tendency for babies to stick out their tongue in response to others doing so was later to become the basis for claims of neonatal imitation (Meltzoff and Moore 1977). McDougall also noted that the emotional responses of others (e.g., happiness, distress) can trigger the same emotions in ourselves, a form of "imitation" that he anticipated was widespread in nature. He also observed how concentrated attention on others could elicit an unlearned tendency to imitate the observed actions, such as when we kick or swing our arms while watching football or tennis. Today, we might describe all of these processes as "response facilitation" (Byrne 1994). A fourth type of imitation occurs when observation of a skilled action excites our admiration and we set out to reproduce it (what we might now call "production imitation"; Byrne 2002). McDougall's fifth type was reproduction of the effects of an action, rather than the movements, which we would now call "emulation" (Tomasello 1990). McDougall was also aware of the process of "stimulus enhancement," or perhaps "local enhancement" (Thorpe 1963),[1] although he did not include it in his categorization. He writes: 'It has been shown . . . that when a monkey, A, has failed to obtain food by manipulating some simple mechanism, it may be aided to learn the manipulation by the example of another, B, which has already learned the trick. It would seem that the example of B merely serves to direct the attention of A to the essential part of the mechanism, which A has then to learn to manipulate by his own series of more or less random efforts' (McDougall 1936, 174).

Drawing on associative learning theory, Neal Miller and John Dollard (1941) proposed a further explanation for social transmission in animals based on operant conditioning. They accounted for cases in which one animal followed or reproduced the actions of another as a result of the leader serving as a "discriminative stimulus" for the follower—that is, it indicated on which occasions the observer would be reinforced for performing the behavior. Their hypothesis was based on experiments in which they trained pairs of rats in a T-maze, where the observer rat was either reinforced for turning in the same, or a different, direction to its demonstrator. They argued that the observer's learning was no different than when an abiotic cue, such as a light, indicated which direction to turn. Skinner (1953, 120) subsequently argued that the appropriate contingencies for the development of matched-dependent behavior often occur in nature: "Thus, if a pigeon is scratching in a leaf-strewn field, this is an occasion upon which another pigeon is likely to be reinforced for similar behavior." Galef (1988) points out that, if the pattern of behavior adopted through matched-dependent behavior is to become a stable part of the observer's repertoire, the behavior must come under the control of stimuli not dependent on the presence of the leader. Church (1968) provided evidence that incidental learning could result in the transfer of stimulus control from an organism to a feature of the abiotic environment. When first trained to follow a demonstrator rat in a T-maze to get a reward, observer rats were then given trials in which they followed the leader into the passage marked by a light; then when tested in the absence of the demonstrators, they chose the arm of the maze marked by the light.

---

[1] From McDougall's description, it is difficult to be certain whether it is the location or features of the stimulus that he envisages to be enhanced.

Ethologist William Thorpe (1963, 134) defined local enhancement as "apparent imitation resulting from directing the animal's attention to a particular object or to a particular part of the environment." Galef (1988, 15) subsequently broadened this to include cases where individuals interact indirectly, via the environment: 'If, for example, rats mark foods they have eaten, thereby increasing the probability that conspecifics will eat the same foods (Galef and Beck, 1985), or leave scent trails as they move about the environment, inducing others to follow the same path (Telle, 1966), the absence of the initiator of the pattern of feeding or movement at the time of acquisition by a second individual does not seem to me to change the nature of the basic process."

Further explanatory terms emerged from experimental work. For example, Curio et al. (1978) found that blackbirds exposed to an arbitrary stimulus, such as a plastic bottle, while listening to the mobbing calls of conspecifics, learned to mob the stimulus. They suggested that this mechanism, which functioned to allow birds to learn to recognize predators, was a form of classical conditioning. Here, observation of a demonstrator exposes the observer to a relationship between stimuli (e.g., between a predator and conspecific fear), and exposure to this relationship effects a change in the observer (e.g., mobbing behavior). This form of social learning has become known as "observational conditioning," and resembles a process identified earlier by Berger (1962), which he called "vicarious instigation."

Galef (1988) suggested that local and stimulus enhancement could be subsumed under a single heading, and he advocated the use of "stimulus enhancement" on the grounds that it was the broader term (although in chapter 4 we make the case for continuing to distinguish between these). Galef identified several other explanatory terms in the literature that he thought represented distinct social learning mechanisms, including "social facilitation" (where the mere presence of conspecifics promotes dominant responses), "contagion" (which resembles C. L. Morgan's "instinctive imitation"), "observational conditioning," "matched-dependent behavior," "copying" (or vocal mimicry), and "imitation" (or "observational learning").

Later, a classification by primatologists Andrew Whiten and Rebecca Ham (1992) condensed these into just four types of social learning—stimulus enhancement, observational conditioning, imitation, and goal emulation—with the last referring to cases in which an observer animal learns from the demonstrator what goal to pursue. In contrast, they categorized cases of "contagion," "exposure," "'social support," and "matched-dependent learning" as instances of "social influence." Shortly afterward, Celia Heyes (1994) proposed a completely different classification, based on associative learning theory. Heyes identified just three classes of social learning: stimulus enhancement (which she viewed as a subset of single stimulus learning); observational conditioning (which she portrayed as a subset of Pavlovian conditioning, or stimulus-stimulus learning, where observation of the demonstrator exposes the observer to a relationship between stimuli); and observational learning (which she categorized as a subset of operant conditioning, or response-reinforcer learning, where observation of a demonstrator exposes the observer to a relationship between a response and a reinforcer). This scheme therefore has the advantage that social learning can be subsumed within the general explanatory framework applied to asocial learning. We (Hoppitt and Laland 2008a) have built upon Heyes scheme; our categorization is described in detail in chapter 4.

In the last 50 years experimental demonstrations of imitation have become increasingly compelling. An important methodological breakthrough came in 1965, with Dawson and Foss' introduction of the "two-action task" as a means to study imitation. The two-action method requires experimental subjects to solve a task with two alternative solutions. Typically, half of the subjects observe a demonstrator solving the task in one way, while the other half the alternative. Subjects are then tested to see which method they use, and if they disproportionately used the method that they observed; this is taken as evidence of imitation. In Dawson and Foss' (1965) original two-action test, budgerigars used their beak or feet to open food containers and the observer birds copied these behavior patterns. This two-action approach has become widely used within the field of social learning, although it remains a point of contention as to whether it can provide definite evidence for imitation (Hoppitt and Laland 2008a). In 1998, Whiten conducted a modified version of the two-action test where the alternative actions were replaced with action sequences. He provided evidence that four chimpanzees could learn by imitation how to open an "artificial fruit," a box containing food. This could only be opened after disabling a number of "defenses," following the performance of the sequence by a human demonstrator. The experiment constituted evidence that an arbitrary sequence of some kind had been learned through observation, although once again interpretations other than imitation are possible (chapter 4).

Back in 1952, Hayes and Hayes famously claimed that a hand-reared chimpanzee called Vicki could "do-as-I-do"—that is, she could be trained to reproduce the actions of the experimenter. Unfortunately, the original study suffered from various methodological deficiencies, rendering its findings difficult to interpret as clear evidence of imitation. However, the method was later repeated in a more rigorous manner by Debbie Custance and colleagues (1995). These researchers trained two chimpanzees over a three-month teaching phase, to copy a human demonstrator performing 1 of 15 different actions in response to the command, "Do this!" They then tested whether the subjects would imitate 48 arbitrary actions in response to the same command, and found that they were significantly more likely to perform an action similar to the demonstrated action than any other. This experiment, and a subsequent one on orangutans (Call 2001), are commonly interpreted as evidence that chimpanzees and orangutans are capable of imitation, although once again, other interpretations remain (see chapter 4).

It is not only apes that have been put forward as imitating animals. Akins and Zentall (1996, 1998) conducted studies of imitation learning in quail (*Coturnix japonica*), finding that in a two-action test, where the observer watches a demonstrator either stepping on or pecking at a treadle, the observer will only imitate the demonstrator if it observes the demonstrator being rewarded for its actions. Dorrance and Zentall (2002) provided evidence for a similar phenomenon in pigeons. While, as before, there are multiple potential explanatory processes (chapter 4), all of these are consistent with Thorndike's notion that when an act is "witnessed [observers] learn to do the act."

Theoretical models of imitation have been categorized as "transformational" or "associative" (Heyes 2002). Transformational theories, such as Bandura's (1986) "social cognitive theory" and Meltzoff and Moore's (1977) "active intermodal matching" (see section 2.5 on child development, below), postulate processes that convert sensory input from an observation into a mental representation that can be used directly to generate matching behavior. In contrast, associative theories state that the ability to imitate is a result of past experience. Such theories usually require that the observer be rewarded for copying the demonstrator, or that the observer requires some form of tutoring for the appropriate associations to form (Holt 1931; Miller and Dollard 1941). Heyes and Ray's (2000) associative sequence learning (ASL) theory specifies that preexisting action units are recognized in a demonstrator's behavior as a result of prior associations between the sensory and motor representations, and are combined into a matching linear sequence. Byrne's (1999) behavior parsing model of production imitation involves the detection of statistical regularities in the action components of demonstrated behavior that allow an observer to extract the underlying program for the demonstrator's behavior.

In light of the above eclectic history, it becomes apparent why the field of social learning suffers from such a bewildering array of often overlapping and inconsistent terms, and why new classification schemes continue to be proposed.

2.3 Animal "Culture"

Over the last century, field researchers have reported many instances of the spread of novel foraging behaviors in natural animal populations. Lefebvre and Palameta (1988) document "possible socially transmitted foraging behaviors" in a variety of vertebrates, going back to 1887, when Carpenter reported the putative, socially transmitted habit of crab-eating macaques cracking oysters with stones. Such behavioral innovations appear to have spread too quickly to be explained plausibly by population genetic, ecological, or demographic factors, and have been assumed to spread through social learning. However, researchers have, in general, rarely been able to substantiate the claim that such diffusions are actually the product of social (as opposed to asocial) learning, thus leaving the assumption that the behaviors are spread socially open to criticism (Laland and Janik 2006; Laland and Galef 2009).

The modern debate over animal culture began in earnest in Japan, a little more than a half century ago. Inspired by Imanishi's (1952) claim that culture is widespread in animals, Japanese researchers began to document traditions in free-living, often provisioned, primate populations. The most famous among these is the washing of sweet potatoes by Japanese macaques. Imo, an 18-month-old, female macaque, was first observed in 1953 to wash a dirt-covered sweet potato in a small freshwater stream on Koshima Islet in the Sea of Japan. A dozen years later, when the first publication appeared in the West describing the pattern of diffusion through Imo's troop of the habit of washing dirt from sweet potatoes before eating them, its author referred to the behavior, and other unique patterns of behavior seen on Koshima, as "pre-cultural" (Kawai 1965). The use of the word "culture," even prefixed, implied some degree of correspondence between monkey and human behavior, be it homologous or analogous. In the years that followed, researchers studying social transmission increasingly began to refer to it as "pre-cultural" (e.g., Menzel 1973b), "protocultural" (Menzel et al. 1972), or "traditional" (e.g., Beck 1974; Strum 1975) behavior. In 1978, McGrew and Tutin claimed that the term "culture" was appropriate to describe the grooming handclasp prevalent in a troop of chimpanzees, together with other population-specific chimpanzee behaviors, arguing that these behaviors satisfied the criteria used to identify cultural patterns in humans. Subsequently, talk of "preculture" and "protoculture" frequently changed to discussion of animal "culture," particularly when speaking of chimpanzees (Goodall 1986; Nishida 1987; McGrew 1992; Boesch 1993; Wrangham et al. 1994; Whiten et al. 1999).

McGrew studied chimpanzee behavioral variation systematically, making detailed comparisons between sites, and reporting a number of different behavior patterns ranging from foraging, to sexual, aggressive, and even medicinal behavior, which varied systematically among chimpanzee populations. This perspective received considerable attention through McGrew's influential book *Chimpanzee Material Culture* (1992), Wrangham et al.'s (1994) edited volume *Chimpanzee Cultures* (1994), and popular science books by Franz de Waal, notably *The Ape and the Sushi Master* (De Waal 2001).

However, some skeptics, notably psychologists Jeff Galef (1992, 2003) and Michael Tomasello (1994, 1999), took issue with claims of animal culture, both by questioning the evidence that the putative traditions were a consequence of social learning, and by suggesting that the parallels between animal and human culture rested on superficial analogies, rather than homologies in cognitive processing. Galef and Tomasello insisted that human culture was supported by imitation and teaching, different psychological mechanisms than those they thought underlay animal traditions. Tomasello (1994) argued that imitation and teaching were critical for cumulative culture, since they alone afforded high-fidelity transmission; he suggested that humans uniquely exhibited cultural transmission with a "ratcheting" (cumulative knowledge-gaining) quality.

Meanwhile, biologists had also started to apply the phrase "culture" to animals. John Tyler Bonner (1980), in his popular book *The Evolution of Culture in Animals*, defined culture broadly as the "transfer of information by behavioral means." Similarly, Mundinger (1980) and Catchpole and Slater (1995) characterized vocal learning in birds as culture. Geographical variation in the songs of many birds had been documented since the 1960s, notably among white-crowned sparrows and chaffinches (Marler and Tamura 1964; Catchpole and Slater 1995). In their book *Genes, Mind, and Culture*, Charles Lumsden and Edward Wilson (1981) adopted an even broader definition. These authors treated any extra-genetic form of acquired information as "cultural," leading to the suggestion that culture could be found in 10,000 species, even including some bacteria.

From the 1970s, evidence began to appear for vocal traditions in mammals too, particularly cetaceans, such as bottlenose dolphins (*Tursiops spp.*), and humpback whales (*Megaptera novaeangliae*) (Caldwell and Caldwell 1972; Janik and Slater 1997). As described by K. Payne and R. Payne (1985), all males in a humpback whale population shared a song that changed gradually through the singing season, a change much too rapid to be explained by changes in genotype. Most striking, off the east coast of Australia, a song changed in two years to one previously heard only off the west coast of Australia, possibly as a result of movement of a few individuals from west to east (Noad et al. 2000). Rendell and Whitehead's (2001) review "Culture in whales and dolphins" directed widespread attention to the sheer number of putative cases of cetacean culture. These authors described a wide range of traits that they claimed might be interpreted as cultural, including killer whales (*Orcinus orca*) beaching themselves during foraging, and bottlenose dolphins using sponges to forage.

The case for chimpanzee culture was strengthened by a major collaborative study involving leading primatologists who collated behavioral information from 7 long-term field studies across Africa; they generated 42 categories of behavior that exhibited significant variability across sites (Whiten et al. 1999). While some of this variation was attributed to differences in the availability of resources (e.g., absence of algae fishing can be explained by the rarity of algae at some sites), most behavior patterns, including tool use, grooming, and courtship behaviors, were common in some communities, but absent in others, and the authors claimed this distribution had no apparent ecological explanation. Whiten et al. (1999) titled their article "Cultures in chimpanzees," and some of these authors went on to argue that chimpanzee and human cultures result from homologous processes (McGrew 2004, 2009; Whiten 2005, 2009).

Whiten et al.'s (1999) analysis triggered the application of similar methods to other species, including orangutans, capuchin monkeys, and bottlenose dolphins (van Schaik et al. 2003; Perry et al. 2003; Krützen et al. 2005). Collectively, these papers implied that differences in the behavioral repertoires of many conspecific, large-brained mammals living in different locales provide evidence that they are cultural beings. Orangutan primatologists Carel van Schaik and colleagues' use of the term "cultural" clearly implied homology with human culture. They wrote (2003, 105): "The presence in orangutans of humanlike skill (material) culture pushes back its origin in the hominoid lineage to about 14 million years ago, when the orangutan and African ape clades last shared a common ancestor."

At about the same time, researchers studying capuchin monkeys published results of a major, long-term collaborative study of white-faced capuchin monkeys (*Cebus capucinus*) revealing behavioral variation in the social conventions of 13 social groups throughout Costa Rica (Perry et al. 2003). Several striking, and often bizarre, social conventions were reported, including hand sniffing, sucking of body parts, and placing fingers in the mouths of other monkeys. However, Perry et al. (2003) carefully avoided describing these traditions as culture.

The animal cultures debate remains unresolved, with a spectrum of views manifest in the literature (for an overview, see Laland and Galef 2009). We will be returning to these issues in chapters 5–7 of this book, and will also present new methods that potentially help to resolve some of the points of contention.

2.4 The Diffusion of Innovations

In contrast to the animal cultures literature, there is little doubt that technological innovations among humans spread through social transmission, and this has been a longstanding topic of interest within several academic fields. The French sociologist Gabriel Tarde is widely known for his emphasis on the importance of imitation in social life; indeed, he claimed that "society is imitation" (1903, 74). For Tarde, imitation was responsible for the stability of social patterns—that is, it was a conservative influence on society—while invention or innovation was the source of novelty. Tarde argued that imitation could lead to a "group mind" or herd behavior. Tarde's view that complex social dynamics could be understood as the product of psychological interactions between individuals is very much in accord with contemporary thinking concerning, for example, social learning strategies (Boyd and Richerson 1985; Laland 2004), collective decision-making (Couzin 2009), and human herding behavior (Raafat et al. 2009).

Tarde (1903) observed certain generalizations about the diffusion of innovations that he called the "laws of imitation." It was Tarde who first observed that the spread of an innovation typically follows an S-shaped curve. He also argued that the upward curve in the S-curve of adoption begins to occur when opinion leaders in a system adopt a new idea, and that its rate of spread depends on its compatibility with established ideas.

Tarde is regarded as the primary European forefather to the American field of diffusion research, which took up Tarde's challenge to understand the rules specifying why some innovations succeed and others fail (E. Rogers 1995). Both Tarde's position and that of the American school, can be contrasted with the British and German diffusionist schools, both of which were influential in the early part of the twentieth century but adopted the rather extreme position that all social change within a society stemmed from the input and diffusion of external ideas. According to E. Rogers (1995), the American tradition is not so much a single intellectual college but rather a disparate set of quasi-independent strands spanning a variety of disciplines, including anthropology, sociology, education, public health, communication, economics, and marketing.

Particularly influential was Ryan and Gross's (1943) hybrid-corn study, which shaped the methodology and interpretations of diffusion studies in sociology and other disciplines. Following the widespread adoption of hybrid corn by Iowa farmers in the 1930s, Ryan and Gross conducted a survey in which they interviewed several hundred farmers to determine when they had adopted the corn, from whom they acquired knowledge of it, how quickly they went from hearing about it to trial cultivation, and from testing it to exclusive use, and so forth. They subsequently analyzed the data, finding that it exhibited many of the patterns anticipated by Tarde; these patterns included an S-shaped diffusion curve, and significant roles of social networks in propagating the innovation. Subsequently, the American diffusionist school was to quantify and elaborate on Tarde's laws, and generate an extensive body of concepts and generalizations specifying the factors affecting the spread of innovations (E. Rogers 1995).

2.5 Child Development

Psychologists, like sociologists, have long argued that imitation is critical to human social behavior. Eminent psychologist William McDougall wrote (1945, 281–2): "Imitation . . . is the great agency through which the child is led on from the life of mere animal impulse to the life of self-control, deliberation, and true volition." McDougall develops the suggestion, which had been made repeatedly over the centuries, that imitation plays a particularly important role in child development.

In 1962, the Swiss developmental psychologist Jean Piaget provided what is perhaps the best-known theory of imitation as a key element of intellectual development. For Piaget, children progressed through various developmental stages, each associated with characteristic imitative capabilities, and through imitation children learned about their worlds. A central idea of his theory was that there are cognitive constraints limiting the imitative competence of children. Accordingly, as infants move through stages of cognitive development they progress through different stages of imitative skill. Piaget believed that infants begin life with independent sensory systems, and in the first year of life are limited to imitating simple vocalizations and manual gestures. These types of imitation are possible because the infant can both hear their own and the other individual's sounds, and can also see their own and the other's hand movements, such that no cross-modal matching is required to reproduce perceived behavior. In essence, the infant imitates only those actions that it already has in its repertoire, and for which the sensory experience of self and other's performance are similar.

From around year one, Piaget believed children become capable of the imitation of facial gestures, which would of course require matching from visual to proprioceptive or kinesthetic feedback. Finally, from around 18 months he suggested children become capable of duplicating a wide range of observed acts, even after the observed individual has gone. This coincides with the ability to mentally represent an object or action that is not present. Sensory-motor imitation was thought to become internalized as a kind of "internal imitation," or mental representation. Thus, for Piaget, imitative ability is a reflection of the child's development of intellect, in particular through reflecting progress in intermodal coordination, but this ability is also a primary means by which the child learns about the world. In sum, it is the developmental precursor to mental representation.

Another major impetus for developmental research into imitative behavior came from Albert Bandura's classic social learning experiments (Bandura 1977). Bandura was a leading and highly influential psychologist at Stanford University, and the architect of what he termed "social learning theory." The latter was the view that people learn new behavior primarily through copying others, and that whether they did so depended on whether the modeled behavior had desirable outcomes. He is famous for his Bobo doll experiments conducted in the 1960s. These seemed to show that aggressive tendencies could be acquired through social learning. Young children were exposed to aggressive or nonaggressive adult demonstrators, the former playing violently with an inflatable doll. Bandura found that children exposed to aggressive models were more likely to be aggressive toward the doll. These experiments are widely thought to have changed the face of psychology, and helped precipitate the shift from behaviorism to cognitive psychology.

Dating back to Preyer (1889), a variety of writers from a number of theoretical traditions had agreed that while some kind of imitation was possible in very young infants, this "early imitation" was qualitatively different from the "later imitation" characteristic of older children and adults. The former was more automatic, less reliable, and less general (Anisfeld 1991). Piaget agreed with these previous writers that genuine imitation was not possible amongst neonates, and did not emerge until the last quarter of the first year. However, this view was to be challenged in the late 1970s.

Meltzoff and Moore's (1977) classic study of neonatal imitation was a milestone in research and theorizing on the development of human imitation. These researchers provided evidence, in the laboratory for the first time, that neonates are able to imitate three facial gestures of adults (tongue protrusion, mouth opening, and lip protrusion). The study was significant because it seemed to rule out both the view that neonatal imitation is the result of reinforcement from the experimenter or the neonate's parents, and the view that it is the result of an "innate releasing mechanism" triggered by the target gesture. Rather, Meltzoff and Moore (1977, 77–8) claimed neonatal imitation is due to "an active matching process . . . mediated by an abstract representational system." Thus, for these researchers, early imitation was not qualitatively different from later imitation. From their perspective, imitation did not disappear as previously thought, and Meltzoff and Moore concluded that it is not reflexive, but a "genuine" form of "later" imitation.

While Meltzoff and Moore's study was valuable in stimulating multiple new lines of research (Meltzoff and Prinz 2002), their basic claims remain contentious. Anisfeld (1996) reviewed and reanalyzed many studies on neonatal imitation and showed that only imitation of tongue protrusion can be observed reliably in the laboratory. This specificity and unreliability means that the distinction between early imitation and later imitation has become tenable again, as has the view that early imitation is not strictly imitation, but rather an unlearned response. This perspective is fueled by the observation that neonatal imitation is observed in rhesus macaques (Ferrari et al. 2006); the species has never been shown to exhibit adult imitation (Visalberghi and Fragaszy 2002). The same controversy surrounds vocal imitation. Some researchers (e.g., Kuhl and Meltzoff 1996) claim infants as young as a few weeks imitate the sounds they hear, while others, going as far back as John Watson (1919), remain unconvinced. Nonetheless, the notion that imitation plays vital roles in child development remains widely accepted by developmental psychologists.

2.6 Cultural Evolution

Alongside teaching, imitation is also central to the idea of cultural evolution. Pre-dating even Darwin's writings, cultural evolution has a very long history. A linear and progressive conception of societal evolution was championed by some late nineteenth and early twentieth century sociologists and anthropologists, notably Herbert Spencer (1857), Edward Tyler (1865), and Lewis Henry Morgan (1877), and still has its advocates today (Caneiro 2003). However, with notable exceptions (Campbell 1960), the development of a Darwinian theory of cultural evolution remained comparatively unexplored until the 1970s.

In *The Selfish Gene* (1976) Richard Dawkins introduced a cultural replicator that he called the "meme." Dawkins suggested that fashions, diets, customs, language, art, and technology evolve over historical time through differential social learning. He coined the terms "replicator" and "vehicle" to distinguish between the "immortal" genes that are replicated in each generation in the transient, vehicular organisms that house them. Dawkins portrayed a meme as a newly evolved, insidious kind of replicator that "infected" humans with catchy ideas. The meme idea was given further attention through a series of highly successful and popular science books, including Daniel Dennett's *Consciousness Explained* (1991) and *Darwin's Dangerous Idea* (1995), and Susan Blackmore's (1999) *The Meme Machine*. In the late 1990s, the first academic conferences were held on the topic of memetics, raising the possibility that memetics might emerge as an active research program. However, in academic circles the meme fell on stony ground. With notable exceptions, such as the philosopher David Hull, anthropologist Bill Durham, and archaeologist Stephen Shennan, the meme concept failed to take off as a serious explanation for cultural phenomena, and interest in memes, at least in the scientific forum, began to wane. See Aunger (2000) for a useful analysis of the strengths and weaknesses of memetics.

In spite of this, a broader science of cultural evolution emerged that spans biology, psychology, and anthropology, with a strong theoretical foundation based on the mathematical modeling of cultural change (Mesoudi et al. 2004, 2006; Richerson and Boyd 2005; Pagel 2012). In the 1970s, geneticists Luca Cavalli-Sforza and Marcus Feldman at Stanford University began to publish the first mathematical models of cultural inheritance and cultural transmission, gradually building up a body of mathematical theory exploring the processes of cultural change and the interactions between genes and culture. This work climaxed in their influential monograph *Cultural Transmission and Evolution* (1981). They frequently took advantage of the parallels between the spread of a gene and the diffusion of a cultural innovation, borrowing or adapting established models from population genetics. University of California anthropologists Robert Boyd and Peter Richerson extended this work. Their monograph *Culture and the Evolutionary Process* (1985) introduced further theoretical methods and stimulating ideas into the field. More recently, phylogenetic methods have been deployed to interpret aspects of human cultural variation, from languages to human mating systems (Gray and Jordan 2000; O'Brien and Lyman 2000; Mace and Holden 2005; Pagel et al. 2007; Currie et al. 2010). These methods, again borrowed from evolutionary biology, reconstruct the history of cultural traits by mapping diverse populations onto a tree of relatedness.

With its technical and explicitly mathematical foundation, the field of cultural evolution initially struggled to spawn experiments or empirical studies. However, in recent years, this has changed, and there now exists a vibrant field of experimental cultural evolution, with many anthropologists, archaeologists, economists, and psychologists investigating aspects of cultural change in the laboratory, frequently testing hypotheses and ideas derived from formal theory (McElreath et al. 2005; Mesoudi and O'Brien 2008; Caldwell and Millen 2008, 2009; Kirby et al. 2008; Fay et al. 2008; T.J.H. Morgan et al. 2012).

Cultural evolutionists have described a number of evolved learning rules, known as "social learning strategies," that specify when an individual should copy and from whom they should learn (Boyd and Richerson 1985; J. Henrich and McElreath 2003; Laland 2004). These include "copy the most successful individuals," "copy in proportion to the demonstrator's payoff," and "conform to the majority" behavior. Each generates a different pattern of cultural change. Several of these command support from theoretical analyses and some from experimentation (see Rendell, Fogarty et al. 2011 for a review).

Several of these ideas also have a long history. For example, English economist Walter Bagehot (1873), who was also interested in societal evolution, seemed to anticipate modern interest in the roles of conformity and strong reciprocity in cultural evolution (e.g., Boyd and Richerson 1985; Fehr and Fischbacher 2003). He believed that individuals in a group were rewarded and punished depending on their adherence to the group's traditions. Bagehot was also the first person to draw attention to the phenomenon of "cultural inertia" (Boyd and Richerson 1985), through which old habits slow down the acquisition of new ones through social reinforcement. Similarly, William McDougall (1945, 290) was amongst the first researchers to draw attention to the operation of a prestige bias in social learning. He referred to the "fundamental law of imitation," the idea that the primary source of adopted innovations is an individual or social group enjoying prestige. He was also aware of how the socially transmitted traditions that characterize societies could act as ethnic boundaries, suggesting this resulted from "contra-imitation"—a tendency *not* to copy those in out-groups (members of different groups or societies). He wrote: "Most Englishmen would scorn to kiss and embrace one another or to gesticulate freely, if only because Frenchmen do these things" (McDougall 1945, 297). Interest in social learning strategies remains a major growth area within social learning research (see chapter 8).

2.7 Conclusions

It will be apparent from the above review that scientific interest in social learning research stems from a number of distinct sources spread across a broad range of disciplines, including evolutionary biology, comparative and developmental psychology, anthropology, sociology, and economics. While some of the afore-mentioned historical traditions have petered out, many continue to shape the intellectual landscape. Social learning remains central to attempts to understand the evolution of mind and cognition, child development, the diffusion of innovations, and humanity's place in nature. We encourage social learning researchers to consider the history of their field, as doing so frequently helps make sense of the present. For example, it is only in the light of history that the eclectic terminology of social learning researchers can be understood.

Chapter 3

Methods for Studying Social Learning in the Laboratory

At least from the nineteenth century, it has been apparent that social learning and imitation cannot reliably be deduced from the casual observation, and that experimentation is the most straightforward means to investigate the underlying mechanisms. The designs implemented in social learning experiments reflect the research questions and conceptual frameworks of the scientists involved. As outlined in chapter 2, primary concerns have been to identify the different processes through which social learning can occur, to establish methods for distinguishing between these in the laboratory, and thereby to establish which nonhuman animals, if any, were capable of what were thought to be the most sophisticated processes. Of long-standing interest has been the experimental investigation of "true" imitation, both in animals and in humans (particularly children), and of the role it plays in cognitive development. Methods for distinguishing between alternative social learning processes are discussed in chapter 4.

More recently, with the recognition that simple processes can underpin complex behavior (Shettleworth 2010), this research agenda has broadened, In part, this reflects the belief that all of the psychological processes supporting social learning are worthy foci for experimental research. Over and above investigations into mechanism, in recent decades laboratory experimental studies have been conducted to explore the population-level aspects of social transmission; an example is the investigation of aspects of tradition, diffusion, and innovation. Recent years have also witnessed a rush of interest in social learning strategies, and many experiments have been conducted to isolate factors that affect when individuals copy, what they learn, and from whom they acquire knowledge.

The primary advantages of the laboratory experimental approach are control and reliability. In the laboratory, extraneous and potentially confounding variables can be removed or counterbalanced, precise manipulations of the variables of interest can be conducted, control conditions can be deployed, and procedures can be reproduced on replicate animals or populations. Such methodologies generate reliable and replicable investigations and potentially robust outcomes. This is important for a topic like social learning, where two individuals can come to exhibit the same behavior following social interaction through a variety of means other than social learning or imitation (Galef 1988; Whiten and Ham 1992), and hence experimentation is important to determine the developmental processes involved. First and foremost, the experimentation can establish whether social learning actually occurs, and it can go on to isolate learning processes.

One disadvantage of the laboratory approach is a potential lack of ecological validity. If, as we describe below, quail are shown to learn by imitation to press a treadle with their feet in the laboratory (Akins and Zentall 1996), we do not know whether, or how frequently, quail acquire behavior through imitation in their natural environment, whether other animals do so, or whether learning in general occurs in a similar manner in nature. At the extreme, some findings may be an artifactual consequence of impoverished and unnatural conditions in the laboratory. This concern has led to the development of methods for studying social learning in more naturalistic circumstances (R. L. Kendal et al. 2010) to allow laboratory findings to be validated in the field (see chapters 5–7). Nonetheless, a positive result in a laboratory experiment is sufficient to show that a species has the capacity for a particular type of social learning. Laboratory studies remain the primary modus operandi for social learning researchers, as they are for cognitive scientists in general.

However, in spite of being largely a laboratory science, understanding of the biological bases of social learning is primitive. Little is known about the neural, endocrinological, or physiological processes underlying social learning, and the vast majority of studies are exclusively behavioral. This is starting to change, however, and a small number of neuroendocrinological studies of social learning have been conducted, as well as investigations of the neural mechanisms of observational learning; the latter are particularly focused on the mirror neuron system and imitation. This development is very welcome, because the neural and physiological mechanisms that support social learning, and their relationship to the neural mechanisms of asocial learning, have hitherto been contentious, underinvestigated, and poorly understood.

In this chapter, we review the primary laboratory experimental approaches to the study of social learning. We begin by considering the traditional social learning experiment, with its trained demonstrator and naïve observer design. We also consider extensions of this methodology designed to investigate the transmission properties of social learning, including transmission chain and diffusion experiments. We discuss the experimental study of behavioral innovation, an important but neglected aspect of the diffusion of novel behavior (Reader and Laland 2003b). Finally, we end with a brief review of what is known about the neuroscientific basis of social learning from experimental studies.

3.1 Traditional Social Learning Studies

The classical design to a social learning experiment involves pairing a set of experimental subjects, termed "observers," each with a single "demonstrator" animal that has been trained to be proficient at a target behavior. The subjects' performance is assessed, first during the observation phase while in the presence of their demonstrator who performs the target behavior. Then, after the demonstrators have been removed, their preformance may be evaluated again in order to ascertain whether the observers have acquired the behavior, or whether its acquisition has been improved or accelerated as a result of the observational experience. Frequently, the performance of experimental subjects is compared to that of individuals in one or more control conditions, for example, that have no demonstrator, an untrained demonstrator, or a demonstrator unable to perform the target behavior.

Demonstrator-observer pairings have been central to the experimental study of social learning from the work of Thorndike onward, and for the most part have attempted to address the question: "Can animals imitate?" (see chapter 2). While efforts to demonstrate imitation in animals are no longer the primary focus of social learning experimentation, this tradition has proven enormously valuable in at least three ways. First, it has left a legacy in which experimental rigor is appreciated and encouraged within the field, and social learning researchers generally set high standards for experimental design compared with research on many other aspects of social behavior. Second, there now exist clear methods for establishing through experimentation whether animals exhibit social learning, and the list of taxa known to be capable of learning from others is extensive as a result. Such experimentation has helped establish the sheer pervasiveness of social learning in nature as well as its fundamental importance to the acquisition of adaptive behavior in countless species. These studies are the "bread and butter" of the field, without which the topic of social learning would surely not have reached its current state of prominence. Third, researchers are now in a position to identify, with some degree of precision, the psychological processes underlying a specific instance of social learning. Though in practice, researchers are frequently content to conclude that the process underlying the social learning of their animal might be X, Y, or Z, this does not negate the fact that such methods are available.

A significant advance in the study of imitation came with the introduction of the "two-action method" (Dawson and Foss 1965). Researchers seeking to establish whether a particular motor pattern can be acquired through observation were able to deploy a design in which demonstrator animals performed one of two alternative actions in different conditions to determine whether observers acquired the method to which they were exposed. Observers were subsequently tested in the absence of their demonstrators to establish whether they disproportionately performed the action of their demonstrators. The two-action method was first deployed by Dawson and Foss (1965), who allowed budgerigars to observe conspecifics remove a lid from a food container either by twisting it off with

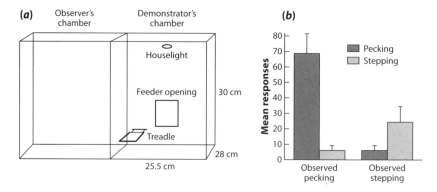

Figure 3.1. A schematic diagram (*a*) of the apparatus used, and the mean (SE) frequency (*b*) of pecks and steps in the first five minutes of the test involving quail that observed pecking and stepping conspecifics in Akins and Zentall's (1996) experiment. Based on figures 1 and 3 in Akins and Zentall (1996).

their beak or by grasping it with their foot. The subjects tended to remove the lid in the same manner as their demonstrator. Such a finding is of interest because it cannot easily be explained by a variety of alternative social learning processes, such as local enhancement, and accordingly a positive result may imply imitation. (In chapter 4 we suggest that a time delay between observation and test phase, or evidence of context specificity, is necessary to exclude an alternative explanation.)

Akins and Zentall (1996) provide an example of the classic social learning experimental design (fig. 3.1). These authors placed a young Japanese quail ($n = 8$) in the observer compartment of an experimental chamber (fig. 3.1a). Over a 10-minute observation period, the quail witnessed a demonstrator conspecific pressing a treadle to receive a food reward, either by pecking it ($n = 4$) or stepping on it ($n = 4$). Immediately afterward, the demonstrator was removed and the observer was placed in the demonstrator's chamber for a 30-minute test period. Akins and Zentall found that the subjects' first action was always the same as their demonstrators' (i.e., quail that watched pecking, then depressed the treadle through pecking, while those that observed stepping used the step method). Moreover, in the first five minutes of the test session, subjects tended to perform more of the action that they had observed than the alternative (fig. 3.1b).

The study provides reasonable evidence for contextual imitation. By testing the animals in the absence of the demonstrators, the authors are able to rule out other explanations for the subjects' performance (i.e., motivational or facilitatory effects) and show that some sort of learning has occurred. By deploying a two-action design, allied to a single manipulandum (in this case, the treadle) that moves in the same manner irrespective of the motor pattern used to depress it, the authors exclude local and stimulus enhancement explanations, and are able demonstrate that the subjects' learning is related to the demonstrators' movement. Probably the subjects learn an *S-R* (stimulus-response) association through observation, and associate the context of the treadle and demonstrator's compartment with a particular motor pattern (peck or step), although the precise nature of the learning in this experiment is unclear. What is incontrovertible, however, is that social learning has occurred. This is the real strength of the traditional demonstrator-observer pairing design—it provides an effective means for establishing that social learning has taken place, and sheds light on the underlying process.

This basic design can be extended in a number of useful ways. In addition to the two experimental conditions, it can often be helpful to also set up a control group that does not observe any demonstration. Comparison of experimental and control conditions potentially reveals whether the rate of acquisition of the task, and not just the motor pattern deployed, is affected by the observation of an experienced conspecific. Moreover, in principle there is no need for the number of demonstrators to be restricted to a single individual. Indeed, there is evidence from several species that the likelihood of social learning, or the magnitude of the effect, increases with the number of demonstrators (Sugita 1980; Beck and Galef 1989; Lefebvre and Giraldeau 1994); this is also true when testing for conformity effects in humans (Ash 1955; Coultas 2004; T.J.H. Morgan et al. 2012). For example, naïve guppies exposed to a single demonstrator taking one of two routes to reach food show no evidence of social learning (fig. 3.2), while those exposed to multiple demonstrators tend, when subsequently tested alone, to take the same route as their demonstrators (Laland and Williams 1997). Moreover, attention to rates of demonstrator action, or the number of demonstrators at the task, can provide information about the underlying process. For example, the rate at which chickens initiate bouts of preening is strongly related to the number of birds already preening in the same aviary; however, this relationship was not apparent in an adjacent aviary visually isolated from the first aviary. The experiment is suggestive of a response facilitation mechanism (Hoppitt et al. 2007). Likewise, designs attempting to isolate imitation and emulation often deploy "ghost control" or "end-state" conditions, in which the apparatus is seen to move (i.e., treadle moves downward) or is seen in its final state (i.e., treadle is depressed), in the absence of a demonstrator (Heyes et al. 1994; Fawcett et al. 2002; Subiaul et al. 2004; Call et al. 2005; see Hopper 2010 for a review). We discuss the strengths and weaknesses of these control conditions in detail in chapter 4.

In addition to studies of mechanism, laboratory experimental studies are now increasingly arranged to investigate social learning strategies (chapter 8). Such studies go beyond exploring whether, or how, social learning occurs, to determine

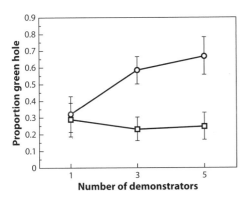

Figure 3.2. The proportion of trials in which observer fish tested singly took the green hole in a partition to get to a food source, given that their demonstrators had been trained to take the green (*circle*) or red (*square*) hole, and that they had one, three, or five demonstrators. The vertical bars give standard errors. It can be seen that the probability of social learning increases with the number of demonstrators. Based on figure 1 in Laland and Williams (1997).

the contextual variables affecting reliance on social learning. Some of these reflect the state or characteristics of the learner; for example, young or unsuccessful individuals may disproportionately rely on copying (Kirkpatrick and Dugatkin 1994; Duffy et al. 2009; Grüter et al. 2011). Others pertain to the characteristics of the demonstrator; for example, individuals seemingly copy high-status demonstrators more than low-status ones, and also copy successful more than unsuccessful demonstrators (Horner et al. 2010; T.J.H. Morgan et al. 2012). In such studies, the potential biases affecting reliance on social and asocial sources of information are manipulated, with conditions that, for example, render asocial learning more or less costly, or allow the behavior of demonstrators to vary in their consistency.

Laboratory and captive-facility experiments have also been run to investigate adaptive specializations in social learning. One familiar example concerns acquisition of a fear of snakes among monkeys (Mineka and Cook 1988). A rhesus monkey (*Macaca mulatta*) that has had no contact with snakes will typically not show fear when exposed to the animal. However, if the monkey is exposed to another individual reacting fearfully to a snake, or even a toy snake, it will also display fear, and show fear when later presented with the snake (Cook et al. 1985). The animal has been conditioned through observation of fearful conspecifics. Of particular interest here is the observation that these monkeys cannot be so conditioned through observation to develop a fear of just any arbitrary object, such as a flowerpot. Rather, they seem to have been "prepared" by a history of selection to form some fear-related associations through social learning more readily than others, a mechanism that may be both ancient and phylogenetically widespread (Olsson and Phelps 2007). Other examples of adaptive specializations in social learning include the ability of some songbirds to learn conspecific song more readily than heterospecific song (Catchpole and Slater 2008) and the ability of some sticklebacks to acquire public information about food-patch quality (Coolen et al. 2003).

3.2 Alternative Experimental Approaches

Historically, the study of social learning has been dominated by two distinct research traditions: (*i*) traditional laboratory experiments with the above dyadic demonstrator-observer design, primarily investigating psychological mechanisms, and (*ii*) field studies of the natural behavior of animals. The latter report behavior thought to be socially transmitted; examples are behaviors that follow natural diffusions or account for inter- or intrapopulation variation. Both traditions have their advantages. Laboratory experiments introduce a scientific rigor and control of key variables that generate reliable, replicable findings, while field studies shed light on the natural context and social processes leading to social learning. However, each tradition also has disadvantages. In the field it is usually difficult to establish through observation alone that the spread of a behavior results from social learning, as opposed to asocial learning (an exception is the spread of new songs in birds and cetaceans, where asocial accounts are implausible; e.g., Noad et al. 2000). It is also difficult to investigate the processes or learning strategies underlying diffusion. At the same time, traditional laboratory experiments sacrifice validity for reliability, and tell us little about how, or even whether, social learning operates in natural populations. Nor do they shed light on any of the population-level aspects of social transmission.

One possibility is to conduct experiments on natural animal populations, as ethologists and behavioral ecologists do routinely (Slater 1985; J. R. Krebs and Davies 1991). Those few experiments investigating social learning in natural populations are among the most informative and exciting studies of recent decades (Helfman and Schultz 1984; Warner 1988, 1990; Lonsdorf et al. 2004; Thornton and McAuliffe 2006; Slagsvold and Wiebe 2007; Davies and Welbergen 2009; see Reader and Biro 2010 for a review). However, not all species are suitable for such procedures (Laland and Hoppitt 2003). Moreover, as we will find in chapters 5–7, mathematical, statistical, and experimental procedures are starting to be devised that enable the investigation of social learning, and its underlying processes, in natural social conditions.

Similarly, researchers investigating social learning in the laboratory, while innovative in devising clever controls and experimental procedures, have nonetheless been rather conservative in use of the basic (demonstrator-observer) design structure. This design can be represented in abstract terms as "$A \rightarrow B$," since information passes from the demonstrator (individual A) to the observer (individual B) (see fig. 3.3a). In contrast, there have been comparatively few social learning experiments employing designs with multiple demonstrators (many: one designs; fig. 3.3b). These create opportunities to investigate the role of alternative forms of social interaction (e.g., scrounging, social rank, aggression), as well as directed social learning and model-based biases. Designs with multiple observers (one: many designs; fig. 3.3c) have also been relatively scarce. These afford similar

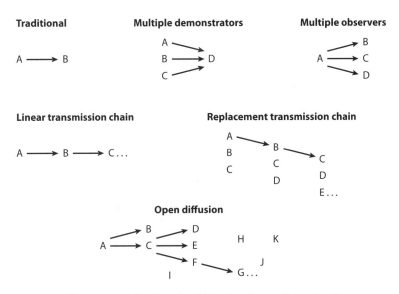

Figure 3.3. Designs for experimental studies of social learning. See text for explanation.

opportunities to explore state-based (i.e., *copy-when-uncertain*) or observer-based biases, as well as aspects of social interaction (see chapter 8 for discussion of these biases). Equally rare are studies that allow for multiple sequential bouts of social learning, with transmission of information along a chain of individuals (transmission chain designs; fig. 3.3d, e) or propagation through a population (diffusion studies; fig. 3.3f). Given that social transmission, tradition, and culture, by their very nature require multiple transmission episodes, and may exhibit population-level properties, various researchers have been concerned that examining social learning only at the dyadic level falls short of the methodology that is needed (Laland et al. 1993; Galef and Allen 1995; Whiten and Mesoudi 2008).

Fortunately, matters have begun to change in recent decades, and a variety of alternative experimental approaches have begun to emerge. Here we highlight two such innovations in the form of "transmission chains" and "controlled diffusion studies."

3.2.1 Transmission chain studies

The goal of transmission chain studies is to investigate whether the behavior exhibited by a set of original demonstrators, or "founders," can be propagated along a chain of individuals, and to explore the factors that enhance or degrade transmission. The distinguishing feature of this design is that the "observer" individuals are given the opportunity to act as demonstrators to other individuals in a structured chain. The "linear transmission chain" (fig. 3.3d) can be represented in the same abstract terms as above (as "A → B → C . . . "), and in structural terms resembles the familiar party game known as "Chinese whispers" or "telephone." "Replacement transmission chains" extend this approach to small groups of individuals, the composition of which are gradually replaced over time (i.e., "ABC → BCD → CDE . . . "; fig. 3.3e). Like the linear chains, in replacement transmission chains the experimenter imposes a systematic series of steps or transmission episodes, but at each step a naïve individual replaces a more experienced individual. Early in the chain the replaced individual may be one of the original founder members, while later naïve individuals replace nonfounders who have interacted directly with founders. In long chains, naïve subjects may replace individuals with no form of contact with founders.

The replacement transmission chain design has been characterized as intermediate in structure between the linear transmission chain and open diffusion approaches (Whiten and Mesoudi 2008). The approach benefits from some of the experimental control that characterizes the linear chains but also allows restricted forms of social interaction, which enhances validity (Galef and Allen 1995; Whiten and Mesoudi 2008). A disadvantage is that large numbers of experimental subjects may be required.

3.2.1.1 Linear transmission chains

The transmission chain approach was developed by the psychologist Frederic Bartlett (1932), and originally involved a text or picture being passed along a linear chain of participants. The first participant in such a chain read or viewed the stimulus material and later recalled it. The resultant recall was then given to the next participant in the chain to recall, the result of which was given to the third, and so on. Studying how the material changes as it is transmitted, and comparing the degradation rates of different types of material, can reveal specific biases in cultural transmission (e.g., Mesoudi and Whiten 2004; Mesoudi et al. 2006).

Linear transmission chains have all of the benefits of traditional social learning experiments (e.g., control of extraneous variables) but with the additional advantage that they allow investigation of factors influencing social transmission (e.g., those affecting whether information is lost or gained in transmission, or which mechanisms perpetuate learned knowledge). The chains allow researchers to track what happens at each step of the transmission process; for example, researchers can determine at what point a particular corruption occurs, without the complications, noise, and confounding effects of other forms of social interaction (Whiten and Mesoudi 2008). Replicate chains can be set up in each condition to enhance reliability.

Bartlett's primary interest was in memory rather than social transmission. Nonetheless, he established some key patterns that emerge in transmission, including a loss of detail such that only the broad features of the stories or generalizations remained, and the distortion of the material to render it more consistent with the subjects' prior knowledge. The years immediately following Bartlett's original work witnessed a series of transmission chain studies in human subjects generating broadly consistent findings (Maxwell 1936; Northway 1936; Allport and Postman 1947; T.H.G. Ward 1949; Hall 1951).

In the 1950s, the transmission chain method seemingly fell out of favor, and was not widely used within experimental psychology until reemerging as an experimental tool during the last decade (Bangerter 2000; Kashima 2000; Barrett and Nyhof 2001; Mesoudi and Whiten 2004; Mesoudi et al. 2006; Kalish et al. 2007; Griffiths et al. 2008). The recent studies also support Bartlett's general conclusions. For example, Mesoudi and Whiten (2004) found that descriptions of everyday events were given in gradually more general terms as they were passed along chains. They interpreted this as implying that information is stored in the brain in a hierarchically organized manner, such that the lower-level detail could be lost while the higher-level broad structure of events was retained. Likewise, Kashima (2000) found that transmitted information converged on cultural stereotypes, consistent with Bartlett's reported influence of prior knowledge. Barrett and Nyhof (2001) found that descriptions of living creatures or physical objects that were "minimally counterintuitive"—that is, they contained a small number of counterintuitive features. These descriptions were passed along transmission chains with more accuracy than either entities with no such counterintuitive elements or extremely counterintuitive (i.e., bizarre) material. A small amount of counterintuitive information can apparently render a story more memorable, but as Norenzayan et al. (2006) have shown, too many counterintuitive elements decrease the memorability of narratives, presumably because unfamiliar material is vulnerable to being lost.

Another advantage of the transmission chain design is that it provides a sensitive means of detecting transmission biases. Content biases are tendencies to copy certain kinds of material more readily than other kinds (see chapter 8). Mesoudi and Whiten (2004) argue that transmission chain studies have identified several content biases, including the aforementioned biases toward transmitting stereotypes and broad structure, and a bias toward social information. The latter emerges in a study by Mesoudi et al. (2006) that found information concerning third-party social interactions was transmitted with higher fidelity than equivalent nonsocial information, a finding interpreted as supporting social intelligence hypotheses (Byrne and Whiten 1988; Dunbar 2003).

Animal researchers resurrected the transmission chain approach after its use waned in the 1950s. Curio and colleagues (1978) carried out laboratory experiments demonstrating that mobbing of artificial stimuli could be socially transmitted along a chain of blackbirds. Observer birds learned to mob a novel nonraptorial bird, and even a plastic bottle, as a consequence of witnessing another transmitting bird exhibit the behavior. As a result of this learning, observers were themselves able to act as effective transmitters of the mobbing behavior to other naïve observers, allowing transmission to be effective along a chain of six birds.

Laland and Plotkin (1990, 1992) found that foraging information could be transmitted along replicate chains of Norway rats digging for buried food. These authors established a linear transmission chain design, with conditions that did or did not start with a trained demonstrator. There was a decreasing trend in foraging efficiency along the chains that started with a trained demonstrator, while chains that began with an untrained demonstrator exhibited an increasing trend. All trends became stable at a performance level significantly above those of individuals that foraged alone. This elevated level of performance may reflect a balance reached between the loss of information that occurs with each transmission episode, and the enhancement of the observers' performance by the sum of the demonstrators' social and asocial learning. The observed patterns of gain and loss of material parallel that reported in human transmission studies (e.g., Bartlett 1932).

In a second set of studies, Laland and Plotkin (1991, 1993) built on the finding that excretory deposits surrounding food sites can facilitate the social learning of food preferences in Norway rats (Galef and Beck 1985); they investigated whether dietary preferences could be transmitted along chains of animals by this means. The authors found that when the transmitted information conflicted with individual dietary preferences, transmission broke down rapidly, but when the rats worked in concert, transmission was enhanced and stable. They concluded that previously learned, experiential, or unlearned predispositions may significantly affect the stability of socially transmitted traits. Once again, there is a nice parallel here with the aforementioned human transmission chain studies of Bartlett (1932) and Kashima (2000) where prior knowledge was found to also bias transmission. Laland and Plotkin (1993) found that the stability of the transmitted preference could be bolstered by the introduction of a second social learning mechanism, the learning of food preferences via gustatory cues (Galef and Wigmore 1983), resulting in much slower decay of transmitted information.

3.2.1.2 Replacement transmission chains

The replacement transmission chain method was pioneered in a classic study by Jacobs and Campbell (1961), although it was first proposed by Gerard et al. (1956). Building on a study showing conformity in response to a perceptual illusion (Sherif 1936), Jacobs and Campbell exposed human subjects in small groups to "confederates" (individuals in league with the experimenters who were pretending to be subjects). The confederates were required to estimate the distance that a stationary point of light had moved in a dark room. Each gave their estimate, and subsequently subjects typically gave similar estimates to their confederates. They then gradually replaced group members and found that the biased judgment instigated by the confederates could be transmitted along the chain, although the initial bias was typically lost after a few iterations. Zucker (1977) repeated Jacob and Campbell's study with the addition that participants were given instructions emphasizing that they were part of an organization or institution, which substantially increased transmission stability. A similar approach was taken by Insko and colleagues who used the replacement method to explore traditions for cooperative and coercive interactions in humans, as well as the emergence of leadership (Insko et al. 1980; Insko et al. 1982; Insko et al. 1983).

Insko and colleagues found that groups comprised of trading parties improved their efficiency over time, as manifested in increasing productivity and earnings along the chain; for example, groups introduced division of labor and bargaining rules (Insko et al. 1980; Insko et al. 1983). This cumulative improvement along the chain was also a feature of similar transmission chain studies by Caldwell and Millen (2008); they offered groups of participants the goals of constructing high towers from uncooked spaghetti and modeling clay, as well as paper airplanes that could fly far. Similar increases in efficiency along a transmission chain are reported for arbitrary language learning by Kirby et al. (2008).

Menzel et al. (1972) set up a replacement transmission chain to investigate the gradual propagation of habituation to anxiety-inducing objects (novel toys) within a small group of chimpanzees. He observed that anxiety gradually diminished along the chain. Unfortunately this early study suffers from the weakness that with only a single chain there is no replication, and general patterns cannot be separated from individual idiosyncrasies. Replicate chains per condition are vital if transmission chain methods are to generate reliable findings.

One of the strongest examples of this approach is provided by Galef and Allen's (1995) studies of the transmission of dietary information in rats (fig. 3.4). Here two populations of four demonstrator rats were conditioned to prefer one of two flavored diets, with demonstrators in the same population having the same preference. The rats were gradually replaced (one a day) by naïve subjects who had the opportunity to acquire preferences from conspecifics by picking up dietary cues on their breath. Galef and Allen found that this design allowed for the stable transmission of two arbitrary dietary preferences for 14 days, with significant differences remaining in the feeding behavior of the two transmission groups. Moreover, Galef and Allen established that the stability of transmission was substantially enhanced by restricting feeding to just three hours. They point out that

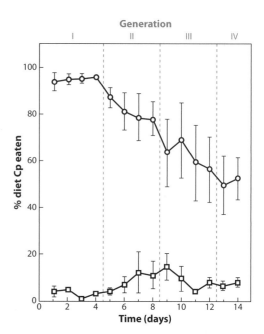

Figure 3.4. The mean (SE) amount of cayenne pepper (Cp) flavored diet, as a percentage of total food intake, eaten by Norway rats housed in enclosures. These enclosures contained founder colonies that ate either the Cp-flavored diet (*circles*) or an alternative Japanese horseradish (Jp) flavored diet (*squares*). Based on figure 5 in Galef and Allen's (1995) study of the transmission of dietary information in rats, which used a replacement transmission chain procedure.

free-living rats are most active at dawn and dusk, with many individuals feeding for only a few hours each day, implying that under natural conditions the stability of transmission may be similarly enhanced.

The findings from transmission chain studies in rats and humans have been reinforced by an investigation of the transmission of foraging information in guppies (*Poecilia reticulata*) again using the replacement design (Laland and Williams 1997, 1998). In this study, untrained guppies fed with trained conspecifics, and in the process learned a route to a food source. Fish in small founder populations were trained to take one of two alternate routes to a food source, and then founder members were gradually replaced with untrained conspecifics. Three days after all founder members had been removed, populations of untrained fish still maintained strong preferences for the routes of their founders. A follow up study determined that even maladaptive traditions could be established for taking long routes to the feeder when a shorter alternative was available (Laland and Williams 1998). While those traditions gradually eroded with replacement, there was still evidence of the social transmission of route preferences, and authors argued that arbitrary and even maladaptive migratory traditions were likely to be more stable in a naturalistic context with greater spatial separation between the alternatives. Interestingly, Reader et al. (2008) report similar behavior in a study of route learning in humans. The experimenters led adult human subjects into a room by one of two routes. Participants followed the demonstrated route choices, and later remembered and preferred this choice even when it was demonstrably suboptimal (longer, and not preferred by control subjects). Moreover, route preferences were stably maintained over multiple transmission episodes. In this instance, it would seem humans and guppies behave in strikingly similar ways.

Stanley et al. (2008) demonstrated that foraging techniques, and not just migratory routes, can be maintained as traditions in laboratory populations of guppies and platies (*Xiphophorus maculates*). Demonstrator fish were trained to collect food from inside vertical tubes, requiring them to swim in a vertical orientation not normally observed. Demonstrators were initially trained to feed by swimming into horizontal tubes to collect food, and over a period of several exposures these tubes were gradually rotated to vertical. This was a foraging task that the fish could not solve by themselves without training; while trained fish reliably fed from these tubes, no naïve fish presented with a vertical tube learned to feed from it on its own. When placed in groups with experienced demonstrators however, naïve fish readily learned to feed from the vertical tubes, establishing the social transmission of a novel feeding behavior. When the experienced demonstrators were gradually removed and replaced with more naïve fish, other group members continued to exploit the feeding tubes. Larger groups of fish showed more stable transmission of this behavior than smaller groups, although this was found to be related to their slower rate of turnover, rather than a direct effect of group size.

3.2.1.3 Insights from transmission chain studies

Transmission chain studies have led to a number of insights concerning social transmission, tradition, and culture (box 3.1). Unlike traditional social learning experiments, the chain studies can potentially establish whether or not social transmission processes exist that might result in the diffusion of learned behavior through populations of animals. The danger of assuming that such dissemination would follow on the basis of the single demonstrator-observer pairings characteristic of traditional social learning studies is illustrated by experiments that reveal the social learning observed in the first pairing may quickly peter out after a small number of steps of the chain, where each successive demonstrator proves less effective than its predecessor (Laland and Plotkin 1993). Equally, and conversely, social transmission effects may only become visible after multiple links in the transmission chain, as they gradually accumulate. For example, habituation to novel anxiety-inducing objects was only manifest after four to eight steps in a chimpanzee transmission chain (Menzel et al. 1972).

In general, in the same way that the behavior of populations cannot always be predicted by the behavior of their constituent individuals, the transmission process has population-level properties that the study of single demonstrator-observer pairings may not uncover. Transmission chain experiments have revealed that social transmission can distort the consumption of resources (e.g., diet composition) and use of environments (e.g., migration routes) away from the pattern of behavior predicted by individual preferences or patterns of reinforcement (the rewards received by individuals for their behavior). This can generate a historical contingency to population-level behavior. In other words, the equilibrium mean proportion of the diet represented by food items for which there is socially transmitted information cannot always be predicted from animals' consumption of such food items in the absence of social information. Diet composition may depend on historical factors and cannot always be predicted

---

Box 3.1

Insights from transmission chain studies

1. Stories, foraging skills, dietary preferences, migratory route preferences, and antipredator behavior can all be socially transmitted along chains of individuals in the laboratory, and maintained as short-term laboratory traditions (Bartlett 1932; Jacobs and Campbell 1961; Menzel et al. 1972; Curio et al. 1978; Insko et al. 1980; Insko et al. 1982; Laland and Plotkin 1990, 1992, 1993; Galef and Allen 1995; Laland and Williams 1997, 1998; Mesoudi and Whiten 2004; Mesoudi et al. 2006; Stanley et al. 2008).

2. Information can be gained as well as lost in transmission, and can lead to cumulative improvements over time (Bartlett 1932; Insko et al. 1980; Insko et al. 1983; Laland and Plotkin 1990, 1992; Caldwell and Millen 2008; Kirby et al. 2008).

3. Innovation does not necessarily require "clever" individuals, but can accrue in social transmission through the accumulated activities of many individuals (Laland and Plotkin 1990, 1992).

4. There may be an equilibrium of socially transmitted information that can be stably transmitted through a population (Laland and Plotkin 1990, 1992).

5. The opportunity to reinforce socially acquired knowledge and skills through further asocial learning and experience allows individuals to maintain an elevated level of information in transmission and reduces decays in performance (Laland and Plotkin 1992).

6. Content biases distort the stability of transmission. Social transmission may be more stable when it reinforces a prior preference (e.g., for a more palatable diet) than when it conflicts with one, and certain kinds of information (e.g., social information of a broad structure rather than fine detail, social stereotypes) may be acquired and transmitted more readily than other types (Laland and Plotkin 1993; Kashima 2000; Mesoudi and Whiten 2004; Mesoudi et al. 2006).

7. An experimental demonstration of social learning employing the traditional laboratory design (A → B) is not sufficient grounds for the conclusion that diffusion of a behavior through a population will inevitably result (Menzel et al. 1972; Laland and Plotkin 1993).

8. Social transmission can distort the consumption or use of resources away from that predicted by individual preferences, generating a historical contingency to population-level behavior. The mean proportion of the diet represented by food items for which there is socially transmitted information cannot always be predicted from animals' consumption of such food items in the absence of social information. Diet composition and migratory pathways may depend on historical factors and cannot always be predicted from patterns of reinforcement, profitabilities, or resource distributions (Laland and Plotkin 1993; Galef and Allen 1995; Laland and Williams 1997, 1998).

9. Two or more social learning processes may add up to reinforce the stability of social transmission (Laland and Plotkin 1993).

10. The stability of the transmission process is positively affected by the number of individuals in a social group and the amount of time available for individual behavior, but the stability is negatively influenced by the rate of change of group composition (Laland and Plotkin 1992; Stanley et al. 2008). Institutionalization reinforces transmission stability (Zucker 1977).

---

from patterns of reinforcement, profitabilities, or resource distributions (Laland and Plotkin 1993; Galef and Allen 1995). The same holds for migratory traditions (Laland and Williams 1997, 1998). The findings of transmission chain experiments also draw attention to a number of factors that bolster the stability of transmission.

Replacement transmission chain studies also show that simple traditions can be established in the laboratory. While these lab traditions are short-term, lasting days or weeks rather than years, such experiments suggest potential mechanisms for longer-term natural traditions. For example, the Laland and Williams studies highlight a simple process that may plausibly underpin the migratory traditions that occur in many natural fish populations (Helfman and Schultz 1984; Warner 1988, 1990). Such traditions are stable because the natural shoaling tendency of some fishes brings about a simple form of conformity. Fish prefer to join and follow large compared to small shoals (Lachlan et al. 1998), thereby lending them a tendency to adopt the majority behavior, and causing migration behavior to become stabilized in the group. This effect, where individuals prefer to copy the majority, is called positive frequency dependence, and generates the "cultural inertia" that can characterize traditions (Boyd and Richerson 1985). This process is sufficiently powerful that it can maintain arbitrary and even maladaptive traditions (Laland and Williams 1998), which helps explain why the mating and schooling sites of natural populations cannot always be predicted from features of the environment (Warner 1990).

3.2.2 Diffusion studies

In diffusion studies, the opportunity to exploit novel resources (e.g., to solve a puzzlebox to get food) is introduced into replicate populations of animals, and the spread of the novel behavior is monitored. Sometimes the populations are seeded with one or more demonstrators previously trained to perform the target behavior, and groups in different conditions may receive demonstrators that utilize alternative solutions. Control conditions may experience no demonstrator. While controlled diffusions are usually carried out in laboratories or in captivity, it has been possible to conduct some diffusion studies on habituated or feral natural populations, such as pigeons and meerkats (Lefebvre 1986; Thornton and Malapert 2009). This latter type of diffusion study, where the target behavior is free to diffuse through the whole population or group, is known as an "open" diffusion.

Aside from the initial introduction of the task, there is usually no structure of social interaction imposed on the population in a diffusion study. Individuals may or may not acquire the novel behavior in a naturalistic manner, and there are a wide range of forms of social behavior possible. In these respects, the method offers greater ecological validity than transmission chains or traditional laboratory studies (Whiten and Mesoudi 2008).

Open diffusion studies initiated by the experimenter also offer considerable advantages over the monitoring of natural diffusions as they occur in the field. First, and most importantly, replicate populations can be seeded with the novel behavior, greatly enhancing the reliability of any researcher inferences made as a result of the diffusion. Second, researchers can usually be confident that they witness the behavior from the diffusion's outset, that they collect all of the data (i.e., no missing events), and can make clear predictions as to precisely which behavior will spread, and under which conditions. Third, the researchers typically have knowledge of the relevant history of the animals, and can ensure that the task is novel.

However the data from open diffusion studies may be less easy to interpret than in other, more structured, experimental designs—for example, it may not be apparent who learns from whom (Whiten and Mesoudi 2008), and it is not always feasible to establish large numbers of replicate populations. Nonetheless, fine-grained analyses of whether copying or innovation has taken place, and exploration of some of the characteristics of these processes, is feasible (e.g., Whiten and Flynn 2010). Accordingly, we see diffusion studies playing a vital role in helping to integrate and better interpret the findings of laboratory and field research, as well as furnishing researchers with novel insights into the dynamics of social transmission.

The work of Lefebvre, Giraldeau, and colleagues stands out for pioneering the combination of laboratory and diffusion studies to investigate social transmission; they did this in studies of food-finding behavior in pigeons (Palameta and Lefebvre 1985; Lefebvre 1986; Giraldeau and Lefebvre 1986, 1987). Having isolated the mechanism involved in the social learning of a paper-piercing behavior (for a food reward) in pigeons through laboratory experiments, Lefebvre and colleagues went on to investigate the diffusion of this behavior through aviary and urban flocks. They found that scrounging of the food reward prevented the socially transmitted behavior from spreading to more than a few pigeons in an aviary flock (Giraldeau and Lefebvre 1986, 1987). Most aviary birds were able to take advantage of a few individuals that acquired the paper-piercing behavior. Lefebvre (1986) found that the diffusion was more rapid in an urban than a laboratory flock, with a greater total number of the urban birds learning the piercing behavior. He interpreted this finding as a consequence of the migration that occurs in open urban flocks, which may destabilize a frequency-dependent equilibrium between paper-piercing "producers" and scroungers. As paper-piercing individuals leave the open city flocks, scroungers exploit information gained from them to become piercers in their own right. Laboratory or field studies alone would not have led to the same level of insight.

In recent years, controlled diffusions have become an important tool for social learning researchers, and have led to a number of valuable insights. The most obvious of these is that repeated bouts of social learning can propagate novel learned behavior through animal populations in nature or captive groups kept in naturalistic conditions. Considerable progress has been made through field experiments (e.g., Warner 1988, 1990) and applying statistical and mathematical methods to infer social learning from natural diffusions and repertoires (e.g., R. L. Kendal, J. R. Kendal, et al. 2009; Franz and Nunn 2009; see chapters 5–6). Nonetheless, diffusion studies provide some of the most compelling hard evidence that animal social learning can plausibly lead to social transmission, tradition, and culture, and can result in behavioral differences between populations.

Some diffusion studies, particularly in primates, establish that observational learning processes underlie the diffusion (Whiten et al. 2005; Whiten et al. 2007). For example, Whiten et al. (2005) presented captive groups of chimpanzees with a "pan-pipes" task from which food could be extracted using either a lift or poke (i.e., two-action) method (fig. 3.5a). One group was seeded with a demonstrator trained to access food by lifting a T-bar with a stick, a second group had a demonstrator trained to get food by poking the stick past a flap to release a mechanism, while a third (control) group had no demonstrator. The authors found that in each experimental group the demonstrated behavior, more so than the alternative, spread to the other chimpanzees (fig. 3.5b), while no diffusion occurred in the control group. A similar pattern of findings that tested different tasks was

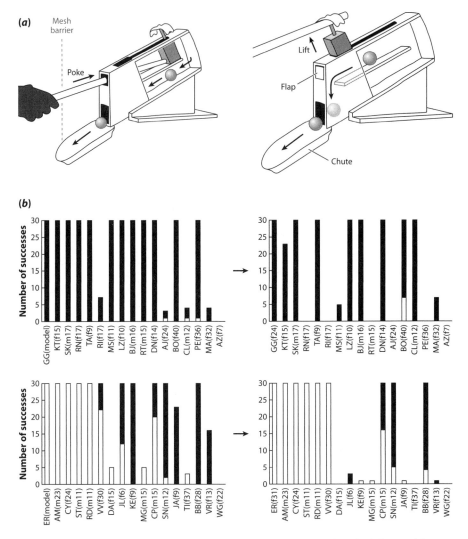

Figure 3.5. Pan-pipes apparatus (*a, top*) used by Whiten et al. (2005) to explore the diffusion of the poke (*left*) or lift (*right*) solutions in captive chimpanzee populations to acquire food. At bottom (*b*), the number of successes using the poke (*black*) and lift (*white*) methods in two groups, where a bar represents the performance of a single chimpanzee. The top and bottom graphs show chimpanzees exposed to poke and lift demonstrators, respectively, while the left and right figures give performance after 10 days and 2 months, respectively. Based on figures 1 and 2 in Whiten et al. (2005).

reported in a later study, which, in addition, reported between-group transmission (Whiten et al. 2007). The findings strongly suggest that the chimpanzees acquired the demonstrated behavior through some form of observational learning, perhaps through contextual imitation. That imitation may underlie captive diffusion enhances the plausibility of similar claims made for natural diffusions (Whiten et al. 1999; Whiten 2005).

Other diffusion studies have identified individual differences (related to sex, age, neophobia, social rank, etc.) that affect whether and how rapidly information spreads through a population (Reader and Laland 2000; Boogert et al. 2006; Boogert et al. 2008). Others have shown that transmitted information spreads more readily among familiar than unfamiliar individuals (Swaney et al. 2001) and determined that (at least in starlings) asocial learning ability predicts the rate of acquisition of a novel behavior in a group context (Boogert et al. 2008).

Diffusion experiments have the advantage that they can often be conducted on groups in their natural environment, though this can also introduce some complications. We discuss these in section 7.2.1.

3.3 Innovation

For the majority of the period in which social learning has been subject to scientific investigation, researchers studying animal (including human) cognition have regarded innovation as central to an understanding of individual problem-solving capabilities, but not key to the study of social learning and diffusion. Social learning researchers have taken an interest in behavioral innovation only in the last few years, and are now starting to view it of direct relevance. We suspect that there are two primary reasons for this change in perspective. Renewed interest in the experimental study of diffusion naturally leads researchers to ask which individuals, or classes of individual, initiate the diffusion. Researchers increasingly appreciate that diffusions could not occur without innovation, and they seek to understand what social, ecological, state (e.g., how hungry an individual is), or personality characteristics facilitate it.

Reinforcing this experimental perspective is the discovery that the incidences of innovation and social learning appear to be correlated across species. Not all species are equally innovative, and those most innovative appear disproportionately to be those that are most reliant on social learning (Reader and Laland 2002, 2003b). For example, across nonhuman primates, reported frequencies of innovation and social learning covary positively (fig. 3.6), after correction for research effort, phylogeny, brain size, and other potential confounding factors. Innovation frequencies covary with brain size in both birds and primates (Lefebvre et al. 1997; Reader and Laland 2002), leading to the view that the frequencies may be a useful, naturalistic, and pragmatic measure of a species' behavioral plasticity.

Further analyses have brought home the adaptive significance of innovation. There are multiple descriptions of specific animal innovations apparently facilitating survival in changed circumstances (Sol 2003; Reader and Laland 2003).

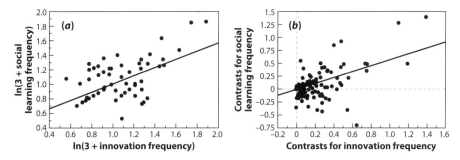

Figure 3.6. Frequencies of innovation and social learning, corrected for research effort, covary across nonhuman primates. The raw data (*a*), with each point representing one species, and independent contrast data (*b*). Based on figure 2 in Reader and Laland (2002).

Innovation is thought to be important to endangered or threatened species forced to adjust to impoverished environments (Greenberg and Mettke-Hofman 2001). Also, species that exhibit more behavioral innovation have been found to be more likely to survive and establish themselves when introduced to new locations, even new continents (Sol 2003; Sol and Lefebvre 2000; Sol et al. 2005). Evidence is mounting that innovation plays an important role in ecology (e.g., range expansion), evolution (e.g., subspecies diversification), cognition (as the first step of social transmission), and cultural diversification (Laland and Reader 2009). Overington et al. (2009) report that in birds the diversity of innovation is a better predictor of brain size than innovation frequency, suggesting that the cognitive capacity required to perform a wide variety of novel behavior underpins the relationship with brain size. These observations also suggest that innovation is not only widespread in nonhumans, but is functionally important.

Moreover, it has long been suggested that innovation and social learning might be linked, and could coevolve, and that these processes may play a role in driving brain evolution (Wyles et al. 1983; Wilson 1985, 1991; Boyd and Richerson 1985; Lefebvre et al. 1997; Reader and Laland 2002; Whiten and van Schaik 2007; Reader et al. 2011). This theoretically derived link feeds back to stimulate interest in innovation among social learning researchers.

How, though, can innovation be studied experimentally? In many researchers' minds, innovation is a relatively rare event, so how could it be investigated? Kummer and Goodall (1985, 213) offered a suggestion: "Systematic experimentation (such as the introduction of a variety of carefully designed ecological and technical 'problems') both in free-living and captive groups would provide a new way of studying the phenomena of innovative behaviors and their transmission through and between social groups." In the last decade, this approach has started to be implemented in animal studies. Innovation can be studied experimentally in captive animals, and even in the field (Morand-Ferron and Quinn 2011; Morand-Ferron et al. 2011), by presenting the animals with novel challenges, such as foraging-puzzle boxes and by exploring the factors influencing innovation (sex, age, social rank) (see Ramsey et al. 2007 for an alternative approach). Below we present two examples.

The prevailing assumption in the primate literature for many years was that young or juvenile primates are more innovative than adult individuals. This putative innovative tendency among the young was thought to be a side effect of their increased rates of exploration and play. However, a review of the primate-innovation literature noted a greater reported incidence of innovation in adults than in nonadults, which may reflect the greater experience and competence of older individuals (Reader and Laland 2001). By presenting novel "puzzle-box" foraging tasks to 26 zoo populations of callitrichid monkeys (marmosets and tamarins), R. L. Kendal, Coe, and Laland (2005) sorted between these conflicting views and recorded the first individual to approach, contact, and solve each task. Their study revealed systematic age differences in callitrichid innovation, with older monkeys significantly more likely than younger monkeys to be the first to solve the tasks. Whereas younger monkeys were disproportionately represented among those first to contact the box, adults were disproportionately the first to solve the tasks, and older individuals were significantly more likely than younger individuals to turn manipulations into successful manipulations. Seemingly, greater experience and competence allowed older individuals to solve novel problems more effectively than younger individuals, although other developmental factors, such as improvements in manipulative skills, increased strength, and maturity with age, may also be important.

Boogert et al. (2008) investigated to what extent the pattern of spread of innovations in captive groups of starlings (*Sturnus vulgaris*) could be predicted by knowledge of individual and social-group variables, including association patterns, social-rank orders, measures of neophobia, and asocial-learning performance. Small groups of starlings were each presented with a series of novel extractive foraging tasks, and the latency (time taken) for each bird to contact and solve each task as well as the orders of contacting and solving were recorded. Object neophobia and social-rank measures characterized which animal was the first of the group to contact the novel foraging tasks, and the time at which other individuals subsequently contacted the tasks was associated with latency to feed in a novel environment (a possible indicator of vigilance). However, asocial-learning performance measured in isolation, predicted the first solvers of the novel foraging tasks in the group. In other words, one can predict how innovative a starling will be on the basis of its previously measured asocial learning performance.

More generally, it is increasingly the pattern that researchers conducting diffusion studies pay attention to the characteristics of the first solver in a social group, and/or to the social contexts that facilitate or hinder innovation.

3.4 The Biological Bases of Social Learning

Throughout its long history, social learning has almost exclusively been investigated solely at the behavioral level. However, within the last 10 to 15 years a small number of studies have emerged investigating the biological underpinnings of social learning. We briefly summarize this work here, organizing the data into three clusters concerned with the (1) neuroendocrinology of social learning, (2) social learning of fear, and (3) neural mechanisms of observational learning and imitation. One area of social learning research for which the neural basis of learning is well developed is birdsong, which we also present (see box 3.2).

3.4.1 Neuroendocrinological studies

In an extensive series of studies spanning several decades, Galef and collaborators have explored the social transmission of food preferences in rats via olfactory cues carried on the breath of conspecifics (see chapter 4 for details). The same process has been reported in other mammals, including bats and several other rodents. This "social transmission of food preferences in rodents" model system has now begun to be used as a vehicle to explore the biological basis of social learning.

A number of brain regions are known to be involved in this form of social learning, including the hippocampus, subiculum and dentate gyrus, basal forebrain, and the frontal, piriform, and orbitofrontal cortices (Choloris et al. 2009). A specialized olfactory subsystem, which includes olfactory sensory neurons (OSNs) expressing the receptor guanylyl cyclase (GC-D), the cyclic nucleotide-gated channel subunit CNGA3 (channels that transmit information about smell from sensory cells to the brain), and the carbon anhydrase isoform CAII (which promotes the hydration of CO_2) is required for the acquisition of socially transmitted food preferences in mice (Munger et al. 2010). Using electrophysiological recordings from genetically altered mice, Munger et al. (2010) showed that GC-D + OSNs are highly sensitive to carbon disulphide (CS_2), a component of rodent breath and a known social signal mediating the acquisition of socially transmitted food preferences (Galef et al. 1988). Hence GC-D + OSNs may detect the chemo-signals that facilitate this form of social learning.

Choloris and colleagues (2009) provide a useful review of the roles of the hormones oxytocin (henceforth OT), arginine-vasopressin (AVP), and estrogens in social learning. Both OT and AVP have been found to improve performance of male rats in the previously mentioned social transmission of food preferences via olfactory cues carried on the breath of conspecifics, such that treated animals are more likely to acquire the demonstrated preference than controls (Popik and Van Ree 1993). However, experimental studies reveal an interaction with task difficulty, in which AVP improves performance on difficult tasks (where control subjects show no preference), but impairs performance on simpler tasks (Strupp et al. 1990); this is plausible because simpler tasks are typically learned asocially (see chapter 8).

Endogenous changes in hormones in female animals also appear to modulate social learning. Postpartum animals outperform virgins in other phases of the estrous cycle (Flemming et al. 1994). Though they did not publish this data, Choloris and colleagues reported in a portion of Choloris et al. 2009 that the same is true of animals in the postestrous phase of the estrous cycle. Moreover, activation of some estrogen receptors (ERβ) improves performance on this task, while activation of other estrogen receptors (ERα) inhibits it (Clipperton et al.

2008). These results point to a role for estrogens in the social transmission of food preferences. Recent evidence also suggests a role for dopamine receptors (Choleris et al. 2011).

Oxytocin may also be involved in mate choice. Studies in mice have established that females will develop a preference for a male whose odor cues are associated with another estrous female. However, females who do not produce oxytocin, and females treated with an oxytocin antagonist, were impaired in their ability to use this social information, and did not exhibit mate-choice copying (Kavaliers et al. 2008). Oxytocin also seems to be important in helping females to avoid parasitized males, and hence may facilitate the transmission of public information about mate quality (Kavaliers et al. 2006). These findings are consistent with behavioral studies suggesting that hormone levels can affect reliance on social information. Webster and Laland (2011) report that female stickleback in reproductive condition copied less than nonreproductives, and male sticklebacks in reproductive condition copied more.

3.4.2 Social learning of fear

Another system that has received some attention is the social learning of fear. Many animals, including birds, mice, cats, cows, and primates, can learn fears by observing a conspecific (Olsson and Phelps 2007), often through an observational conditioning process (see chapter 4). One productive experimental paradigm has been the social learning of stress and avoidance responses to biting flies in mice; this learning was investigated by Martin Kavaliers, Douglas Colwell, Elena Choleris, and colleagues. Mice exposed to biting flies respond with both increased plasma corticosterone levels and active self burial. When subsequently exposed to flies whose biting mouth parts were removed, the mice display the same conditioned responses. Interestingly, this conditioned increase in corticosterone and burial can also be acquired through social learning, without direct experience with the intact biting flies (Kavaliers et al. 2001a, 2003). Kavaliers et al. (2001b) established that an NMDA (N-Methyl-D-aspartate) receptor antagonist (a drug that inhibits reponsiveness to NMDA) given to observers prior to, but not after, presentation of fly-attacked demonstrators blocked the socially determined, conditioned analgesia and self-burying. Kavaliers and colleagues argue that this supports NMDA involvement in the mediation of the social transmission of conditioned fear.

Research on the neurobiology of fear conditioning has focused on the amygdala in the medial temporal lobe, a key structure in the brain's fear circuitry. Amygdala lesions in monkeys, and imaging studies in humans, confirm that this structure is also critical to the observational learning of fear (Amaral 2003; Olsson et al. 2007). Unlike mice, the strength of a primate demonstrator's fear response during its learning is correlated with the observer's expressed fear when tested; this indicates that there may be a greater reliance on emotional expression during the learning process in primates than other mammals. In monkeys and humans, facial fear is a reliable unconditioned stimulus. Olsson and Phelps (2007) suggest that two interacting pathways mediate the social learning of fear (fig. 3.7). First, the demonstrator's expression of fear can be intrinsically aversive, presenting the possibility that sensory representations may be primed by mere observation of another individual's emotional display, perhaps via a mirror-neuron system (see below). Second, factors related to the social context can be involved in the regulation of emotional responding and the learning that results (fig. 3.7b). In addition, humans uniquely possess the additional ability to obtain emotional information through language

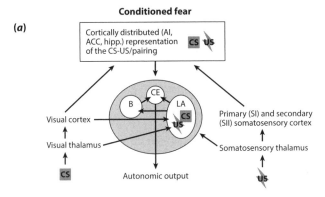

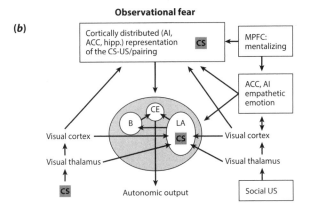

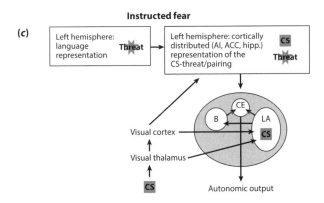

Figure 3.7. Olsson and Phelps' (2007) neural model of nonsocial and social fear learning in humans. The arrows describe the flow of information between different functional brain regions. Fear conditioning (a) occurs by associating the visual representation of the CS (conditioned stimulus) with the somatosensory representation of the aversive US (unconditioned stimulus). The lateral nucleus (LA), in which sensory representations of the CS and US converge, is believed to be the site of learning. The amygdala also receives input from the hippocampal memory system (hipp.), anterior insula (AI), and anterior cingulate cortex (ACC), the latter which contains secondary representations of the CS and US, information about the learning context, and the internal state of the organism. In observational fear learning (b), the visual representation of the CS is modified by its association with a representation of the distressed other, serving as the US. As in fear conditioning, it is hypothesized that representations of the CS and the US converge in the LA. The strength of the US may be modified by the medial prefrontal cortex (MPFC) input related to the interpretation of the other's mental state, as well as cortical representations of empathic pain through the ACC and AI. Instructed fear learning (c) occurs by modifying the processing of the visual representation of the CS through its association with an abstract representation of threat. Instead of being coded in the amygdala, the CS-"threat" (US) contingency is likely to be represented in a cortically distributed network, critically depending on the hippocampal memory system. Based on figure 2 in Olsson and Phelps (2007).

(fig. 3.7c). Olsson and Phelps (2007) describe how observational and instructed fear is distinguished by involvement of additional neural systems implicated in social-emotional behavior, language, and explicit memory, and propose a modified conditioning model to account for social fear learning in humans (fig. 3.7).

3.4.3 Neural mechanisms of observational learning

Many neuroimaging studies have investigated the brain regions involved in observational learning, particularly imitation (Iacoboni et al. 1999; Iacoboni et al. 2001; Buccino et al. 2004; Williams et al. 2006; Williams et al. 2007; fig. 3.8). While they have not always generated completely consistent findings, a limited number of areas are strongly implicated. These include the inferior frontal gyrus, the dorsal and ventral premotor cortex, the inferior parietal cortex, the superior parietal lobule, and the posterior superior temporal sulcus (see Brass and Heyes 2005 for a summary). For example, Williams et al. (2007) conducted a study in which human participants observed one of two actions and were instructed to imitate the action they observed, or to perform the alternative nonmatching action. The researchers witnessed the greatest contrast between matching and nonmatching actions in the insula, intraparietal sulcus, dorsal premotor cortex, and superior temporal gyrus, but imitation was also associated with activity in the prefrontal cortex and lateral orbitofrontal cortex, amygdala, and other regions. Williams et al. suggest that the lateral orbitofrontal cortex responds to action-perception mismatch. The cerebellum is also thought to be involved in the internal representation of action and the learning of skills through observation (Petrosini et al.

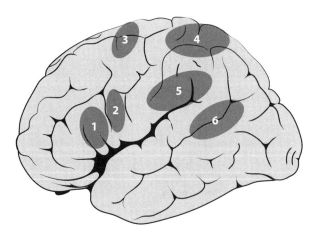

Figure 3.8. Areas of shared activation during movement observation and imitation. A schematic, lateral view of the human cortex showing areas that have consistently been found to be active during imitation tasks and during passive observation of biological motion: (*1*) the pars triangularis and the pars opercularis of the inferior frontal gyrus, (*2*) the ventral premotor cortex, (*3*) the dorsal premotor cortex, (*4*) the superior parietal lobule, (*5*) the inferior parietal cortex, and (*6*) the posterior superior temporal sulcus. Based on figure 2 in Brass and Heyes (2005).

2003), and lesion studies suggest that it plays a critical role in the observational learning of spatial tasks (Leggio et al. 2000).

Burke et al. (2010), again based on a brain imaging (fMRI) experiment, concluded that human subjects' observational learning could be explained by two previously uncharacterized forms of prediction errors—that is, differences between predicted and obtained outcomes. The first they describe as "observational action prediction errors"—the actual minus the predicted choice of others—which is associated with activity in the dorsolateral prefrontal cortex. The other they call "observational outcome prediction errors"—the actual minus the predicted outcome received by the demonstrator being observed—was associated with activity in the ventromedial prefrontal cortex. The two error signals contain the information required for the observer to update their knowledge of what choices should be performed in a particular context, and what the consequences of such actions are likely to be. Feasibly, this process might underlie the observational R-S learning described in the previous chapter.

Klucharev et al. (2009) report that conflict with group opinion triggered a neuronal response in the rostral cingulated zone and the ventral striatum, similar to the prediction error signal in reinforcement learning. Moreover, the amplitude of the conflict-related signal predicted subsequent conforming behavioral adjustments, while the individual amplitude of the conflict-related signal in the ventral striatum correlated with differences in conformity across subjects. Plausibly, aspects of conformity may be mediated by activity of the rostral cingulate zone and ventral striatum, and these findings may help to characterize the biological underpinnings of the conformist transmission (the majority) strategy described in chapter 8.

Behrens et al. (2008) found that two neighboring regions of the anterior cingulated cortex were central to learning about social and (asocial) reward-based information, but that when these sources of information were combined in decision making this excited activity in the ventromedial prefrontal cortex. Other regions appear to be instrumental in mediating social influences on value (on the extent of our esteem) in the human brain. For example, Campbell-Meiklejohn et al. (2010) found that when someone agrees with two "experts" on music choice, activity is produced in a region of ventral striatum that also responds when receiving a valued object. It is apparently psychologically rewarding to agree with others, or at least to have one's judgments endorsed by authorities. These authors also found that the anterior insula cortex/lateral orbitofrontal cortex uniquely responds to the unanimous opinions of others regardless of whether or not the opinion is in agreement or disagreement with the subject. This finding is suggestive of an evolved neural sensitivity to consistency in demonstrator behavior; it is consistent with an economics experiment that suggests people are more reinforced by following social information than otherwise expected by a payoff (i.e., financial reward) alone (Biele et al. 2009).

No discussion of the neural basis of social learning would be complete without consideration of the role of mirror neurons. Considerable attention has surrounded the discovery of a class of neurons in area F5 of the primate prefrontal cortex that fire both when a monkey performs a specific action and when the monkey observes the same action being performed by a second individual (Galesse et al. 1996; Rizzolati et al. 1996; Gallese and Goldman 1998; Rizzolati and Craighero 2004; Brass and Heyes 2005). These cells, termed "mirror neurons," appear to function as a bridge between higher visual processing areas of the brain and the motor cortex. Actions most effective in triggering mirror neurons are hand actions such as grasping, manipulating, and placing, although some mirror neurons respond when an object is grasped in the mouth. Mirror neurons studied thus far typically respond to goal-related motor acts: when the same actions are observed or performed in the absence of a target object, the relevant mirror neuron does not respond. Transcranial magnetic stimulation and functional imaging of the human motor cortex provide circumstantial evidence for the existence of a similar system with the properties of mirror neurons in our own species (Fadiga et al. 1995; Grafton et al. 1996); however, controversy remains over whether the posterior part of the inferior frontal gyrus, which is thought to be the human

---

Box 3.2

Birdsong and the song system

Both male and female songbirds produce vocalizations all year round, but the term "song" is reserved for the long, complex vocalizations produced, typically by males, during the breeding season (Catchpole and Slater 2008). Usually male songbirds learn their song from an adult male tutor, such as a neighbor. There are two phases of songbird song learning, including a (1) memorization phase, in which the vocal information of the tutor song is stored in long-term memory (usually during an early sensitive period), and (2) a sensorimotor phase later in life, when the bird's own vocal output is compared with the memorized information (Bolhuis and Gahr 2006). Initially the young bird produces subsong, which does not closely resemble that of the tutor, but this output is gradually refined to resemble the memorized template.

The neural basis of birdsong has been investigated over many decades. An extensive network of interconnected brain nuclei, known as the song system, has been identified as being involved in the perception, learning, and production of song (see fig. 3.9). Lesion and immediate early gene studies in adult and young songbirds have identified two pathways in the songbird brain. A caudal pathway, involved in the production of song, encompasses HVC, which sends projections to RA (robust nucleus of the arcopallium), which in turn projects to other nuclei controlling motor output to the syrinx (nXIIts). A rostral pathway, involved in song learning, involves HVC and LMAN (lateral magnocellular nucleus of the anterior nidopallium), which each send projections to a nucleus known as Area X, which in turn projects to DLM (dorsal lateral nucleus of the medial thalamus).

Studies in canaries and sparrows have established that the size of brain nuclei responsible for song production (HVC and RA) can be dramatically larger (up to 100%) during the breeding season, resulting from increases in the size, connectivity, or number of neurons (Nottebohm 1981; Tramontin and Brenowitz 1999, 2000). As many birds learn new songs each year, or modify their existing songs, new neurons may be required in nuclei that control song output to store instructions. Alternatively, seasonal fluctuations may reflect energy savings made in the nonbreeding season. For a recent summary of research into the neural mechanisms of birdsong, see Bolhuis and Gahr (2006), and Ziegler and Marler (2008).

homologue of the premotor area F5 in monkeys, is necessary for imitation (Brass and Heyes 2005). Recently, auditory mirror neurons have also been discovered in the telencephalic nucleus HVC, in the brains of male swamp sparrows (*Melospiza georgiana*) (Prather et al. 2008; see box 3.2). These neurons show nearly identical patterns of firing when the bird hears another bird singing as when it sings the same note sequence itself.

There has been much debate regarding the functional significance of mirror neurons, with suggested roles for them in the theory of mind (Gallese and Goldman 1998), language (Rizzolatti and Arbib 1998), and complex imitation (Williams et al. 2001). However, there is little evidence for any such abilities in monkeys, which implies their initial function may be more general. Perhaps, rather than being the underlying mechanism that *enables* imitation and other cognitive processes to occur, mirror neurons are instead more appropriately regarded as the *outcome* of imitation and other forms of social learning, or indeed,

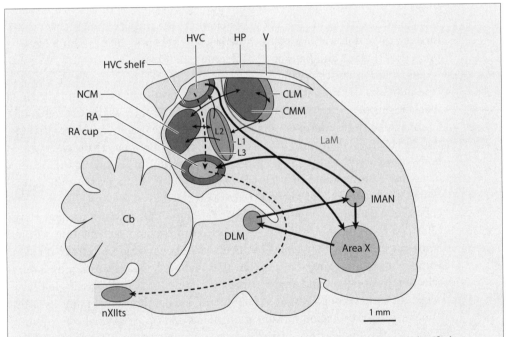

Figure 3.9. An extensive network of interconnected brain nuclei (the song system) has been identified as being involved in the perception, learning, and production of birdsong. This schematic diagram of a songbird brain gives approximate positions of nuclei and brain regions involved in song. Lesion studies in adult and young songbirds led to the distinction between a caudal pathway (*dashed arrows*), considered to be involved in the production of song, and a rostral pathway (*thick black arrows*), thought to have a role in song acquisition. Abbreviations: Cb, cerebellum; CLM, caudal lateral mesopallium; CMM, caudal medial mesopallium; DLM, nucleus dorsolateralis anterior, pars medialis; HP, hippocampus; HVC, a letter based name; L1, L2, L3, subdivisions of field L; LaM, lamina mesopallialis; IMAN, lateral magnocellular nucleus of the anterior nidopallium; NCM, caudal medial nidopallium; nXIIts, tracheosyringeal portion of the nucleus hypoglossus; RA, robust nucleus of the arcopallium; V, ventricle. Based on figure 3 in Bolhuis and Gahr (2006).

of general associative processes (Laland and Bateson 2001; Keysers and Perrett 2004; Brass and Heyes 2005).

It is well established that learning processes result in changes in strength of synaptic connections between neurons. Hence it is certainly plausible that, given appropriate sensory experience and reinforcement, networks of connected neurons will form in appropriate regions of the brain that exhibit properties like mirror neurons (Brass and Heyes 2005; Catmur et al. 2009); that is, neurons in networks will fire when an individual perceives both herself and another performing an action. A number of recent studies have suggested that the human mirroring system can be modified by experience (see Catmur et al. 2009 for a review), and may acquire its properties as a side effect of general associative processes (Keysers and Perrett 2004; Brass and Heyes 2005). For example, mirror neurons for tool use can develop during ontogeny (Ferrari et al. 2005), while observation of piano playing is associated with stronger motor activation in the brains of pianists than musically naïve controls (Haslinger et al. 2005). These findings are consistent with our own examination of learning processes using simple unsupervised neural networks; these neural network models imply that forms of social learning resulting in a match of demonstrator and observer's behavior will forge synaptic connections between neurons in networks that exhibit mirroring properties (Laland and Bateson 2001; Hoppitt 2005).

Generalist theories of imitation propose that the correspondence problem (the difficulty of translating perception into action) can be solved by general mechanisms of associative learning and action control (Brass and Heyes 2005). Consistent with the aforementioned neural net modeling, they assume that imitation is achieved by activation of motor representations (representations of body movements, in the brain) through observation of action. Several studies have now demonstrated that passive observation of action leads to the activation of a set of brain regions known to be involved in movement execution (Iacoboni et al. 1999; Iacoboni et al. 2001; Buccino et al. 2004). Brass and Heyes (2005, 489) argue compellingly: "Mirror neurons could *do* imitation without being *for* imitation; they could be involved in generating imitative behaviour without imitation being the function that favoured their evolution by natural selection." We agree, and suggest that mirror neurons may equally underpin social learning through response facilitation, local enhancement, and a variety of other processes, including contagion effects and observational conditioning. If correct, this implies that mirror neurons will be found in a wide range of taxa capable of simple forms of social learning, rather than being restricted to large-brained or cognitively sophisticated animals. Of course, mirror neurons might, over recent evolutionary time, have been co-opted to play an instrumental role in various "higher" cognitive procedures. Evidence is starting to grow suggesting that in humans the mirror system is involved in solving the correspondence problem for imitation (Brass and Heyes 2005; Catmur et al. 2009). However, that evidence implies that the mirror system, like the imitative capability in general, depends on learning perceptual-motor links through the application of very general processes.

3.5 Conclusions

Recent years have witnessed a blossoming of experimental research into social learning, and with it a diversification in methodology. The traditional laboratory experimental design, with its demonstrator-observer pairing, has been supplemented with transmission-chain and diffusion studies, as well as investigations into behavioral innovation. These allow some of the population aspects of social transmission to be explored. Researchers have also begun to investigate the underlying biological bases of social learning. While this latter body of research is still in its infancy, in certain domains, notably endocrinological studies of the social transmission of food preferences, the observational learning of fears, and imitation, real progress has been made. The experimental tools are now available for researchers to gain a good understanding of the neural and physiological underpinnings of social learning, and we anticipate considerable expansion of this aspect of the research in the near future.

Chapter 4

Social Learning Mechanisms

As the preceding chapters bring to light, there are many fields of scientific enquiry on which social learning research impinges. The psychological mechanisms by which individuals learn from one another are of particular interest to these fields. Over the years, a number of authors have attempted classifications of social learning mechanisms (Galef 1988; Heyes 1994; Whiten and Ham 1992; Whiten et al. 2004; Zentall, 1996, 2001), but in practice, there is only partial consensus over terminology. Typically classifications have been devised by researchers studying animal social learning, at least partly with the goal of cataloguing all plausible imitation-like mechanisms that must be ruled out if instances of "true" imitation are to be recognized. It is often assumed that the purported "simplest" social learning processes, such as local enhancement (Thorpe 1956), are in operation in animal populations in the wild, inferred by a process of parsimony. However in reality, several processes could account for most instances of animal social learning, including imitation, and it is a subjective enterprise to decide which is the "simplest" (Roitblatt 1998). In order merely to validate the existence of a particular social learning process, researchers require data that can discriminate one process from another (Byrne 2002).

We have previously attempted to provide a classification system with this goal in mind (Hoppitt and Laland 2008a). In section 4.1 we provide a summary of this classification, though we extend it to include two additional mechanisms; "coaching" and "opportunity providing." Both are likely to be important mechanisms

---

The text in this chapter is based on two journal articles, which have been substantially edited from their original forms: Hoppitt and Laland 2008a; Hoppitt, Brown, Kendal, Rendell, Thornton, et al. 2008.

of social learning (Caro and Hauser 1992; Hoppitt et al. 2008). We also shift the focus slightly to discuss the circumstances under which each mechanism can result in social transmission. In our original classification, we also reviewed the evidence for each mechanism in nonhuman animals. In this chapter we begin by defining all mechanisms in the classification (section 4.1) before going on to discuss the methods needed to distinguish between them (section 4.2).

In our previous review (Hoppitt and Laland 2008a), we found that in most published social learning studies it is very difficult to determine exactly which mechanisms are operating. This is because experiments are often not designed with this primary purpose. Nonetheless, we recognize that in such cases a researcher may still wish to draw some inferences about the process underlying a particular case of social learning. Consequently, in section 4.3 we suggest an alternative way of classifying the processes involved in a case of social transmission. We relate this system to that presented in section 4.1, allowing a researcher to establish the sets of processes, or combinations of processes, that might account for the case in question. We do not attempt to provide an exhaustive review of the evidence for each process, but only give examples of cases to illustrate the methodology required for the detection of each process. In addition, here we place more emphasis on the evidence required to show that a particular process is not only operating, but is resulting in social transmission. Finally, in section 4.4 we discuss how teaching fits into our classification.

4.1 A Classification of Social Learning Mechanisms

Our scheme builds on the classifications of other authors, in particular that by Heyes (1994), which emphasized the utility of definitions based on observable criteria. However, Heyes was concerned only with mechanisms by which learning occurred as a direct result of social interaction. In contrast, we include a number of mechanisms where the behavior of the demonstrator causes a transient change in the observer's behavior that can lead to its learning by asocial means.

We discuss twelve mechanisms, which are listed with definitions in table 4.1. In each case we specify whether the mechanism results in social learning directly or indirectly, and under what circumstances it results in social transmission. The classification has a number of limitations; like others previously devised, it exhibits overlapping and nonhierarchical categories (e.g., local vs. stimulus enhancement), as well as some categories that can lead to learning by both direct and indirect means (e.g., stimulus enhancement). This reflects the history of the terms involved; if one were to devise a new classification scheme from scratch, conceivably such problems might be avoided. However, our primary goal here is not to provide a new classification scheme but rather to provide a guide to how the terms are currently used in the literature. In our previous classification on which the current one is based (Hoppitt and Laland 2008a), we attempted to use terms in a similar manner to the original definition, but allowed for the fact that some mechanisms had been redefined (e.g., "observational conditioning"),

Table 4.1. Social learning processes

Process	Definition	Source
Stimulus enhancement	*Stimulus enhancement* occurs when observation of a demonstrator (or its products) exposes the observer to a single stimulus at time t_1, and single stimulus exposure effects a a change in the observer detected, in any behavior, at t_2.	Heyes (1994)
Local enhancement	*Local enhancement* occurs when, after or during a demonstrator's presence, or interaction with objects at a particular location, an observer is more likely to visit or interact with objects at that location.	Thorpe (1963)
Observational conditioning	*Observational conditioning* is a subset of stimulus-stimulus learning in which observation of a demonstrator exposes the observer to a relationship between stimuli at t_1, and exposure to this relationship effects a change in the observer detected, in any behavior, at t_2.	Heyes (1994)
Response facilitation	*Response facilitation* occurs if the presence of a demonstrator animal performing an act (often resulting in reward) increases the probability of an animal that sees it doing the same.	Byrne (1994)
Social enhancement of food preferences	*Social enhancement of food preferences* occurs when after being exposed to a demonstrator carrying cues associated with a particular diet, the observer becomes more likely to consume that diet.	Galef (1989)
Social facilitation	*Social facilitation* occurs when the mere presence of a demonstrator affects the observer's behavior.	Zajonc (1965)
Contextual imitation	*Contextual imitation* occurs when, directly through observing a demonstrator perform an action in a specific context, an observer becomes more likely to perform that action in the same context.	Byrne (2002)
Production imitation	*Production imitation* occurs when, after observing a demonstrator perform a novel action, or novel sequence or combination of actions, none of which are in its own repertoire, an observer then becomes more likely to perform that same action or sequence of actions.	Based on Byrne (2002)
Observational *R-S* learning	*Observational R-S learning* is a subset of response-reinforcer learning (*R-S*) in which observation of a demonstrator exposes the observer to a relationship between a response and a reinforcer at t_1, and exposure to this relationship effects a change in the observer detected, in any behavior, at t_2.	Heyes (1994)
Emulation	*Emulation* occurs when, after observing a demonstrator interacting with objects in its environment, an observer becomes more likely to perform any actions that have a similar effect on those objects.	Tomasello 1990; and Custance et al. 1999
Opportunity providing	*Opportunity providing* occurs when the products of the behavior of the demonstrator provide the observer with an opportunity to engage in operant learning that would otherwise be unlikely to arise—for example by providing an easier, less dangerous or more accessible version of the task.	Hoppitt et al. (2008), based on Caro and Hauser (1992)
Inadvertent coaching	*Inadvertent* coaching occurs when the response of a demonstrator to the behavior of the observer inadvertently acts to encourage or discourage that behavior.	Hoppitt et al. (2008), based on Caro and Hauser (1992)

or their meaning had evolved somewhat. We also introduced refinements in the interests of defining processes according to testable criteria. Here, we take the same approach, but shift the focus slightly to discuss the circumstances under which each mechanism can result in social transmission.

4.1.1 Stimulus enhancement

Stimulus enhancement (Spence 1937) occurs when a demonstrator's behavior increases the probability that an observer is exposed to a stimulus, resulting in an increase in the observer's rate of interaction with stimuli of the same type. Heyes (1994) argued that the process could be regarded as a subset of single-stimulus learning, where exposure to a stimulus results in a change in the subject's responsiveness to that stimulus. Here we adopt Heyes' (1994, 216) definition:

> Stimulus enhancement occurs when observation of a demonstrator (or its products) exposes the observer to a single stimulus at time t_1, and single stimulus exposure effects a change in the observer detected, in any behaviour, at t_2.

Stimulus enhancement could be regarded as a case of sensitization, where subsequent responsiveness to the stimulus is increased, but there is no reason to suppose that there might not instead be habituation. Nonetheless, it is the sensitizing form of stimulus enhancement that remains of most interest, since it can lead indirectly to further learning about the stimulus in question, through the observer's subsequent interactions with that stimulus (e.g., reinforcement learning). In particular, it could result in social transmission. In this case, when a demonstrator performs the target behavior, it exposes another individual to a stimulus, causing an increase in the observer's responsiveness to that stimulus. This is likely to cause an increase in the rate at which the observer acquires and performs any behavior associated with that stimulus, including the behavior performed by the demonstrator. Note, however, that this process of social transmission by stimulus enhancement may *not* result in any change in the relative rates of performance or acquisition of alternative traits associated with the same stimulus.

The above definition does not specify the manner in which the observer is exposed to the stimulus—it could have its attention drawn to the stimulus by the demonstrator, or the demonstrator could act in some other way to make the stimulus more accessible to the observer. An example of the latter is a mother bringing prey to her young. Critical to stimulus enhancement is that there is a degree of stimulus generalization that means the observer's response to similar stimuli in different locations may be affected.

4.1.2 Local enhancement

The term "local enhancement" was introduced by Thorpe (1956), who later defined it as "apparent imitation resulting from directing the animal's attention to a particular object or to a particular part of the environment" (Thorpe 1963, 134). It has been suggested that local enhancement be regarded as a subset of stimulus enhancement (Galef 1988; Heyes 1994), where the stimulus in question is a particular location. However, stimulus enhancement, as defined in the previous section, requires that the animals exhibit single-stimulus learning, whereas here we stress that local enhancement can occur without learning.[1] Such nonlearning processes are of interest to social learning researchers because they can indirectly lead to social learning. For example, a demonstrator may attract individuals to its location merely as a consequence of a tendency for individuals to aggregate, and the effect may not last once the demonstrator has moved. This would not constitute stimulus enhancement by Heyes' (1994) definition. Rather than invent a new term, we suggest that local enhancement be retained to refer to all such location effects, irrespective of whether they result in learning. More importantly, it is a relevant distinction to ascertain whether the subject is acquiring information about the demonstrated object or locality, and we have previously argued that evidence for the former occurring in nature is far more extensive than for the latter (Hoppitt and Laland 2008a). An increase in interaction with identical stimuli at different locations, following observation of a demonstrator interacting with a particular stimulus at a particular location, may be surprisingly rare in nature.

Hence we define local enhancement as follows:

> Local enhancement occurs when, after or during a demonstrator's presence, or interaction with objects at a particular location, an observer is more likely to visit or interact with objects at that location.

Our definition differs from Thorpe's in that no reference is made to "attention," a faculty that can be difficult to measure reliably (Heyes 1994), and does not require that the interaction result in social transmission. By our definition, a number of specific processes could result in local enhancement, including (1) stimulus enhancement of a specific location, (2) an aggregation effect (e.g., Waite 1981), (3) a tendency for individuals to move around as a group (Laland and Williams 1997), and (4) the products of a demonstrator's behavior (e.g., scent marks and feces)

---

[1] Some researchers (e.g., Galef 1988) have suggested the term "local enhancement" be used to refer to the entire process of attraction to a location and the subsequent learning, if and only if learning happens to converge on the same trait exhibited by the demonstrator. We suspect this difference in interpretation reflects different readings of Thorpe's (slightly ambiguous) phrase "apparent imitation." However, a close reading of Thorpe's text makes it clear that he did not intend "local enhancement" to intrinsically involve learning. He (1963, 134) writes: "A special form of social facilitation, still possible where only a very slight degree of social behaviour exists, is that known as local enhancement." He contrasts social facilitation with social learning (133): "Imitation as a whole, then, might be described as 'social learning,' and social facilitation can be described as 'contagious behaviour,' where the performance of a more or less instinctive pattern of behaviour by one will tend to act as a releaser for the same behaviour in another or in others, and so initiate the same lines of action in the whole group." This implies that the "apparent imitation" to which Thorpe referred is the act of doing something at the same location as the demonstrator, rather than subsequently learning the same trait that attracted it to the location. The fact that many researchers read Thorpe as we do is illustrated by the common tendency to speak of a case of social transmission as resulting from "local enhancement *followed by* asocial, or trial-and-error, learning," which would obviously not make sense if the term "local enhancement" intrinsically included learning.

attracting other individuals to a location (e.g., Telle 1966). We suggest that the definition of local enhancement be kept broad; once local enhancement has been detected, the precise mode of operation can be investigated.

In principle, any of these routes can result in both social learning and transmission. If local enhancement occurs through aggregation or following effects (2 and 3, above), then social transmission of a behavioral trait can occur if acquiring the trait causes a demonstrator to visit locations at which others can acquire the trait. Alternatively, or in addition, performance of the trait might act to directly attract observers to a specific location at which the trait is performed, causing them to acquire and/or perform it at a higher rate. Products of trait performance could act in the same way. Therefore, the scale at which local enhancement occurs might plausibly dictate how strong the effect of social transmission is, and whether the relative rate of acquisition of different traits is affected. If two variant traits are performed at very similar locations, local enhancement will only cause an observer to be more likely to acquire the same variant if its effects are very spatially specific.

We also note the possibility that uninformed or naïve "demonstrator" individuals may inadvertently act in such a way as to increase the rate at which other "observer" individuals are exposed to, or interact with resources in, a particular locality, and that such social interaction may facilitate learning in the observers. Local enhancement can occur whether or not the demonstrator has previously learned about the locality. For example, Atton et al. (2012) found that by shoaling together, sticklebacks increased the rate at which other shoal members were exposed to a food site, because they all discovered the food location at a similar time; as a result, they learned individually that the location constitutes a reliable source of food. This is an instance of social learning via local enhancement but not of social transmission, because the latter requires that the demonstrators have previously acquired knowledge about the location.

Thus, even if local enhancement did not result in social transmission, it could still cause groups of animals to acquire similar behavioral repertoires. If individuals move around together and different resources are patchily distributed, individuals will encounter the same features of the environment and then acquire similar behavioral repertoires, without social transmission occurring (Van der Post and Hogeweg 2006, 2008). In reality, we suspect that this mechanism operates rarely without an additional social transmission component. If individuals tend to return to locations at which a trait is performed and naïve individuals follow them, the necessary causal link for social transmission is almost inevitable. Galef (1976) and Sterelny (2009) have noted that these types of effects are likely to be highly important in development for species that have an extended period of parental care. The young of such species often accompany their parents, and thus the environment in which they learn is structured by their parent's routines. More generally, several researchers (Galef 1976; Laland et al. 2000; Sterelny 2009; Fragaszy, 2012b) note that parents structure the environment in which their offspring learn; this is a case of developmental "niche construction" (Odling-Smee et al. 2003). As we will see, "opportunity providing" is also likely to play a part in this process (section 4.1.9).

4.1.3 Observational conditioning

Observational conditioning traditionally refers to Pavlovian conditioning in which an unconditioned response (*UR*) to a stimulus on the part of the demonstrator acts as an unconditioned stimulus (*US*), eliciting a matching response on the part of the observer. Consequently, the stimulus to which the demonstrator is responding becomes a conditioned stimulus (*CS*) for the observer, to which it will later respond in the same way (Cook et al. 1985). For example, a rhesus monkey (*Macaca mulatta*) that has had no contact with snakes will not show fear if presented with a snake. However, if it is exposed to another individual reacting fearfully to a snake, it will also display fear, and show fear when later presented with the snake.

Heyes (1994) criticizes the original definition of observational conditioning on the grounds that is too restrictive, and suggests that the term should refer to any case where observation of a demonstrator's behavior increases the probability that an observer will be exposed to a stimulus-stimulus (*S-S*) relationship. Once again, we adopt Heyes' (1994) definition:

> Observational conditioning is a subset of stimulus-stimulus learning in which observation of a demonstrator exposes the observer to a relationship between stimuli at t_1, and exposure to this relationship effects a change in the observer detected, in any behaviour, at t_2.

Our use of the term "observation" is not intended to restrict this process to visual cues. The type of *S-S* learning involved depends, as with asocial learning, on the relationship between the two stimuli (*S1* and *S2*), which can be positively correlated (*S1* predicts *S2*) and result in excitatory conditioning; alternatively, it can be negatively correlated (*S1* predicts absence of *S2*) and result in inhibitory conditioning. The nature of the second stimulus also affects the observed behavioral change, whether it is appetitive (attractive to the animal) or aversive (Heyes 1994). In the former case, observational conditioning could be excitatory-appetitive, with the first stimulus becoming more likely to evoke a response, or inhibitory-appetitive, with the first stimulus becoming less likely to evoke a response.

Therefore, observational conditioning will result in social transmission if (*a*) performance of the trait causes the observer to be exposed to the excitatory-appetitive relationship between *S1* and *S2*, and (*b*) the response to *S1* is the trait in question, or if responding to *S1* *otherwise* causes the observer to acquire the trait at a higher rate. For example, an observer might observe a demonstrator lift the covers on blue cups to find food, resulting in excitatory-appetitive observational conditioning and causing the observer subsequently to respond to the blue cups. The same observer might also observe a demonstrator lifts the covers on red cups for no reward, resulting in inhibitory-appetitive observational conditioning and causing the observer not to respond to red cups. This offers a plausible process by which animals can learn socially where to find food (Galef and Giraldeau 2001; Zentall 2001).

Perhaps the best studied of any social learning process is the social enhancement of food preferences in rats (e.g., Galef 1996). Galef and Wigmore (1983) found that an observer rat (*Rattus norvegicus*) would prefer to eat a novel diet after having been exposed to a demonstrator conspecific that had recently eaten that food. The authors found that physical contact between the demonstrator and observer is not necessary for social enhancement to occur, but instead depends on odor cues on the demonstrator's breath.

In our previous review (Hoppitt and Laland 2008a), we justified social enhancement of food preferences as a category of its own right on the grounds that it appears to be a specialized case that is taxonomically widespread, at least within mammals. However, even if it did involve a specialized learning mechanism (this is now doubtful; see section 4.2.3), it is clear that it fits the definition of observational conditioning given above, and so we include it here as a subset. Nonetheless, social enhancement of food preferences in rats provides an exemplary case study for how the mechanisms of social transmission can be isolated, and we provide a discussion in section 4.2.3. It is also one of the few social learning processes for which the underlying biological processes are comparatively well explored (see chapter 3).

4.1.4 Response facilitation

Byrne (1994, 237) introduced the term "response facilitation" to refer to instances when

> the presence of a demonstrator performing an act (often resulting in reward) increases the probability of an animal which sees it doing the same.

It is generally assumed that such a process would have a transient effect on behavior, because the effect may be the product of priming of brain records corresponding to an action (Byrne 1994). Such priming could be accounted for by residual neural activity remaining for a short time after observation of the demonstrator's actions. Alternatively, transience may be due to the fact that as other, mutually exclusive, actions are observed and the probability of their performance by the observer increases, the probability of the first action being performed by the observer will necessarily reduce.

Several related terms must be considered. "Contagion" was used by Thorpe (1963) to refer to the unconditioned release of an instinctive behavior in one animal triggered by the performance of the same behavior in another animal. Contagion can therefore be seen as a subset of response facilitation that requires no prior learning. "Social facilitation," has been used in a similar manner. For example, Visalberghi and Adessi (2000) define social facilitation as "the increased probability of performing a class of behaviours in the presence of a conspecific performing the same class of behaviours" (69). One further apparent difference between response facilitation and contagion or social facilitation (in the sense of Visalberghi and Adessi 2000) is that the latter two are usually assumed to apply to whole classes of behavior, such as feeding or running (Zentall 1996), whereas response facilitation may affect specific individual actions (Byrne 1994, 1999). However, response facilitation could be viewed as potentially operating at a number of different levels of "resolution," much like imitation (see below).

Furthermore, response facilitation provides another plausible mechanism by which social transmission of a trait might indirectly occur. By synchronizing behavior, response facilitation might act to inform animals when and where to perform certain actions. For example, a frugivore might learn that the fruit of a particular tree is good to eat if it is present in the tree with other individuals who are eating and it is primed to eat as a consequence. In this manner, response facilitation could result in learning that is functionally equivalent to contextual imitation (see below), by a similar process to Suboski's (1990) releaser-induced recognition learning.

4.1.5 Social facilitation

In addition to being regarded a synonym of "contagion," the term "social facilitation" has been used in cases where the mere presence of another animal affects behavior (Zajonc 1965). While Zajonc described this effect as mediated by an increase in arousal, we suggest that some types of behavior may be facilitated whereas others may be unaffected or inhibited by the presence of others. This leads us to the following definition:

> Social facilitation occurs when the mere presence of a demonstrator affects the observer's behavior.

This process could potentially lead indirectly to social learning if, for example, an individual is more likely to exhibit exploratory behavior in the presence of other individuals, perhaps through a reduction in neophobia, allowing the individual to learn about novel objects. Social facilitation could, in theory, play a role in social transmission in natural situations, perhaps in concert with local enhancement, because performing a location-specific trait requires an individual to visit the location at which that trait is performed. The demonstrator's mere presence at that location could then make other individuals more likely to acquire and/ or perform the trait. However, there is no general reason to think that its mere presence would increase the rate of performance of the trait relative to other traits that could be performed at that location. The main interest in social facilitation is as a process that must be ruled out as an explanation for apparent social transmission; to be sure that social transmission is occurring, one must rule out the explanation that the mere presence of a conspecific is responsible for the increased rate of acquisition or performance of a trait (see section 4.2).

4.1.6 Imitation

Imitation is among the most contentious of social learning processes, with seemingly little consensus on exactly what "imitation," or "true imitation," is, and how it should be defined. There seem to be three main issues underlying differences of opinion (Byrne 2002). First, some researchers (e.g., Tomasello 1996) have suggested that imitation should be seen as evidence of a capacity for intentionality, and hence researchers should only consider cases of social learning to be imitation if there is evidence of intentional copying. However, intentionality is not directly observable, and although it may be inferred, it is not easy to study experimentally (Zentall 1996). In addition, we see no reason why an animal should not be able to learn about behavior through observation without a capacity for intentionality (Byrne 1999), nor why an animal that possesses a capacity for intentionality (i.e., humans) should necessarily apply it every time that they imitate.

Second, there is disagreement over how accurately the observer must copy the demonstrator to constitute a case of imitation (Nehaniv and Dautenhahn 2002). There are a number of factors that might determine how accurate an imitative match will be. One reason why the match might not always be perfect is that imitation might be used in concert with other factors in skill learning, such as unlearned predispositions and asocial learning (Byrne 2002). Nonetheless, if researchers can reliably detect a level of correspondence between the actions of the demonstrator and observer that is lacking in suitable control subjects, this could be taken as evidence of an imitative influence on the learning of behavior.

Third, some researchers maintain that only novel acts, not already in the observer's repertoire, can be imitated (Byrne and Tomasello 1995). This requirement stems from the concern that if actions are already in the observer's repertoire, then they might be subject to contagion, which could be mistaken for a general imitative capacity. While this may be a concern when assessing anecdotal reports of imitation from the field, in the laboratory other social learning processes can be controlled for experimentally (see section 4.2). At the same time, the processes by which an individual acquires topographically novel behavior for its repertoire are of great interest to anyone investigating how an animal adapts to its environment. The problem of novelty is addressed neatly if we adopt Byrne's (2002) definitions of two types of imitation, "contextual imitation" and "production imitation," a distinction stemming from one made in the vocal learning literature (Janik and Slater 2000). A related argument is that only novel actions can be used to study imitative learning, since one cannot know whether familiar actions have been imitated. Whether this is correct or not, we see no reason why the criteria necessary to isolate a mechanism should be regarded as a part of the definition of that mechanism. Equally, novelty can be regarded as central to definitions of processes irrespective of whether that makes it easy or hard to detect those processes in practice. For example, we argue below that novelty is an important criterion for production imitation.

4.1.6.1 Contextual Imitation

Byrne (2002, 82) defines contextual imitation as "learning [by observation] to employ an action already in the repertoire in different circumstances, not learning its form." Here we adopt a variant definition that does not exclude the possibility that an animal might learn a novel behavior by production imitation while simultaneously learning the context in which to use it:

> Contextual imitation occurs when, directly through observing a demonstrator perform an action in a specific context, an observer becomes more likely to perform that action in the same context.

The effect of contextual imitation is "direct," rather than an indirect result of another social learning process, such as stimulus enhancement followed by trial and error learning. For example, an observer could learn through contextual imitation to perform an action in a specific location, at a specific time of day, or in response to a specific stimulus, making it a potentially powerful means of learning.

Contextual imitation can be regarded as a case of stimulus-response (S-R) learning. Although somewhat discredited for asocial cases of instrumental learning, S-R learning is gaining credibility in the field of social learning (e.g., McGregor et al. 2006; Saggerson et al. 2005). Our definition of contextual imitation differs from Heyes' (1994), who defined imitation as a subset of response–reinforcer (R-S) learning. Here we recommend that researchers be open to the possibility that observers might imitate "blindly," in the absence of demonstrator reward (Shettleworth 1998). There is empirical support that "blind" imitation can occur (McGregor et al. 2006; Moore 1992, 1996), and it has rarely been shown that the observer must see the demonstrator being rewarded for contextual imitation to occur (Akins and Zentall [1998] is an exception). This mechanism is consistent with recent neural network models of contextual imitation, based on Hebbian and Stentian learning principles (Hoppitt 2005; Laland and Bateson 2001).

The conditions for contextual imitation to occur are similar to those for observational conditioning, in so far as the observer is "presented" with two stimuli: the contextual stimulus and the experience of observing a conspecific perform an action. However, observational conditioning involves the formation of an S-S association, whereas contextual imitation involves the formation of an S-R association between the contextual stimulus and the observed response. Contextual imitation requires some kind of neural connection between the experience of observing an action and the performance of that action. Contextual imitation differs from response facilitation in that the observer's response is context specific.

4.1.6.2 Production Imitation

For Byrne (2002), "production imitation" refers to imitation where the form of a novel action is learned through observation. This is different from R-S learning (section 4.1.7) where an individual learns about the consequences of a response. Production imitation involves learning by observation to perform the response itself—for example, learning how to hit a golf ball.

To a degree, the requirement of novelty falls foul of our endeavor to define social learning processes according to observable criteria, because one can rarely be sure which actions are already in an individual's repertoire. Even if a complete history of the individual's behavior is available, it is difficult to exclude the possibility that an individual has an unlearned predisposition for an action they have never performed. However, the problem seems unavoidable if researchers are to investigate the addition of novel actions to an individual's repertoire, and in experimental contexts, the problem can be finessed by presenting animals with actions that they are unlikely to have previously performed.

Novelty presents a further, paradoxical problem. Every action any individual performs is likely to be novel to some degree as it adapts its motor output to the exact environmental conditions and the orientations of the objects with which it is interacting (Schmidt 1991). For example, a tennis player never plays exactly the same shot twice as they adapt their motor output to allow for differences in height, speed, and spin of the ball; their own position in the court; the opponent's position; and so forth. This, of course, does not mean the tennis player has to learn how to play every shot from scratch. Instead, a player has a limited repertoire of different shots (e.g., slice, lob) which they have learned to adjust to the circumstances. Conversely, no action is ever completely novel, because it is always built up from component actions that have been performed before.

A solution would be to conceptualize production imitation to involve the learning of a novel sequence of simpler actions (termed action units by Heyes and Ray 2000). Such actions need not themselves be novel because all novel actions can be seen as a sequence of familiar actions at some level (Byrne 2002). A number of models and reviews have observed that novel skill learning appears to occur through sequence learning (e.g., Adams 1984; Schmidt 1991). Recent models of imitation reflect this, hypothesizing that production imitation occurs through the learning of novel sequences of actions or action units (consider the string parsing model by Byrne [1999] and associative sequence learning model by Heyes and Ray [2000]) (see fig. 4.1 for the latter). In theory, novel actions could be learned as a combination of existing actions because an individual could learn to combine two action-units simultaneously rather than sequentially (e.g., Moore 1992). We propose the following definition to account for these recent conceptual advances:

> Production imitation occurs when, after observing a demonstrator perform a novel action, or a novel sequence or combination of actions, none of which are in its own repertoire, an observer then becomes more likely to perform that same action or sequence of actions.

A number of qualifying statements are required. To be regarded as imitation, the novel action must be acquired by the observer directly through observation, rather than through the indirect effects of other social processes that are followed by asocial learning. The requirement that an observer should become "more likely" to perform the novel action or sequence is necessary, because there is always a chance that animals will spontaneously "invent" an action without first observing it.

If production imitation does occur through the learning of novel sequences of action units, this has important implications for how accurately the observer imitates the demonstrator. The closeness of the match will depend critically on the "size" of the action units that make up the sequence being copied. It follows that

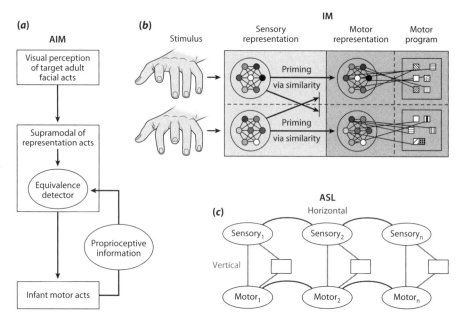

Figure 4.1. Three models addressing the correspondence problem of imitation (based on fig. 1 in Brass and Heyes 2005). (*a*) Active intermodal mapping (AIM) theory assumes that visually perceived acts are actively mapped onto motor output via a supramodal representation system. (*b*) Ideomotor theory (IM) assumes that actions are represented in the form of the sensory feedback they produce. A perceived action leads to priming of the corresponding motor representation because of the representational overlap of the sensory and motor representation. IM assumes that the visuo-motor translation required for imitation results from the general organization of motor control. (*c*) The associative sequence learning (ASL) model assumes that visual (sensory) representations of action become linked to motor representations (encoding somatosensory information and motor commands) through Hebbian learning. In environments where the same action is simultaneously seen and executed, links are formed between visual and motor representations of the same action. These contiguity-based "matching vertical associations" mediate priming of motor representations by action observation.

imitation can occur at a number of levels. For illustration, imagine an observer learning by production imitation how to pick fruit from a tree after observing a demonstrator grasp a fruit and then pull back to detach it from its stem. The novel action could be copied as a sequence of two simpler actions: (1) grasp fruit, and (2) pull back; or it could be copied as a longer sequence of smaller actions, where each movement around each finger joint is copied to create the grasping movement, followed by the movement around the elbow and shoulder joint to create the pulling back movement. In the first case, the imitated fruit-picking action might vary considerably from that of the demonstrator if the observer's "grasp" and "pull back" actions happen to differ from the demonstrator's versions of the same action. Yet if every movement of each digit and the arm is copied, then the imitated version will closely match that of the demonstrator.

At this stage, researchers cannot be certain that production imitation does operate through sequence learning. However, following Heyes and Ray (2000) and

Byrne (1999), we suggest that it is likely to prove worthwhile to conceptualize the process as the acquisition of novel sequences, because this allows for the possibility that imitation might not occur at the exact-movement level and allows researchers to quantify the probability that a given sequence will arise by chance (Whiten and Custance 1996).

Byrne and Russon (1998) claim that an imitator might not directly copy the actions of a demonstrator, but through repeated observation, instead extract and copy the underlying organization of the demonstrator's behavior. They term this process program-level imitation, as oppose to action-level imitation. Program-level imitation is not the same as merely imitating at a low level of resolution because it requires inferences concerning which aspects of a sequence are important, which parts of a sequence form different subroutines, and the rules by which those subroutines are applied (Byrne 2002). In other words, the observer can extract the program for behavior from a demonstrator, perhaps by detecting statistical regularities in demonstrators' behavior, a method that Byrne (2003) calls "behaviour parsing."

The term "over-imitation" has been used for production imitation where the observer faithfully copies all aspects of the demonstrator's behavior, regardless of whether they are necessary for achieving the assumed goal of the behavior pattern. This is contrasted with rational imitation, where irrelevant aspects of the behavior are not copied, or are replaced with alternatives that are easier to perform (e.g., Gergely et al. 2002; Range et al. 2007). An in-depth discussion of such distinctions is beyond the scope of this book, and we refer the reader to recent reviews of the imitation literature by Huber et al. (2009) and Whiten at el. (2009). However, in passing, we urge caution in assuming that imitation necessarily requires an information-gaining function, because there are good reasons for thinking that it may sometimes possess a social function, such as strengthening social relations (Carpenter 2006). Where imitation has a social function, the term "over-imitation" is inherently misleading, since aspects of the task that are deemed functionally irrelevant may in fact be socially relevant. That is, there would be no sense in which individuals are imitating too much; rather they may be imitating the optimal amount.[2] An alternative explanation is also provided by work on observational causal learning (see box 4.1).

Finally, we acknowledge that there are additional ways in which an observer might be said to have learned by imitation, without acquiring a novel sequence or learning to perform an action in a novel context. Whiten and Custance (1996) note that other components of imitatable form include the shape, extent, speed, laterality, and orientation of the demonstrator's movements. Nonetheless, we feel the terms contextual and production imitation delineate the phenomena of primary interest to social learning researchers.

---

[2] Equally, the term "over-imitation" would be misleading if it were interpreted as excessive imitation that reduced fitness. We anticipate that in natural circumstances the heavy reliance on imitation observed in humans is highly adaptive (Boyd and Richerson 1985), and it is only the artificiality of the experimental context that creates the appearance of unnecessary imitation.

Box 4.1

Imitation of causal action sequences in children

Recent research into child development has found evidence that human children learn about networks of causal relationships in the world by using observed statistical patterns (Gopnik and Tenenbaum 2007)[1]. This also appears to be the case for social learning, where the causal relationships between actions and outcomes are inferred from observation of a demonstrator. For example, Buchsbaum et al. (2011) ran an experiment on 81 preschool children, in which the children observed repeated demonstrations of different action sequences directed toward a toy. Some action sequences were followed by a positive outcome, a musical effect. Demonstrations were designed such that the pattern of action sequences and outcomes was consistent with a different causal structure in each of three experimental conditions. The children were able to imitate the actions necessary to produce the music, and in some conditions tended not to imitate components of the action sequence that were not causally necessary. This fit closely with the predictions of a Bayesian model that inferred causal relationships using the statistical information the children were hypothesized to be using.

How does observational causal learning fit into our scheme? In one sense, such learning is a more sophisticated version of observational *R-S* learning, allowing children to learn the outcomes of different actions within a causal network. In another sense, it is a type of production imitation, because novel actions sequences are acquired on the basis of whether they achieve desired outcomes. One possible advance for the study of social learning mechanisms might be an investigation of whether observational causal learning also occurs in nonhuman species.

A second experiment conducted by Buchsbaum et al. (2011) also sheds light on the phenomenon of "over-imitation" and its relationship to teaching. In this experiment, adult demonstrators were presented as informed tutors who were attempting to teach the children how the toy works (as opposed to naïve demonstrators who did not know how the toy worked). Buchsbaum et al. hypothesized that under such circumstances children might expect the teacher to choose actions with the goal of helping them infer the correct causal hypothesis. These authors adapted their Bayesian model to include this prior information about the goals of the tutor, and found that the new model tended to select demonstrated action sequences with actions that were inferred to be causally irrelevant by the first model. The children's behavior also showed this pattern in the second experiment, where they were more likely to include actions that seemed causally irrelevant in the first experiment (i.e., "over-imitation"). This suggests that, far from imitating blindly, children attend to statistical evidence in deciding which actions to imitate.

Buchsbaum et al. (2011) suggest that their Bayesian perspective helps reconcile the seemingly conflicting findings of "rational imitation" (e.g., Gergely et al. 2002) and "overimitation" (e.g., Lyons et al. 2007). Indeed, the studies suggest a "rational" mechanism for the phenomenon of "over-imitation." It is a natural outcome of the statistical learning that children undertake to reproduce parts of a causal sequence that are not actually demonstrably necessary for the effect, given their priors and the statistical data available.

[1] More specifically, children appear to test different causal theories tacitly using statistical information in a very similar way to the method we suggest for studying repertoire-based data on social learning (section 6.4).

4.1.7 Observational *R-S* learning

While our definitions of contextual and production imitation do not specify that the observer must witness the demonstrator being rewarded in order to imitate, it could conceivably be adaptively advantageous for an observer to be sensitive to demonstrator reinforcement; this would enable the observer to bias copying towards copying of advantageous actions. In addition, if an observer could learn *R-S* contingencies through observation, they could potentially learn what not to do in a specific context, as a result of watching other animals make mistakes (Heyes 1994; Want and Harris 1998). Such learning does not fulfill the definition for contextual imitation, but could perhaps be regarded as a case of observational *R-S* learning that overlaps with contextual imitation in cases where it results in social transmission. We avoid using the term observational learning favored by Heyes (1994), because it is a term that has been used in a number of different ways, but we adopt her definition for the phenomenon:

> Observational *R-S* learning is a "subset of response-reinforcer learning (R-S) in which observation of a demonstrator exposes the observer to a relationship between a response and a reinforcer at t_1, and exposure to this relationship effects a change in the observer detected, in any behaviour, at t_2." (Heyes 1994, 225)

Observational *R-S* learning could potentially be combined with production imitation to ensure that an observer disproportionately acquires novel action sequences that it sees being rewarded.

4.1.8 Emulation

The term emulation has acquired a number of different meanings since it was first introduced to animal learning (Tomasello 1990; see Byrne 2002 for a discussion). In general terms, it refers to instances where the observer copies the results of a demonstrator's behavior in the environment, rather than the demonstrator's actions themselves. There are a number of ways in which this could occur. Goal emulation (Whiten and Ham 1992) was initially used to refer to cases where the observer not only understands that the demonstrator's behavior has certain consequences but also recognizes that it can achieve the same goal in a different way. The process could be interpreted as a case of low-resolution imitation (Whiten and Ham 1992), with the observer substituting their own functionally equivalent action. Observational conditioning could result in a similar phenomenon. For example, if an observer sees a demonstrator interacting with a manipulandum for a food reward, then it might form an association between the manipulandum and the food, causing it to find its own way of interacting with the manipulandum.

There are other ways in which an observer could learn to copy the results of a demonstrator's behavior. For one, an observer could try and recreate the movements of objects with which the demonstrator interacted; this is termed object movement re-enactment (Custance et al. 1999). Alternatively, an observer could try and recreate the final state resulting from a demonstrator's behavior; Custance et al. (1999) call this final state recreation, though in theory, an observer could recreate a number of intermediate states to emulate the demonstrator's actions. According to our definition:

> Emulation occurs when, after observing a demonstrator interacting with objects in its environment, an observer becomes more likely to perform any actions that bring about a similar effect on those objects.

R-S learning could be implicated in emulation in the same way as in imitation, if an observer disproportionately emulates demonstrators it observes being rewarded. However, there is also the possibility that an observer will "blindly" emulate regardless of its consequences.

It has been noted that if one considers the results of an individual's behavior, such as the movement of objects being manipulated, as an extension of that behavior, there is little distinction between imitation and emulation (Whiten and Ham 1992). In this light, it seems arbitrary to limit "behavior" to the movement of body parts; if limited in this way, the line between imitation and emulation becomes somewhat fuzzy, especially given that an observer could potentially combine aspects of a demonstrator's body movements and aspects of the movements of objects in its environment (or other results). This is especially likely if imitation occurs at the program level (see section 4.1.6.2), when the organization of behavior may depend on monitoring the state of objects being manipulated (Byrne 2002). Consequently, it might be more fruitful to consider body movements as just one type of action- dependent cue that an observer could use in order to imitate the demonstrator (Call and Carpenter 2002).

4.1.9 Opportunity providing

In their review of the evidence for teaching in nonhuman animals (see section 4.4), Caro and Hauser (1992) suggested two mechanisms by which teaching could occur. One of these was "opportunity teaching," where the tutor puts the pupil in a "situation conducive to acquiring a new skill or knowledge." In a broad sense, this could refer to a number of more specific processes—for example, if local enhancement attracts an observer to a location at which a trait can be acquired. However, most of Caro and Hauser's examples of opportunity teaching refer to cases where the products of the tutor's behavior offer a new opportunity for the pupil to practice their skills, through operant learning (i.e., learning the connection between aspects of their behavior and reward). For example, domestic cats bring live prey back to their kittens, allowing them to acquire predatory skills earlier in life (Caro 1980a, 1980b). Caro and Hauser (1992) were primarily concerned with candidate cases of animal teaching, where the evolved function of the tutor's behavior is to provide the pupil with the opportunity to learn, and hence qualify as instances of animal teaching. However, there is no reason why a demonstrator might not "inadvertently" provide such opportunities for other individuals (Hoppitt et al. 2008). The mechanism by which black rats (*Rattus*

rattus) in Israel learn how to strip pinecones for seeds may be an example (Terkel 1996). Although pups raised by a "stripper" mother or foster mother learn to strip cones, naïve adult rats do not. However, adults can learn to strip if provided consecutively with pinecones with decreasing numbers of stripped scales (Terkel 1996). It seems that rat pups learn to open cones through "stealing" partially opened cones from their mother, providing them with the opportunity to learn the task backward, like the above adult rats.

Consequently, we add the mechanism of "opportunity providing" to our classification of social learning processes:

> Opportunity providing occurs when "the products of the behaviour of the demonstrator provide the observer with an opportunity to engage in operant learning that would otherwise be unlikely to arise, for example by providing an easier, less dangerous or more accessible version of the task" (Hoppitt et al. 2008, 490).

This process could result in social transmission if the behavioral trait responsible for producing the relevant products is the same as that which the "observers" acquire through operant learning and interaction with those products. The term "opportunity teaching" is then reserved for cases where the tutor's behavior functions to promote learning in the pupil (see section 4.4).

4.1.10 (Inadvertent) coaching

Caro and Hauser (1992) proposed a second mechanism by which animal teaching could occur: "coaching," where a tutor directly "alters the behaviour of [a pupil] by encouragement or punishment." The maternal display of the domestic fowl (*Gallus gallus domesticus*) is a candidate case of "coaching." The display includes food calls and pecks directed at food and at the ground, which act to attract chicks toward palatable, and away from unpalatable, food (Nicol and Pope 1996). A similar example is provided in a study by Clarke (2010) on ptarmigans. However, there are also cases that we label "inadvertent coaching" (Hoppitt et al. 2008), where the demonstrator's response to the observer's behavior either punishes or reinforces that behavior, even if the demonstrator's response is probably not an adaptation for that purpose. For example, take the aforementioned example of brown-headed cowbirds (*Molothrus ater*). When females hear a preferred song from a male cowbird, they respond with a "wing-stroking" display, which encourages performance of that song in the male (West and King 1988). Wing stroking has an alternative function related to courtship, and hence may not be an adaptation designed specifically and solely to promote learning in the males. If this were the case, then it would not qualify as teaching; rather, female cowbirds inadvertently shape male behavior. This leads us to the following definitions:

> Coaching occurs when "the response of the tutor to the behaviour of the pupil functions to encourage or discourage that behaviour" (Hoppitt et al. 2008, 490, modified from Caro and Hauser 1992).

Inadvertent coaching occurs when "the response of a demonstrator to the behaviour of the observer inadvertently acts to encourage or discourage that behaviour" (Hoppitt et al. 2008, 490).

Inadvertent coaching is likely to be a highly important mechanism for any social species learning to interact with conspecifics. In addition, Sterelny (2009, 2012) suggests that this process ("socially structured or guided trial-and-error learning") is likely to be an important process for young learning from their parents. Here, parents monitor the behavior of their young, and the young are responsive to their parents' emotional state, treating such cues as stand-ins for the world's actual (and costly) error signals.[3] However, cases in which inadvertent coaching results in social transmission might be limited. This is perhaps most likely to occur if acquisition of a trait by one individual causes it to act in a way that encourages the development of the same trait in others. Possible examples of this might be the social traditions in chimpanzees (Whiten et al. 2001), such as handclasp grooming. However, indirect chains of transmission might also occur by inadvertent coaching, such as the example of social transmission of female courtship preferences and male song types in cowbirds (see section 2.2).

4.2 Distinguishing Social Learning Mechanisms

We focus here on experimental laboratory techniques to distinguish different social learning processes. In chapter 8 we discuss briefly how equivalent field experiments might be used for the same purpose.

4.2.1 Stimulus enhancement

To show that stimulus enhancement is occurring, researchers need to establish that a demonstrator's actions toward a particular stimulus caused single-stimulus learning on the part of the observer. To show that stimulus enhancement results in social transmission of trait T or performance of trait T, it must be shown that (*a*) the observer's response rate to the stimulus increases, and (*b*) responding to the stimulus increases the rate of acquisition or performance of trait T, respectively. To do so, researchers must rule out forms of local enhancement that are not a result of stimulus enhancement, such as a tendency to aggregate, or a demonstrator's products attracting observers to a particular location. Local enhancement could be ruled out by stimulus generalization, where the observer is more likely to interact with an object of the same type as the one with which the demonstrator is interacting, rather than the exact same object. Alternatively, researchers could show that the increase in responsiveness persists for more than a short

---

[3] By our definitions, if the cues provided by parents are signals evolved for the purposes of promoting learning in offspring, this would be a case of coaching rather than inadvertent coaching. However, it is possible that such signals evolved with the more immediate function of leading offspring away from danger (see section 4.4).

period after the demonstrator has left the area, and all traces of its products, such as feces or odors, have been removed.

There are many cases of social transmission following direct observation of a demonstrator that are consistent with a stimulus enhancement explanation. An example is provided by J. R. Krebs et al. (1972) who showed that great tits (*Parus major*) were more likely to interact with a particular "type" of place for food, after having seen a conspecific find food in a similar location. In order to demonstrate stimulus enhancement, researchers need to rule out observational conditioning and observational *R-S* learning, by ensuring that the observer does not associate the stimulus or a response with an appetitive stimulus. In cases where the demonstrator is not rewarded, observational conditioning is usually ruled out. For example, Warden and Jackson's (1935) study of rhesus monkeys used a duplicate cage method, where the observer and demonstrator were provided with identical sets of objects. They showed that after the demonstrator had been allowed to select an object (without reward), the observer tended to choose one of the same type. While such cases could be examples of stimulus enhancement, they could also be instances of contextual imitation, where the observer learns to direct a particular action at the object (e.g., "grasp" or "attack").

Some cases of "mate-choice copying" (Dugatkin 1996; White 2004) provide among the clearest evidence for stimulus enhancement. For example, Galef and White (1998) found that female quail (*Coturnix japonica*) preferred to associate with males they have seen in proximity to another female (see also White and Galef 1999). This preference is not location specific (White and Galef 1999) and generalizes to other males with similar traits (White and Galef 2000), ruling out local enhancement. Observational conditioning seems unlikely, because there are no clear *S-S* associations that could have been formed. Contextual imitation is ruled out because the effect depends only on observed female-male proximity and not on the demonstrator's behavior, because a stuffed female demonstrator will achieve the same effect (Akins et al. 2002). However, if the "proximity condition" is not as effective as observation of trait performance, this does not rule out stimulus enhancement, since the performance of the trait might be more effective at exposing the observer to the stimulus than a demonstrator not performing the trait.

4.2.2 Local enhancement

It has long been appreciated that many species of animal aggregate and move as a group, so there is a large body of evidence suggesting that local enhancement occurs on a relatively large scale (Hoppitt and Laland 2008a). To show that such large-scale local enhancement can result in social transmission of a trait, a researcher must show that individuals acquiring trait *T* tend to lead naïve individuals to locations at which trait *T* is acquired at a higher rate than other locations. To show that large-scale local enhancement can result in social transmission of trait performance, a researcher must show that performance of a trait attracts informed individuals to the location at which trait *T* is performed at a higher rate than other locations.

In practice, local enhancement on a large scale might result in a very small social transmission effect, if there are a number (N) of traits that can be performed at the locations in question. In such cases large-scale local enhancement cannot account for the differences in the relative rate of acquisition or performance of the N traits. Furthermore, traditional demonstrator-observer experiments will not detect such effects since the observer is constrained in close proximity to the task apparatus. Consequently, the more germane question is not whether local enhancement occurs, but on what scale—for example, whether individuals are attracted to a specific part of an object, such as a lever, that has to be pressed for food. There are a number of laboratory experiments showing social transmission, following observation, of a foraging task solution consistent with small-scale local enhancement (e.g., Heyes and Saggerson 2002). These cases can usually be explained by observational conditioning, where the observers have formed an *S-S* association between the location and the food, though this explanation can be ruled out if the demonstrator is not rewarded for their actions (e.g., Fritz et al. 2000).

Leadbeater and Chittka (2007) provide an invertebrate example of local enhancement that cannot be accounted for by observational conditioning. In their experiment, bumblebees (*Bombus terrestris*) were presented with an arena containing "artificial flowers" of two types, blue unrewarding flowers and yellow rewarding flowers. Bees learned to reverse an initial unlearned preference for blue flowers, and did so faster in the presence of informed demonstrators who had learned the discrimination. Leadbeater and Chittka found that artificial model bees placed on yellow flowers were sufficient to increase the speed at which the preference was learned. A model bee cannot engage in feeding, so cannot trigger feeding through response facilitation and also cannot reveal a reward to the observer, thereby ruling out observational conditioning. Leadbeater and Chittka also found that dead bees would attract naïve foragers to specific yellow flowers, and found no evidence that the effect generalized to unoccupied yellow flowers, suggesting that local enhancement is due to conspecific attraction rather than stimulus enhancement. In this case, it is clear that local enhancement acts to attract naïve foragers to specific yellow flowers, causing them to learn the association with reward earlier. As far as we are aware, this is the only case where visual local enhancement directly involving a conspecific demonstrator has been shown to operate on such a small scale.

There is more evidence that individuals can be attracted to a highly specific location by a demonstrator's products, even in the absence of a demonstrator. For example, many species of ant are well known to follow pheromone trails left by other individuals in the colony in order to find a food source (e.g., Denny et al. 2001). Also, rats are attracted to locations by conspecific scent marks, and will follow scent trails left by other individuals (e.g., Telle 1966). To test whether products are necessary for social transmission to occur, they can be removed before the observer is exposed to the task apparatus. To test whether products are sufficient for social transmission to occur, they can be presented to the subject during the test phase, without a prior observation phase.[4] Campbell and Heyes (2002) provide evidence that odor cues can result in social transmission by local enhancement in rats. They found that when rats were exposed to a lever that a conspecific had recently pushed either up or down while out of sight of the "observer," they preferred to push the lever in the same direction, due to the odor cues deposited on the lever.

4.2.3 Observational conditioning

Observational conditioning occurs when observation of a demonstrator results directly in the observer forming an *S-S* association. There are numerous examples of social transmission that could be accounted for by observational conditioning. These include, for example, any cases when an observer learns where to find food after observing a demonstrator foraging (e.g., J. R. Krebs et al. 1972; McQuoid and Galef 1993). In such a case, the observer could be learning to associate environmental cues with food. However, such examples are also consistent with other processes such as stimulus enhancement.

If observational conditioning were involved in such cases of social transmission, we would expect social transmission only to occur if the demonstrator is rewarded.[5] This also means that if the demonstrator is rewarded for performing trait *A* but not trait *B*, the observer should acquire and/or perform *A* at a higher rate than *B*. Palameta and Lefebvre (1985) describe such an example. They presented pigeons (*Columba livia*) with boxes of grain covered with a sheet of white paper marked with a red spot, and showed that birds who had seen a demonstrator peck through the paper for a reward of grain were more likely to do so than birds that saw a demonstrator peck through the paper for no reward. One explanation is that the birds that had observed the demonstrator being rewarded had formed an association between the red dots and food, and those that had seen the demonstrator not being rewarded had not formed the association. However, an alternative explanation is that the first group had their attention attracted to the cups more effectively than the control group, because the sight of a feeding demonstrator might be a more effective cue for stimulus enhancement than the sight of a nonfeeding demonstrator.[6]

The paradigm cases of observational conditioning are the social transmission of snake fear in monkeys (Cook et al. 1985) and studies of the social transmission

---

[4] Note that this only provides evidence of social transmission by local enhancement if the products do not provide an easier version of the task for the subjects to solve; that would be opportunity providing (see section 4.1.9).

[5] Strictly speaking, we would expect the social transmission effect to be stronger than when the demonstrator is not rewarded, because observational conditioning might normally act in concert with another process that operates even when the demonstrator is not rewarded.

[6] Such ambiguities could be resolved using a devaluation procedure similar to that used by Balleine and Dickinson (1998) to study asocial instrumental learning in rats (see Hoppitt and Laland 2008a for a suggestion on how to modify this procedure). However, this is unlikely to be feasible for researchers who wish to infer which mechanisms are responsible for a given case of social transmission.

of predator recognition in blackbirds (*Turdus merula*) (Curio 1988). In the former, naïve rhesus monkeys acquired snake fear as a result of observing a demonstrator reacting fearfully to a snake. It was noted that the observer monkeys, while seeing the demonstrator react fearfully to the snake, also showed an emotional fear response. This leads Mineka and Cook (1988) to hypothesize that the fear response of the demonstrator acts as a *US* releasing a matching *UR* in the observer. Since the *US* is paired with another stimulus, the snake, the second stimulus becomes a *CS*, which elicits a conditioned response, or a fear response, through classical conditioning.

This example (and other similar cases) fit in with the general definition given for observational conditioning, because the observer has been exposed to an *S-S* contingency; that is, snakes are associated with an observed fearful response. The transmission of matching behavior in such cases is reliant on the fact that the demonstrator's response releases matching behavior in the observer (i.e., they are reliant on response facilitation). Similar cases of observational conditioning have been shown in several species of fish (Brown and Chivers 2006) and New Zealand robins (Maloney and MacLean 1995). Learning is not always reliant on visual cues. Curio (1988) found that playback of mobbing calls, from a conspecific or other species, was enough to elicit a mobbing response in the observer, and thereby result in social transmission. In fish, the modality of transmission is usually chemical; observer fish respond to alarm substances released by other fish in response to danger.

Social enhancement of food preferences occurs when after being exposed to a demonstrator carrying cues associated with a particular diet, the observer becomes more likely to consume that diet. The classic case is that of the Norway rat; if rats smell a particular food type on a conspecific's breath, they will show a subsequent preference for that food (Galef and Wigmore 1983; Galef et al. 1984). The evidence obtained by Galef and his colleagues shows that social enhancement of food preferences fits within Heyes' (1994) definition of observational conditioning given in section 4.1.3.

Theoretically, stimulus enhancement could result in a process similar to social enhancement of food preferences if the demonstrator acted merely to expose the observer to the odor of the food, thus sensitizing it to that smell. However, Galef et al. (1985) found that exposure to a food type in the absence of a demonstrator was not sufficient to induce a preference for that food. In addition, Galef et al. (1997) found that the mechanism could reverse a learned aversion to a particular food type. This suggests that exposure to a recently fed demonstrator changes the observer's perception of the "palatability," or some other property, of the food that the demonstrator has eaten, rather than acting to sensitize the observer to that stimulus.

Galef et al. (1988) went on to show that if food-related cues were paired with carbon disulfide, a compound present in rat breath, then this was sufficient to induce a comparable effect on diet preference. The process therefore satisfies the definition of observational conditioning. Galef and Durlach (1993) found evidence that the process does not conform to standard associationist phenomena such as overshadowing, blocking, or latent inhibition, suggesting it may be reliant on specialized learning mechanisms. However, Ray (1997) used a slightly different procedure and found that these phenomena were manifested. Heyes (2011) discusses this issue further.

4.2.4 Response facilitation

In order to isolate response facilitation, researchers must first show that the rate of performance of a particular action is increased by exposure to other individuals performing the same action. There are numerous cases where groups of animals seem to be synchronized in their behavior, but synchrony does not have to be a product of response facilitation. If external stimuli or conditions are affecting a number of individuals in the same way, this could cause individuals to do the same things at the same time, in the absence of any social effect. One way to rule out this explanation is to ensure that the cues causing the demonstrator's behavior are known, but also unlikely to be the direct cause of similar behavior in the observer. Ideally, one would want to experimentally trigger the target behavior in the demonstrator. Such convincing cases of response facilitation come from the literature on observational conditioning. Cook et al. (1985) showed that an observer rhesus monkey, when seeing a demonstrator responding fearfully to a snake (characterized by piloerection, grimacing, and vocalizations), would show similar behavior itself. Because the observers in pretests had not responded to snakes fearfully, it is highly unlikely that the observer and demonstrator were responding to an external cue in the same way, and indeed, the observers responded fearfully even when they could not see the snake.

Another possibility is to replace the cue with one provided by the experimenter, such as video footage, or when appropriate, by direct human demonstration. For example, Byrne and Tanner's (2006) experiments on a western lowland gorilla (*Gorilla gorilla*) suggest that response facilitation might operate on specific actions. They found that the subject was spontaneously more likely to execute a hand gesture similar to one she had observed the experimenter performing during the period following the demonstration, rather than during the control period preceding it. There is no question of external cues affecting demonstrator and observer in the same way, because the demonstrator was a human experimenter, making arbitrary hand gestures.

Where one of these forms of experimental manipulation is not possible, reasonably strong inferences might be made by comparing the level of synchrony between individuals who are free to interact and observe one another; a first group's level of synchrony could be compared with a second, isolated group that is, nonetheless, likely to be subject to the same external cues. For example, Hoppitt et al. (2007) found evidence of a response facilitation effect on the rate at which domestic fowl initiated bouts of preening. The rate at which chickens initiated bouts of preening was more strongly related to the number of birds already preening in the same aviary than it was to the number of birds preening in an adjacent, visually obscured aviary. This rules out the possibility that any plausible external cues could be wholly responsible for the behavioral synchrony.

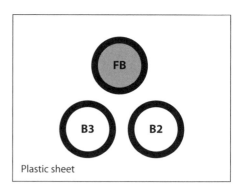

Figure 4.2. Experimental apparatus for the three-bowl test for response facilitation of drinking (not drawn to scale). FB: food bowl; B2/B3: water bowls. The apparatus for the three-bowl test for response facilitation of feeding was identical except with two food bowls (B2/B3) and one water bowl (WB). Based on figure 1 in Hoppitt and Laland (2008b).

Behavioral synchrony could also be accounted for by local enhancement; if individuals move around together, they will experience the same locations in their environment at the same time, and perhaps respond to stimuli at those locations in the same way. This is especially problematic if behavior is highly location-specific. For example, feeding might often be directed toward a specific location, such as a food bowl. We addressed this issue by using a "three-bowl test" to investigate whether there is a response facilitation effect on feeding and drinking in domestic fowl (Hoppitt and Laland 2008b). The experimental setup consisted of three identical bowls, placed adjacent to each other, two of which contained food, one of which contained water, or vice versa (see fig. 4.2). We found that the rate at which one individual initiates bouts of drinking from a given water bowl is increased if another individual is already drinking from an adjacent bowl. Conversely, there was no evidence that an individual drinking increased the rate at which other individuals fed from the adjacent food bowl. This suggests the effect cannot be wholly accounted for by local enhancement, because there was an action-specific effect. We take this finding as evidence of a response facilitation effect on drinking. Similar methodology could be applied in other cases where the target actions can be performed at a number of different locations.

Response facilitation is perhaps capable of generating social transmission of trait performance, where the trait in question is the action that is subject to response facilitation, provided it results in observer learning. Such effects may be short lived, though this in an empirical question (e.g., see Hoppitt and Laland 2008b), so it is necessary to distinguish between transient facilitatory effects and more lasting cases that support social transmission. Though response facilitation, in isolation, can only apply to actions already in the repertoire, it could also result in social transmission where the trait is performance of a familiar action A in a novel context C, similar to contextual imitation (see section 4.1.6.1). In such cases, response facilitation could also result in permanent social transmission of trait performance if response facilitation results in a short-term increase in performance rate of A in C, and the effect is perpetuated by reinforcement learning. We discuss how response facilitation and contextual imitation can be distinguished in such cases in the following section.

4.2.5 Contextual imitation

Contextual imitation is, by definition, social transmission. It is social transmission of trait acquisition if the focal behavior is an action (possibly familiar) that A performed in a *novel* context C. However, contextual imitation might also operate to produce social transmission of trait performance if observing others perform A in C causes the observer to perform A in C at a higher rate once it has acquired the trait (i.e.,the observer has already performed A in context C).

The '"two-action method"' is widely regarded as the most successful method for testing imitative ability (Dawson and Foss 1965). The experimental subjects must solve a task with two alternative solutions. Typically, half of the subjects observe a demonstrator solving the task in one way, the other half the alternative. Subjects are then tested to see which method they use, and if they disproportionately used the method that they observed, this is taken as evidence of imitation. Whether other social learning processes can be ruled out as explanations for matching behavior depends on the nature of the actions demonstrated. If the two actions are directed to different parts of the experimental apparatus, then local enhancement cannot be ruled out.

Likewise, if the actions are both directed to the same location, but there is no evidence that the action used by the demonstrator is being copied, this does not rule out local or stimulus enhancement as an explanation for task acquisition (e.g., Nagell et al. 1993). If the two actions result in different movements of the experimental apparatus (e.g., Bugnyar and Huber 1997), then emulation cannot be ruled out, unless it can be shown that the movement is copied in egocentric space (relative to oneself), rather than allocentric space (relative to the external environment). The latter method was used by Heyes et al. (1992), who showed that a rat would push a joystick to its left or its right, as it had seen a demonstrator do, and this effect remains when the joystick is moved so that it is operated in a plane perpendicular to the demonstration. However, it was later shown that odor cues left on the joystick could act as a cue for the observers, which could result in social transmission of trait acquisition through local enhancement (Campbell and Heyes 2002; Mitchell et al. 1999).

Others have attempted to control for emulation by providing a control group that observes a "disembodied" movement (Whiten and Ham 1992) or "ghost" control, where the manipulandum moves by itself; this may be followed by a reward for an inactive "demonstrator" (e.g., Fawcett et al. 2002; Heyes et al. 1994). If the "disembodied-movement" control group fails to move the manipulandum in the way they observed, but those subjects who saw a full demonstration do copy the movement, it is taken as evidence that the subjects are imitating and not emulating the demonstrator. However, Byrne (2002) notes that spontaneous movements of objects could be extrinsically of less interest to animals than the actions of a demonstrator, so one cannot rule out emulation on the strength of this evidence. Even if an animal does attend to the movement of a manipulandum followed by presentation of food, it may be more likely to learn that the spontaneous movement of the manipulandum signals food presentation (observational conditioning) than it is to try to reproduce those movements.

The two-action test does not inherently test for production imitation because it does not show that the alternative actions are novel. However, it provides evidence consistent with contextual imitation, so long as the context-action combination is novel. The method has been widely used to test for an imitative ability in primates, rodents, and birds. Among the best evidence of contextual imitation is the work of Zentall et al. on quail and pigeons (Akins and Zentall 1996; Kaiser et al. 1997; Zentall et al. 1996). Observers watched demonstrators step or peck on a treadle for a food reward and were more likely to solve the task using the method they had seen. As both actions were directed to the same location and resulted in the same movement of the treadle, local enhancement and emulation are ruled out.

However, Byrne (1999, 2002) argues that response facilitation provides an alternative explanation for the data generated by two-action tasks. For example, this explanation may account for the finding of Zentall et al. (1996) that the pigeons who observed the demonstrator pecking the treadle also tended to peck, whereas those that observed the demonstrator step on the treadle were more likely to step on it. It may be that birds that saw a demonstrator peck were just more likely to peck when they were tested on the task, purely because of a transient increase in pecking rate (or the probability of pecking) in relation to stepping, and vice versa for the other group. This ambiguity is exacerbated by the fact that in many two-action tests, the observer is rewarded for its responses (of either action) during the test phase, and this would act to amplify any initial preference for a specific action, which might otherwise have proved to be transitory.

If researchers are to be sure that they have a case of contextual imitation, rather than response facilitation, it must be shown that the observers have learned to use the target action in that context. One way of eliminating response facilitation as an explanation could be to introduce a delay filled with another observed or performed activity between observation and exposure to the task, so as to let the possible effects of response facilitation wear off. However, it is difficult to know exactly how long a response facilitation effect could last. Alternatively, researchers could introduce control subjects who do not observe demonstrators solving the task, but instead observe demonstrators performing the same actions in a different context. Perhaps the most effective method for ruling out response facilitation is to show directly that an observer has learned to use an action in a specific context during the observation phase.

For example, Dorrance and Zentall's (2002) experiments on pigeons aimed to investigate whether imitative learning of a peck/step two-action task is context specific. Observers first received asocial conditional discrimination training—for example, peck in response to a white light and tread in response to a green light. One group of observers then watched a demonstrator successfully responding to the stimuli in a manner consistent with their own training, whereas another group observed a demonstrator successfully performing the reverse discrimination. All observers then received conditional discrimination training that was a reversal of their initial asocial training. It was found that those birds that had observed demonstrators responding to the reverse discrimination learned the discrimination more rapidly themselves. This case qualifies as social transmission of trait performance by contextual imitation, because observation of a trait performance (e.g., peck in response to white) caused an increase in the relative rate of performance compared to an alternative trait (step in response to white).[7]

4.2.6 Production Imitation

Production imitation is also, by definition, an example of the social transmission of trait acquisition, where the trait is a specific motor pattern or sequence of behavior. Production imitation cannot be invoked for social transmission of trait performance, because once the action itself is no longer novel, the equivalent effect is response facilitation if the effect is not context specific, and contextual imitation if it is context specific. To demonstrate production imitation, it must be shown that an observer has acquired a novel action or sequence of actions not already in its repertoire. In section 4.1.6.2 we suggested that it is likely to prove of utility to view production imitation as the learning of novel sequences of action units (Heyes and Ray 2000). This approach was utilized by Whiten (1998) in sequential imitation experiments on chimpanzees. Whiten investigated whether four chimpanzees could learn by imitation how to open an "artificial fruit," (a box containing food) that could only be opened after disabling a number of "defenses." The defenses consisted of a bolt, a pin, and a handle, each of which could be removed or disabled in two different ways (using one of two actions). They could also be removed in any order, with six possible arbitrary sequences of removal. In Whiten's experiment, each of four observers watched a human demonstrator performing a different sequence of behavior patterns to open the artificial fruit. They were subsequently tested to see if each removed the defenses in the same sequence, and using the same actions, as the demonstrator. This is essentially a modified version of the two-action test, where the alternative actions are replaced with action sequences. Because each sequence is arbitrary, Whiten (1998) rules out the possibility that observers would inevitably converge on the demonstrated solution by trial and error learning. Such an approach also allows one to quantify the probability that an individual will hit on the same sequence as its demonstrator by chance, under the null hypothesis that observers do not imitate the observer. This approach differs from the two-action tests described in section 4.4.6, in that it can test for production imitation. In Whiten's experiment, though it is plausible that each component action is familiar to the observer, it

---

[7] In another experiment, Dorrance and Zentall (2002) found that if pigeons previously observed a demonstrator responding in a manner consistent with their initial discrimination training, there was no evidence that they acquired the discrimination more quickly. Nonetheless, these results seem robust, because Saggerson et al. (2005) were able to replicate the findings of both these experiments. It is not clear why the demonstrator's actions were only imitated when the social information obtained from the demonstration contradicted the observer's first-hand experience.

is improbable that the precise sequence of actions it observes would already be present in their repertoire, if they have not previously had access to the experimental apparatus and do not usually combine the actions in question.

Whiten (1998) found that the order in which defenses were removed was transmitted to the four observers, but interestingly, each sequence took time to emerge. Subjects were tested over three test trials, interspersed with demonstrations, and there was no evidence that the sequence had been imitated on the first two trials, but there was strong evidence on the third trial. However, this cannot be accounted for by trial and error learning, because several arbitrary sequences could lead to reward. The findings could provide insight into the mechanism underlying production imitation; perhaps the chimpanzees formed a template of the demonstrator's behavior and then modified their own behavior to match the template.

The results of Whiten's (1998) experiment constitute evidence that an arbitrary sequence of some kind had been learned through observation. However, subjects did not appear to have used exactly the same actions as the demonstrator in order to make the required manipulations. The chimpanzees only copied one pair of alternative action units used in the task (see also Whiten et al. 1996), poking a bolt versus pulling and twisting it, actions which result in different movements of the bolt itself, suggesting that emulation may be involved. This hypothesis is supported by the fact that one of the subjects was seen to copy the general form of the poking action unit, but showed variation in her performance of the action unit by varying the digit used to poke, and also using her knuckles and finger joints. An alternative explanation is that chimpanzees were imitating at a low level of resolution, using functionally equivalent actions to accomplish the same purpose. Overall, in this case it is clear that observation had a direct effect on skill learning, but the results could be interpreted as imitation or emulation.

Sequential tasks of this kind have great potential to detect production imitation, because if the demonstrator's sequence is copied, most other processes are ruled out. In Whiten's (1998) task, each action in the demonstrated sequence is directed to a different part of the apparatus. Nonetheless, a sequence match could not be explained by local or stimulus enhancement, at least not exclusively, because these processes operating in isolation could not affect the order in which the observer interacts with the different locations. It could be that the observers do not directly learn the sequence of actions, but instead learn the order in which different parts of the apparatus should be interacted with. They may also learn to direct each action unit to a particular location, in a manner analogous to contextual imitation. To rule out this possibility, one could use a sequence of action units that are directed to the same location, or are location independent, such as hand gestures (Custance et al. 1995; Hayes and Hayes, 1952). However, there is no reason why the former process should not be considered to be production imitation.

Response facilitation is unlikely to account for imitation of the sequential structure of behavior, although it might increase the probability that each component action be performed. It would then act to make an observer more likely to converge on the same sequence as the demonstrator than individuals that had seen no demonstration. However, the process is unlikely to account for different individuals using the same actions in a different order, and matching the order they saw demonstrated. One could argue that the actions seen earlier in a sequence might be primed more strongly than later actions, and therefore might be utilized first by the observer. However, if such a process increases the probability the observer will converge on the same sequence as the demonstrator, there is no reason not to label it "production imitation." In fact, conversely, we would expect response facilitation to be strongest on the action seen most recently.

A moot point is how accurate the sequence match must be for learning to be considered a case of production imitation. As was noted in chapter 2, accuracy depends in part on the level of resolution at which imitation occurs. One can regard a sequential task as being set at a certain level, and a positive result implies that the subject is imitating at a resolution equal to the level of the task or higher, whereas a negative result suggests that the subject is not imitating at a level of resolution high enough to be detected, or it is not imitating at all. For example, one could test whether subjects use the same body part as the demonstrator. A positive result would indicate they are copying the body part used, but they might be generalizing between actions that use the same body part, or imitating at a higher level of resolution. An alternative approach to detecting production imitation in a task-solving context is to carry out a detailed motion analysis of the motor patterns used by demonstrators and observers (Voelkl and Huber 2007).

Hayes and Hayes' (1952) famous "do-as-I-do" experiment on chimpanzees is often considered to be strong evidence of production imitation. This method was later repeated in a more rigorous manner by Custance et al. (1995), and has been replicated in orangutans (Call 2001). In the Custance et al. (1995) experiment, two chimpanzees were subjected to a three and a half month teaching phase, where they were trained to copy a human demonstrator performing 1 of 15 different actions in response to the command, "Do this!" Custance et al. then tested whether the subjects would imitate 48 arbitrary actions in response to the same command, and found that they were significantly more likely to perform an action similar to the demonstrated action than any other. Because the actions did not result in the movement of any objects, emulation is ruled out. These experiments are commonly interpreted as evidence that chimpanzees and orangutans are capable of production imitation (Whiten and Custance 1996).

However, it is difficult to quantify the possibility that the actions were already part of the subjects' gestural repertoire. The problem is illustrated by Byrne and Tanner (2006), who ran a similar experiment on a western lowland gorilla, except they did not train the subject to respond to a "Do this!" command as in the previous cases. They found the gorilla spontaneously imitated four of seven arbitrary actions presented, in the absence of any command or reward. On analysis of videotapes documenting the gorilla's past gestures, it was found that it had previously performed all of the actions that had been "imitated," suggesting they were already part of its repertoire. Byrne and Tanner (2006, 226) suggest that this offers an alternative explanation for the apparent production imitation in all three great ape species; after observing the test gesture, subjects became more likely to use whichever action in their repertoire "most resembled the demonstrator's action," making it a case of response facilitation. Similar experiments to Hayes and Hayes' (1952) have been run on bottlenose dolphins (*T. truncatus*, reviewed by Herman 2002). In Hoppitt and Laland (2008a) we discuss the extent to which response facilitation can be ruled out for these cases.

4.2.7 Observational *R-S* learning

To demonstrate observational *R-S* learning, one must show that an observer has formed an association between the action it saw a demonstrator perform and the observed consequences of that action. When the observed consequences of the action are favorable, the observer is more likely to perform the same action in the same context, resulting in social transmission of the trait and/or trait performance. As noted in section 4.1.7, such cases would also be classified as contextual imitation. Therefore, showing that social transmission has occurred by observational *R-S* learning, requires all the same criteria discussed in section 4.2.5. In addition, we would expect social transmission to only occur (e.g., Palameta and Lefebvre 1985), or be more effective (Giraldeau and Templeton 1991), if the observer sees the demonstrator being rewarded. However, in such cases, the observer may not have learned an association between the demonstrator's response and the reward (an *R-S* association), but may instead have learned an *S-S* association between the task apparatus and reward (observational conditioning), making it more likely to respond to the task apparatus. Observational conditioning alone is ruled out if the observers are shown to be using the same action used by the demonstrator (i.e., a positive result in a two-action test), but such results could be explained by observational conditioning in combination with response facilitation or "blind" contextual imitation.

In principle, researchers could distinguish observational *R-S* learning from other processes using a modified version of the two-action test, where the observer sees a demonstrator respond to the same contextual stimulus with two different actions, one of which is rewarded, while the other is not. If observers then preferentially imitate the action they saw rewarded, then this constitutes evidence they have formed an *R-S* association by observation. Blind contextual imitation, response facilitation, and observational conditioning could not account for this result.

An alternative approach was used by Saggerson et al. (2005), who used a devaluation procedure to test for the formation of *R-S* associations in the observer. Observers first watched demonstrators stepping on a treadle to receive a reward of grain that was lit by a red light, and pecking at the same treadle to receive a reward of grain lit by a green light (or vice-versa). Devaluation training consisted of separate presentations of each of the lights, red and green, one of which was paired with food, while the other, devalued stimulus was not. The intention was that the birds would value the devalued outcome less than the alternative, because the former would be a less reliable indicator of food presentation. When observers were subsequently given access to the treadle, they preferentially responded with the action they had observed leading to the non-devalued outcome, suggesting they had indeed formed an association between an action and its outcome through observation. This is social transmission of trait performance of the trait that is observed to lead to the (perceived) preferred outcome.

4.2.8 Emulation

To show that an observer has acquired a novel skill through emulation, researchers need to confirm that it has recreated the results of a demonstrator's actions, rather than reproduced the actions themselves. One way of testing for this is to expose the observer to the results of the demonstrator's actions, without allowing them to observe the actions directly. Here we focus on the evidence that animals can learn an action by recreating the movements of objects that they observe moving when the demonstrator acts on them.

There are cases suggestive of emulation from two-action test experiments, where an observer has recreated the movements of the experimental apparatus, after seeing a demonstrator manipulate that apparatus without using the same exact actions (see sections 4.2.5 and 4.2.6). However, an alternative explanation is that subjects are imitating at a low level of resolution, meaning emulation is practically indistinguishable from contextual imitation. One potential solution is to expose the observer to the "disembodied" or "ghost" movements of experimental apparatus (Whiten and Ham 1992) and to test whether it recreates the movements it observed. Denny et al. (1983) and Denny et al. (1988) showed that after observing a pendulum bar that moved to the right, followed by food presentation, and to the left, followed by no food, or vice versa, rats would tend to push the bar in the direction they had seen followed by food. This is strong evidence of an emulation effect. Klein and Zentall (2003) made a similar finding for pigeons, but the effect has not been replicated in other species (e.g., Fawcett et al. 2002; see Hopper 2010 for a comprehensive review). However, as noted above, spontaneous movements of objects could be of little interest to animals (Byrne 2002). Animals observe many spontaneous movements in their natural environment (e.g., branches moving in the wind) that they do not try and recreate. It may be that animals do emulate object movements in general, but only when certain rules are met (e.g., when that object is touched by a conspecific). If that were the case, it would be unlikely that a disembodied movement task would yield positive results.

An alternative approach is to manipulate the salience of the object movements involved in a task. For example, in a task involving manipulation of a pendulum, one could modify the thickness of the pendulum. If the observers are imitating the demonstrator, then this should have no effect, whereas if they were emulating, we would expect observers to recreate the movements of the more salient, thicker pendulum.

Another alternative would be to use a modified two-action test, where experimental apparatus is devised that can be operated in four different ways, each of which results in one of only two possible movements of the apparatus. For example, one could provide a lever that can be twisted or pushed down in order to obtain a food reward, and each of these movements can be accomplished in two ways, using the teeth or using the hand. Observers would then observe one of four possible demonstrations: twist lever with hand, twist lever with teeth, push lever down with hand, and hold lever in teeth and push down. If observers copied the exact manner in which the demonstrator solved the task, this would be evidence it was imitating the demonstrator. However, if only the movement was copied, this would be evidence that it was emulating the demonstrator. We are unaware of any studies in which the two-action test has been modified to isolate action-based cues and results-based cues in this way.

Elsewhere we (Hoppitt and Laland 2008a) have questioned whether the dichotomy between imitation and emulation is useful, at least with respect to action learning. It is typically assumed that imitation and emulation are the results of different underlying mechanisms. Imitation is seen as the result of a high-resolution mechanism that enables observers to translate the movements of a demonstrator in order to yield the motor program for executing matching movements. Emulation is viewed as the result of a mechanism by which the observer forms a template of the results of a demonstrator's behavior, and attempts, by trial and error, to match that template with its own behavior. However, the latter mechanism could also be used to account for some cases of imitation, because the observer could attempt, by trial and error, to match its own body movements to a remembered template of the demonstrator's body movements. In addition, imitation may not occur at the movement level, but as noted above, might involve learning a novel sequence of action units. If this were true, each action unit would be recognized in the demonstrator's behavior by observable cues characteristic of that action unit; these could include both body movements and environmental effects. The emulation controls in the experiments described above only really act to isolate imitation at the movement level of resolution.

If the above reasoning is correct, there is a strong case to be made that the imitation/emulation dichotomy be dropped. Instead, one could consider the actions and results of a demonstrator's behavior as two different sources of information that can be used by an observer. Call and Carpenter (2002) suggest a similar framework where the observer can use three sources of information: actions, results, and goals. However, because an observer's understanding of a demonstrator's goals is not directly observable, we suggest it might prove profitable to concentrate our efforts on results and actions. In our suggested framework, production imitation can be based on action-based cues, results-based cues, or a combination of both. Other cases, namely Whiten's (1998) artificial fruit experiment on chimpanzees and Moore's (1992) studies of an African gray parrot, are also production imitation, but are probably based on a combination of action- and results-based cues. Recently, Huber et al. have suggested a similar framework classified in terms of movement, action, and result (2009).

4.2.9 Opportunity providing

To show that access to the products of performance of a trait T are implicated in social transmission of T, one could remove the products after the demonstration phase. If the observers deprived of the products show a retarded rate of acquisition or performance of T relative to those who receive the full demonstration, this constitutes evidence of social transmission via products in the latter case. A negative result does not necessarily mean that opportunity providing does not ordinarily occur if there are other processes by which social transmission can occur, and the rate at which these take place is elevated in the absence of products.

An alternative approach is to provide the products of a demonstrator's trait performance (or equivalent, artificially constructed objects) to individuals who do not receive direct access to demonstrators. If the rate of acquisition or performance of T is elevated relative to controls who do not receive products, this constitutes evidence that access to the products are sufficient for social transmission to occur.[8] Terkel (1996) used this approach to study the mechanism of transmission of pinecone stripping from black rat mothers to offspring (see section 4.1.9).

A variant on this approach is to provide some naïve individuals with additional products to supplement those they receive from demonstrators. If the rate of acquisition or performance is increased relative to controls that do not receive additional products, this constitutes evidence that social transmission occurs through the products. A negative result in such cases would not rule out social transmission through products of trait performance, since demonstrators might already "saturate" observers with products. Thornton and McAuliffe (2006) used this approach on meerkats (*Suricatta suricata*) to show that access to live, disabled scorpions, usually provided to pups by adult helpers, caused them to develop more effective scorpion-hunting skills in a subsequent test.[9]

However, showing social transmission via products does not necessarily mean the mechanism is opportunity providing. This is illustrated by a study conducted by Sherry and Galef (1984) on black-capped chickadees (*Parus atracupillus*). They found that observer birds exposed to a demonstrator removing the lid to a pot containing cream were more likely to acquire the trait themselves than controls. They also found that birds that received no demonstration, but instead were presented with pots with the lids removed, were also more likely to acquire the trait than controls. However, in this case the product presented to the birds was already "fully solved," so there was no opportunity for the birds to engage in operant learning about the task. Instead, it is likely that they fed from the open pots, thus learning to associate them with food, and causing them to be more likely to solve the task when presented with lidded pots. Such cases could simply be described as *S-S* learning of an association of the task with food, resulting from the products of the demonstrator's behavior.[10]

---

[8] If the rate of acquisition/performance of T is lower than for individuals who receive a full demonstration, including access to products, this suggests that there is another pathway to social transmission, or at least that there is another social process involved. For example, the demonstrator might attract the observer to the products by local enhancement.

[9] This case also provides strong evidence of teaching on the part of the helpers; see section 4.5.

[10] This is not observational conditioning, since it is not observation of the demonstrator that exposes the observer to the *S-S* relationship.

4.2.10 (Inadvertent) coaching

This is a rather difficult mechanism to detect through experimental manipulation, because it involves feedback between demonstrator and observer. One must establish that the demonstrator responds to the observer's behavior X with response Y, and that Y encourages or discourages X in the observer. In some cases, a researcher might be able to replace Y with an artificially produced stimulus—for example, if Y is a distress call. However, such a case might prove to be exceptional. The best method might often be detailed observations of the demonstrators' reactions to the observer's over time, and the effect this has on the observers' behavior.

It seems inevitable that inadvertent coaching will be common for any social species learning to interact with others, as individuals' behavior will inevitably be shaped by the responses of those around them. Of most interest will be candidate cases of teaching that are thought to operate through coaching (see section 4.4), and whether coaching can account for the social transmission of social traits (e.g., mating behavior of cowbirds; see section 2.2). In the latter case, the researcher must also show that acquisition of the target trait causes an individual to respond to others in a way that results in acquiring the trait at a higher rate.

4.3 A Pragmatic Approach to Characterizing Mechanisms of Social Transmission

The historical development of the terms used for different social learning processes has left researchers with an extremely difficult task when attempting to identify the mechanisms responsible for a given case of social transmission. This is because common classifications systems have overlapping categories, and encompass mechanisms that are almost impossible to tell apart, given the data that are likely to be available. This means that even when researchers possess a lot of information about the processes responsible for social transmission (e.g., the conditions under which they occur, and the effect they have on the observer), it appears that little is known because a unique label cannot be assigned.

However, we believe that there are usually just a few main features that can be used to characterize the mechanisms underlying cases of social transmission. First, one can determine whether observation of trait performance, or else access to the products of trait performance, is required for transmission to occur.[11] If social transmission occurs through products, one can ascertain whether this is a result of opportunity providing, S-S learning of an association between the task and a reward, or local enhancement. If social transmission occurs through observation we suggest a researcher ascertain whether the effect is action-specific, context-specific, and whether it is sensitive to the outcome of the demonstrator's behavior (e.g., whether the demonstrator receives a food reward).

---

[11] In practice, social transmission could operate through both products and observation in a given case. As noted in section 1.2, social transmission can also occur through instructions or descriptions of a trait. Coaching is also possible, though we suspect this is rarely responsible for social transmission outside of humans.

In figure 4.3 we present a decision tree showing how the terminology described in section 4.2 relates to these features. Based on this, a researcher can identify which set of mechanisms, or combinations of mechanisms, could account for their case of social transmission.[12] In the tree the first decision is whether or not the social effect is context specific, and if so, in what respect: Is it specific to a location, or is it specific to a particular class of stimuli? Next, a researcher ascertains whether the effect is sensitive to the outcome of the demonstrator's actions. For example, does the effect only occur if the demonstrator is rewarded? A researcher then determines whether or not the effect is action specific, and finally, if it is action specific, whether that action is a novel action or action sequence.

Note that in all cases, the researcher might not have any evidence on which to base the decision in question. For context specificity, the results might be consistent with both a location-specific effect and a stimulus-specific effect, unless some attempt is made to ascertain whether the effect generalizes to other stimuli of the same type. Likewise, if all observers see the demonstrator being rewarded, a researcher will be unable to ascertain what effect this has on the observer. Finally, in some cases it may not be known whether the effect is action specific or not (e.g., if the test is not a two-action test).[13] In all these cases, the mechanisms obtained following either branch of the tree are consistent with the results of the study.

Even when evidence is available for a specific decision, the decision may not be clear-cut. For example, in investigating context specificity a researcher may find evidence that the effect generalizes to stimuli of a similar kind, but also that the effect is stronger at the demonstrated location. In this case, the effect might be decomposed into location-specific and stimulus-type specific components, with the researcher obtaining estimates of the magnitude of each. Similar logic can be applied to questions about sensitivity to outcome and action specificity. This reasoning suggests an alternative way of thinking about the mechanisms underlying social transmission. Rather than attempting to classify cases into categories, such as local enhancement and observational conditioning, researchers could view context specificity, action specificity, and sensitivity to outcome as three dimensions on which mechanisms can be quantified (e.g., see box 4.2).

There are then clearly further questions one might ask of the effects that are responsible for social transmission, such as: How long do the effects last (box 4.2; Hoppitt et al. 2012)? In the case of action-specific effects, at what level of resolution do they operate? And are action-based cues, results, or both used to identify the action, or action components? However, we feel the four criteria in figure 4.3 are the features that characterize the conditions most commonly presented in

---

[12] Here the case for using parsimony is stronger than for comparing to different mechanisms: For example, we would prefer an explanation involving just contextual imitation to one involving contextual imitation and local enhancement, since one explanation is "nested" in the other. However, it not clear that researchers should a priori prefer an explanation of contextual imitation to one of local enhancement and response facilitation, or vice versa.

[13] Strictly speaking, a negative result in a two-action test only tells us there is little or no evidence of action specificity. However, by quantifying the effect size with confidence intervals, one can establish that an effect, if it exists, is likely to be small.

Figure 4.3. The set of mechanisms or combination of mechanisms responsible for a case of social transmission through observation, based on whether the effect is (*a*) context specific, (*b*) sensitive to the outcome of the demonstrator's actions, and whether the effect is (*c*) is action specific, and if so, whether the action is (*d*) a novel action sequence or a familiar one. PI = production imitation; ORSL = observational *R-S* learning; OC = observational conditioning of an excitatory-appetitive relationship; LE = local enhancement; SE = stimulus enhancement (sensitization); RF = response facilitation; CI = contextual imitation. The scheme allows for the fact that LE and SE might be more effective when the demonstrator is rewarded. Low-res = low resolution, meaning that ORSL or CI might be operating at a low resolution and that action-specificity is consequently not observed. If any of the criteria is unknown, the set of possible mechanisms is found by appropriate pooling of the cells of the table (e.g., if there is a location specific effect that is sensitive to outcome, but the researcher does not know if it is action specific, the result is: CI, RF + LE, LE, RF + OC, OC. Consequently, the researcher can conclude that they minimally have either CI, LE, or OC, and cannot rule out the possibility that RF is also adding to the strength of the effect.

(*a*) Context specificity

(*b*) Sensitive to the outcome of the demonstrator's actions?

(*c*) Action-specific?

(*d*) Novel action/action sequence

experimental studies of social learning mechanisms. If social learning researchers find this system useful, it might even eventually be preferable to drop the current system of terminology (or parts of it), and classify cases by these criteria. Nonetheless, it remains important to establish the link between this system and established terms.

4.4 Teaching

Until recently it was widely believed that animals typically did not teach, that is, it was thought they did not actively facilitate learning in others. Indeed, teaching, together with imitation, have been lauded as the mental faculties that underly complex human culture by promoting the high transmission fidelity deemed necessary for cumulative culture (Boyd and Richerson 1985; Galef 1992; Tomasello 1994). However, recent research challenges these assumptions, revealing strong candidate cases of teaching in a diverse set of species (see reviews by Thornton and Raihani 2008; Hoppitt et al. 2008). Early studies of animal teaching were hampered by an anthropocentric viewpoint. The "type specimen" of a teacher was a traditional schoolteacher, and definitions of teaching stressed the intention of the tutor to educate (Pearson 1989). This stance effectively restricted teaching to

**Decision tree for identifying
social transmission mechanisms**

Social transmission

None   Location-specific   Stimulus-type specific

Yes   No   Yes   No   Yes   No

Yes   Yes   Yes   No   Yes   No   Yes   No   Yes   No

Yes   No   Yes   No   Yes   No   Yes   No   Yes   No   Yes   No

PI + ORSL   Yes   Yes   ORSL   PI + CI   ORSL   PI + CI

ORSL   PI   RF   RF + OC   PI + LE   RF + OC   PI + SE

PI + ORSL   CI   RF + SE   CI
PI + OC   RF + LE   PI + ORSL   RF + SE
PI + LE   PI + OC
PI + SE

Low-res ORSL   Low-res CI   PI + ORSL   Low-res ORSL   Low-res CI
OC   LE   PI + OC   OC   SE
LE   PI + SE   SE

our own species because intentions are difficult to infer in nonhumans. Progress was made when Caro and Hauser (1992) adopted a functional perspective, defining teaching according to observable criteria:

> An individual actor A can be said to teach if it modifies its behaviour only in the presence of a naïve observer, B, at some cost or at least without obtaining an immediate benefit for itself. A's behaviour thereby encourages or punishes B's behaviour, or provides B with experience, or sets an example for B. As a result, B acquires knowledge, or learns a skill earlier in life or more rapidly or efficiently than it might otherwise do so, or would not learn at all. (153)

In cases of teaching, we term individual A the "tutor" and individual B the "pupil," rather than using the terms "demonstrator" and "observer" as for cases of inadvertent social learning. Caro and Hauser's widely adopted operational definition (1992) is designed to minimize the probability of a false report of animal teaching by ruling out behavior that might be adapted for a different function. However, conceptually, we view "teaching" as a class of behavior patterns that are adaptations for transmitting knowledge and skills to others, or for enhancing the fidelity of information transmission (Hoppitt et al. 2008). It is important to keep this distinction between a conceptual definition (what

Box 4.2

Stochastic mechanism-fitting model (SMFM)

We have proposed a stochastic mechanism-fitting model (SMFM) for inferring social learning mechanisms in either the field or naturalistic circumstances in captivity (Hoppitt et al. 2012). We applied the SMFM to data on groups of wild meerkats learning to solve a foraging task. The experiment involved two boxes containing food, both of which could be accessed using either a flap or a tube, giving meerkats four "options" for solving the task (see fig. 4.4a).

The SMFM views individuals as moving between states of not interacting with the task and interacting with the task. Individuals can either terminate bouts of interactions successfully by solving the task, or abandon the task before succeeding. This gives a total of three transition rates (see fig. 4.4b).

The statistical model uses survival analysis methods (similar to those underlying network based diffusion analysis; see section 5.2) to model the transition rates for each individual, i, as a function of

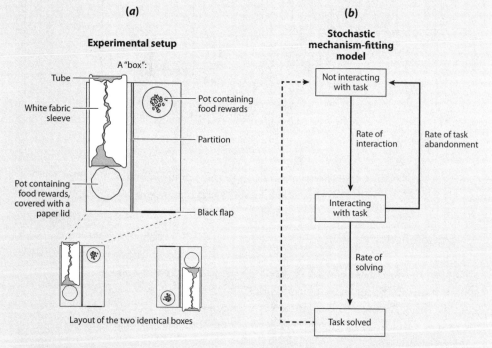

(a)

Experimental setup

A "box":

Tube

White fabric sleeve

Pot containing food rewards

Partition

Pot containing food rewards, covered with a paper lid

Black flap

Layout of the two identical boxes

(b)

Stochastic mechanism-fitting model

Not interacting with task

Rate of interaction

Rate of task abandonment

Interacting with task

Rate of solving

Task solved

Figure 4.4. (*a*) The "Box" apparatus used by Hoppitt et al. (2012) to investigate social learning mechanisms in groups of wild meerkats. The "Flap" technique involved going through a black cat flap to obtain food from a pot; the "Tube" technique involved pushing through a fabric sleeve on the tube and breaking a paper lid to obtain food. The experimental layout of the two identical boxes is shown below. (*b*) A diagrammatic representation of the stochastic mechanism-fitting model (SMFM) showing the three rates of transition that were modeled. In reality, "rate of interaction" involved modeling four "competing" transition rates to each of the four options available: left Flap, right Flap, left Tube and right Tube. Hoppitt et al. recorded an individual as solving the task when it gained access to food inside the box, and as abandoning the task when it terminated a bout of interaction without gaining access to food inside the box. Based on figure 1 in Hoppitt et al. (2012).

both i's previous manipulations (asocial learning), and i's observations of other meerkats manipulations (social effects). The latter was split into (1) direct social learning, where observations had a long lasting effect comparable to that of manipulations made by i, and (2) transient social effects, where a transition rate was increased or decreased temporarily after each observation.

We expanded this model to estimate the effects of sensitivity to the observed outcome of the demonstrator's actions; that is, whether they were observed obtaining food, and whether they were observed gaining entry to the task. We also expanded the model to estimate the extent to which social effects generalized among, or were specific to, options. This allowed us to estimate context specificity; that is, whether each effect was specific to an option (a highly specific location), a box (a less specific location), or to an option type (a type of stimulus—a flap or a tube). The experiment was designed to tease this context specificity apart, but in other cases the options might be designed such that action specificity is also quantified, using a two-action test design.

We found a number of social effects in operation, affecting both the rate of interaction and the rate of task abandonment. Indeed, we were able to identify nine separate learning processes underlying the meerkats' foraging behavior; in each case we could quantify its strength and duration, including local enhancement, emulation, and a hitherto unrecognized form of social learning that we termed "observational perseverance." The dominant social effect was a strong but short-lived (20s half life: time for the effect to halve in magnitude) local enhancement effect, which attracted individuals to interact with the option they had recently observed being manipulated. This effect was strongest for younger observer meerkats (see fig. 4.5). The analysis suggested a key factor underlying the stability of behavioral traditions is a high ratio of specific to generalized social learning effects. The approach has widespread potential as an ecologically valid tool to investigate learning mechanisms in natural groups of animals, including humans.

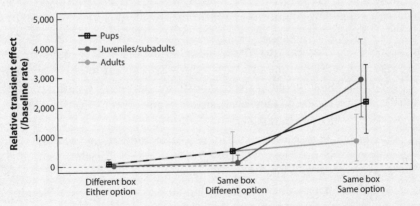

Figure 4.5. Hoppitt et al.'s (2012) estimates of the transient (short-lived) increase in rate of interaction at each option immediately following observation, for different age classes of meerkats. The effect is plotted relative to the baseline (not socially influenced) rate of interaction for naïve individuals. Error bars are 95% credible intervals. Based on figure 3a in Hoppitt et al. (2012).

teaching is) and an operational definition (how we detect it) in mind; behavior is always subject to multiple selection pressures, and this operational definition will miss out on numerous potential cases where a teaching function modifies the selection favoring a behavior that already benefits the performer. Accordingly, we (Hoppitt et al. 2008) previously endorsed Caro and Hauser's definition with one caveat. Their requirement that there be a cost or no immediate benefit to the tutor is only partially successful in ruling out behavior with alternative functions. For example, parental provisioning is costly and can transmit dietary preferences to offspring, but it might have evolved because selection benefits parents that provide nutrition to their young, rather than because provisioning functions as a means to teach. Consequently, for cases to be regarded as teaching, where behavior increases the inclusive fitness of the tutor irrespective of whether knowledge is transmitted to the pupil, we would require evidence that the tutor's behavior has been modified by selection to promote learning. Thornton and Raihani (2010) provide a detailed discussion on the methods for identifying teaching by these criteria.

Leadbeater et al. (2006) argue that "teaching" should be restricted to the passing on of "skills, concepts, rules and strategies," differentiating this from "telling" another individual a fact, such as the location of a particular food source. However, this distinction would require much of human teaching to be recategorized as telling, or we risk imposing stricter criteria for animals than humans (Thornton et al. 2007). Moreover, the distinction between teaching and telling is often unclear; if an individual tells another about the location of a food, and the recipient consequently learns that food of that type is good to eat, it could be said to have acquired a general rule.

Traditionally, teaching was regarded as contributing additional mechanisms of information transfer to the list of social learning processes. As mentioned above, Caro and Hauser (1992) suggested two such mechanisms by which teaching in nonhuman animals might occur: coaching and opportunity teaching. However, we advocate an alternative way to think about teaching (Hoppitt et al. 2008). Social learning mechanisms (section 4.1) relate primarily to psychological processes in the observer (or pupil), whereas teaching processes relate specifically to activities of the demonstrator (or tutor). We suggested that a full description of information transmission requires both. The presence or absence of "active" demonstration (behavior whose function is to facilitate learning in others) can be regarded as orthogonal to social learning processes. Hence, one can categorize instances of teaching as, for example, "teaching through local enhancement" or "teaching through imitation." From the perspective of the pupil, such phenomena would be identical to their inadvertent social learning equivalent, but here a tutor will have actively demonstrated a behavior pattern or actively drawn attention to a location, with the specific function of transmitting information to the pupil. Coaching and opportunity teaching are included in this scheme with teaching and inadvertent equivalents (section 4.1). Table 4.2 shows a number of mechanisms for which there are possible examples of both

Table 4.2. Parallel classification of processes involved in social learning based on "inadvertent"*
information, and social learning based on teaching, with plausible examples

Type	Inadvertent social learning	Teaching
Local enhancement	*A demonstrator inadvertently attracts an observer to a specific location, leading to the observer learning.* Naïve guppies follow informed individuals to food. (Laland and Williams 1997)	*The behavior of the tutor functions to attract a pupil to a specific location, leading to the observer learning.* Tandem running in ants, in which leader ants slow down to ensure followers keep in touch. (Franks and Richardson 2006)
Observational conditioning	*The behavior of the demonstrator inadvertently exposes an observer to a relationship between stimuli, allowing the observer to form an association between them.* Blackbirds learn to recognize predators by observing birds mobbing unfamiliar objects. (Curio 1988)	*The behavior of the tutor functions to expose a pupil to a relationship between stimuli, causing the pupil to form an association between them.* Adult babblers expose nestlings to the relationship between the "purr" call and food. (Raihani and Ridley 2008)
Imitation	*After observing a demonstrator perform a novel action, an observer learns to reproduce that action.* Birds learn to produce novel sounds through vocal imitation. (Janik and Slater 2000)	*The behavior of the tutor functions to demonstrate a novel action, causing the pupil to learn how to perform it.* A human tennis coach demonstrates a shot.
Opportunity providing	*The products of the behavior of the demonstrator provide the observer with an opportunity to engage in operant learning that would arise, for example, by providing an easier, less dangerous, or more accessible version of the task.* Black rats in Israel steal semiprocessed pinecones from their mothers. (Terkel 1996)	*The behavior of the tutor functions to produce products which provide the pupil with an opportunity to engage in operant learning that would arise, for example, by providing an easier, less dangerous, or more accessible version of the task.* Adult meerkats provide pups with dead, disabled, or live scorpions depending on the pups' age. (Thornton and McAuliffe 2006)
Coaching/ inadvertent coaching	*The response of a demonstrator to the behavior of the observer inadvertently acts to encourage or discourage that behavior.* Female cowbirds respond to preferred males songs with "wing stroking," which acts to reinforce that song in the male. (West and King 1988)	*The response of the tutor to the behavior of the pupil functions to encourage or discourage that behavior.* Mother hens attract their chicks away from food the mother perceives to be unpalatable. (Nicol and Pope 1996)

* Use of the term "inadvertent" signifies that the demonstrator's behavior is not adapted to the function of transmitting knowledge, and does not imply that teaching requires any intentionality on the part of the tutor. Other social learning processes might also prove to have a teaching equivalent.

Source: Reproduced from Hoppitt et al. (2008).

teaching and inadvertent equivalents. Potentially, cases could be categorized further on the teaching dimension by specific features of the tutor's teaching behavior, such as response to feedback (Franks and Richardson 2006), or whether the tutor's behavior changes with the competence of the pupil (Thornton and Raihani 2008).

The advantage of adopting a broad functional definition of teaching is that it potentially enables us to use a comparative approach to test hypotheses concerning the conditions under which behavior with a teaching function evolves (Thornton and Raihani 2008; Hoppitt et al. 2008; Fogarty et al. 2011). However, many cases of nonhuman teaching might prove to be specialized adaptations for social transmission of a specific skill, such as meerkats teaching pups how to handle dangerous prey, like scorpions (Thornton and McAuliffe 2006; see section 7.2.2). The conditions favoring the evolution of a generalized capacity for teaching, which can be used for social transmission of any skill a tutor might acquire, presents another question. Theoretical models suggest that this general capability may have coevolved with cumulative culture, since the latter broadens the conditions that favor teaching (Fogarty et al. 2011).

4.5 Summary

In this chapter, we presented a classification of some of the processes that have commonly been considered as mechanisms of social learning, discussed the conditions under which they result in social transmission, and presented methods for distinguishing mechanisms in the laboratory. We have also presented a view of teaching as behavior that has evolved with the function of promoting social learning in others. In subsequent chapters, we go on to discuss methods for studying social learning in the field or naturalistic contexts.

Chapter 5

Statistical Methods for Diffusion Data

This is the first of three chapters focusing on statistical techniques for inferring and quantifying social transmission in groups of animals in the wild, or in "captive" groups of animals in naturalistic social environments. Here we focus on techniques for analyzing time-structured data on the occurrence of a particular behavior pattern, or *behavioral trait*, in one or more groups. For the most part we focus on cases where a novel trait spreads through one or more groups. Following standard terminology in the field of social learning, we refer to the spread of a trait through a group as a *diffusion*, and the resulting data as *diffusion data* (regardless of whether there is evidence for social transmission). Such data may arise if the spread of a naturally occurring trait is recorded, or for diffusions that are initiated by a researcher, by presenting some kind of task (for nonhumans, usually a foraging task) that members of a group must learn to solve (see sections 3.2.2 and 7.2.1). The study of diffusion data is likely to be of crucial importance if researchers are to understand how and when novel innovations spread and give rise to traditions.

The level of detail of diffusion data varies. At one extreme, a researcher might possess a complete history of each individual's performance of the trait, along with a history of its observations of others' trait performances. Such data potentially allow rich inferences to be made about the social learning strategies (box 8.2) and mechanisms (J. R. Kendal et al. 2007; Hoppitt et al. 2012; see box 4.2) being utilized. However, this is perhaps only likely for captive groups, or for the diffusion of the solution to a task in which every manipulation of the task can be monitored closely. More commonly, a researcher might only have an estimate of when each individual first performed the trait, with an associated indirect assay of who is likely to have observed whom. In such cases, one can view individuals as moving from a *naïve* state in which they do not perform the trait, to an *informed* state, through *acquisition* of the trait by means of either *asocial learning* or *social transmission* of the trait from an informed individual (or both). A researcher can then use a model in which the *rate of trait acquisition* is modeled (section 5.2). In other cases, a researcher may not be able to identify individuals in the population at all, and only have an estimate of how many individuals have performed the trait so far (section 5.1). Finally, we look at how social transmission can be inferred from the spread of a behavioral trait through space.

5.1 Diffusion Curve Analysis

A diffusion curve is a plot of the number of individuals observed to have performed a behavioral trait against time.[1] For many years, in both the human and nonhuman social learning literature, it was believed that the shape of the diffusion curve could be used to infer whether social transmission was involved in the spread of a behavioral trait. The idea is that asocial learning proceeds at an approximately constant rate, resulting in an r-shaped diffusion curve. In contrast, if the trait is acquired by social transmission, the per capita rate of acquisition will increase as the number of demonstrators increases, giving an acceleratory curve (see fig. 5.1). If a complete diffusion is documented throughout the entire group or population in question, the diffusion curve is expected to be S-shaped, as it levels out with all individuals having acquired the trait. The reason that social transmission is widely thought to generate an s-shaped curve is that it requires both demonstrators and observers, and when either are rare, as occurs early or late in the diffusion, the rate of spread is constrained; however, when both are common, as in the middle of the diffusion, the spread is at its most rapid. Different functional forms are fitted to the data, and their fit compared (see box 5.3), and social learning is inferred if acceleratory curves fit best.

Diffusion curve analysis has been used extensively to infer social transmission in both humans and nonhuman animals (J. Henrich 2001; Lefebvre 1995; Reader 2004; E. Rogers 1995; Roper 1986). Lefebvre (1995) used this method to analyze 21 diffusions of foraging innovations from the primate literature, including cases from Japanese macaques (e.g., fish eating; Watanabe 1989), vervet monkeys (acacia-pod dipping; Hauser 1988), and chimpanzees (mango and lemon eating; Takahata et al. 1986; Takasaki 1983). Lefebvre found an overall trend for accelerating learning rates, seemingly consistent with models of social transmission.

Unfortunately, recent work suggests that that the shape of the diffusion curve is not a reliable signature of social transmission (Laland and J. R. Kendal 2003; Reader 2004; Franz and Nunn 2009; Hoppitt, Kandler, et al. 2010). First, a number of researchers have pointed out that social learning will not necessarily result

---

[1] Sometimes a proxy for the number of informed individuals is used, such as the number of times a trait is seen being performed during a particular time period.

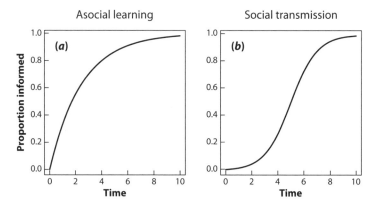

Figure 5.1. Typical diffusion curves traditionally assumed to be characteristic of (*a*) asocial learning (r-shaped) and (*b*) social transmission (s-shaped). Recent theoretical work has cast doubts on these assumptions (see text).

in an S-shaped diffusion curve if the population is structured into subgroups. Laland and J. R. Kendal (2003) and Reader (2004) suggest that directed social learning can result in a step-shaped function, with acceleratory component parts, if the trait spreads more rapidly through closely connected subgroups, such as family units (e.g., Fritz et al. 2000). Furthermore, differences in the rate of acquisition between different subsections of the population might act to obscure any underlying pattern; for example, a strong sex difference might result in a bimodal distribution of latencies to acquire the trait (Reader 2004).

Perhaps even more of a concern, there are good reasons to expect that asocial learning alone can result in an S-shaped diffusion curve, even if populations are homogeneously structured. In general, any process that results in an increase with time in the rate at which individuals acquire the trait will result in an acceleratory diffusion curve. For example, if the trait of interest is the solution to a novel foraging task that is presented to a group of animals, they will often display some neophobia to the task. If the effects of neophobia decrease over time, as we might expect, the rate at which individuals solve the task might also increase with time (Hoppitt, Kandler, et al. 2010). Acceleratory curves can also occur if an individual must move through a number of stages in order to acquire a trait. For example, there may be a number of different steps required to solve a foraging task, such as defenses that need to be removed to access a fruit (Whiten 1998). If the time to complete each step of the task is exponentially distributed, then we would expect the overall time to solve the task to follow an approximate gamma distribution, causing the diffusion curve to become more and more S-shaped as the number of task steps increases (Hoppitt, Kandler, et al. 2010). In conclusion, recent theoretical analyses suggest that researchers cannot reliably infer social learning from the shape of the diffusion curve (Reader 2004 gives further reasons for caution).

5.2 Network-Based Diffusion Analysis (*NBDA*)

Network-based diffusion analysis (*NBDA*), pioneered by Franz and Nunn (2009), infers social transmission if the pattern of spread of a behavioral trait follows the connections of a social network. A social network is a social structure made up of individuals, sometimes called nodes, as well as the connections among them that represent forms of relationship or interdependency, such as patterns of association, interaction, friendship, or kinship (Newman 2010). The assumption here is that a trait will spread sooner between individuals who spend more time together than between less connected individuals. As such, *NBDA* inherently addresses the concern that the pattern of spread of a trait will be influenced by population structure. With social network analysis becoming increasingly popular in both the social (Wasserman and Faust 1994; Newman 2010) and biological (Croft et al. 2008) sciences, the appropriate data for applying *NBDA* to both human and non-human populations is often likely to be available. *NBDA*, therefore, offers a viable alternative for inferring social transmission from diffusion data. Though *NBDA* was developed recently in the field of animal social learning, similar methods had previously been developed in the social sciences. Here we start by describing *NBDA* in detail, before discussing how previous approaches relate to it.

Franz and Nunn's original version (2009) of *NBDA* took as data the *times* at which individuals acquire a behavioral trait, which can be considered the time at which an individual was first observed performing the trait in question. Hoppitt, Boogert, and Laland (2010) introduced an alternative version of *NBDA*, which applies to the *order* in which individuals acquire the trait, but not the exact time. These alternative versions of *NBDA* have become known as time of acquisition diffusion analysis (*TADA*) and order of acquisition diffusion analysis (*OADA*), respectively. In both cases, the researcher fits a model including a social transmission component, in which the rate of transmission between an informed and a naïve individual is proportional to the connection between them. If this model is better than a model without social transmission (see section 5.2.2), then this supports the hypothesis that the trait is transmitted through the social network. In box 5.1, we provide the mathematical and technical details underlying *NBDA*. Code to run *NBDA* in the *R* statistical environment (*R* Core Development Team 2011) is available at http://lalandlab.st-andrews.ac.uk/freeware.html. Here we aim to provide a general guide to using *NBDA* for nonmathematical readers.

Each version of *NBDA* has its advantages and disadvantages. While *TADA* has more statistical power than *OADA* (Hoppitt, Boogert, and Laland 2010), especially when networks are relatively homogeneous, this comes at the cost of stronger assumptions. In its original form (Franz and Nunn 2009), *TADA* assumes that the *baseline rate of acquisition* (the rate of acquisition in the absence of social transmission) remains constant over time. This can result in false positives in the same circumstances as diffusion curve analysis (i.e., if the asocial acquisition rate increases over time, for example, as a result of a reduction in neophobia to a novel task). *OADA* is not vulnerable to such effects, making the weaker assumption that the baseline rate function is the same for all individuals being analyzed. However,

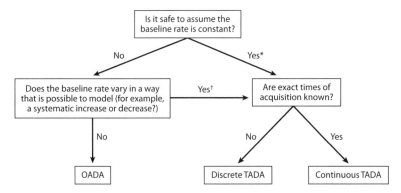

Figure 5.2. Flowchart for selecting the appropriate *NBDA* model. *Researchers should be cautious in assuming the baseline rate of acquisition is constant, because a number of factors can cause increases in the rate (see Hoppitt, Kandler, et al. 2010). †In principle, any function can be used to model the baseline rate. However, the software provided on our website only allows for a systematic increase or decrease. In cases where environmental variables are thought to unpredictably influence the rate of acquisition, but for all individuals in the same way, *TADA* becomes intractable, whereas *OADA* remains appropriate (see Hoppitt, Boogert, and Laland 2010). Based on figure 2 in Hoppitt and Laland (2011).

Hoppitt, Kandler, et al. (2010) extended *TADA* to accommodate a nonconstant baseline rate of acquisition. They suggested use of a baseline rate function corresponding to a gamma distribution of latencies under asocial conditions, which allows for a systematic increase or decrease in the asocial rate of acquisition over time. Nonetheless, *OADA* remains an attractive option if the baseline rate function is thought to follow a form that is difficult to model, or if the researcher does not wish to make any assumptions regarding the shape of the function.

If the researcher chooses to use *TADA*, they also have the choice of treating time as a continuous variable (Hoppitt, Boogert, and Laland 2010), or splitting the diffusion period into a number of discrete units and specifying which individuals acquired the trait in each unit (Franz and Nunn 2009). In practice, the method will fit equivalent models if the number of time units used is large. Typically, computation speeds are faster for the continuous *TADA*, and we recommend that this method be used when exact times of acquisition are known. However, when this is not the case, the discrete *TADA* may be preferable. For example, data might be collected in a series of scans, where, at discrete points in time, the researcher ascertains which individuals are informed. In this case the researcher can only infer a time period during which each individual acquired the behavior, and so the discrete *TADA* is the natural choice.

Even if data is collected individually, in the field it seems likely that there will be observation errors in the recorded time of acquisition. Franz and Nunn (2010) find that this can result in type 1 errors in a discrete *TADA* when the time units are small (and so too, presumably, for the continuous *TADA*), but that if the length of time unit is long enough the problem is alleviated. Franz and Nunn provide a rule of thumb that there should be at least a 50% probability that an individual who has acquired the trait will be observed performing it within any given time unit. A researcher

Box 5.1
Network-based diffusion analysis (*NBDA*)

All forms of *NBDA* can be generalized in the following form:

$$\lambda_i(t) = \lambda_0(t)(1 - z_i(t))\left(s\sum_{j=1}^{N} a_{i,j}z_j(t) + A\right),$$ 5.1.1

where $\lambda_i(t)$ is the rate at which individual i acquires the trait at time t; $\lambda_0(t)$ is a baseline acquisition function determining the distribution of latencies to acquisition in the absence of social transmission; $z_i(t)$ gives the status (1 = informed, 0 = naïve) of individual i at time t; $a_{i,j}$ gives the network connection, or association, between i and j; s is a fitted parameter determining the relative strength of social transmission; and the predetermined form of A determines whether asocial learning of the trait is assumed to occur.

The $(1 - z_i(t))$ and $z_j(t)$ terms ensures that the trait is only transmitted between informed and uninformed individuals. The model assumes that the rate of transmission between such individuals is proportional to the connection between them, and this rate is scaled by the parameter s. Social transmission is inferred if a model including s is better (see box 5.3) than a model where $s = 0$. One strategy for doing this is to compare a model of "pure" social transmission, where all acquisition is through social transmission ($A = 0$) with a model of asocial learning ($s = 0$, $A = 1$) (Franz and Nunn 2009). However, this only works if the analysis starts with informed individuals in the group (e.g., trained demonstrators). An alternative strategy is to assume there is always the chance that an individual can acquire the trait asocially. In this case, the most intuitive way to parameterize the model is to set $A = 1$[1]. This means that s gives the rate of social transmission relative to the rate of asocial acquisition. We prefer this approach, and henceforth replace A with 1.

The difference between *TADA* and *OADA* is the way in which the baseline rate function, $\lambda_0(t)$, is treated. In *OADA*, $\lambda_0(t)$ is unspecified, with the assumption that it is the same for all individuals being modeled, whereas in TADA, $\lambda_0(t)$ takes a specified form that is fitted to the data.[2] In both cases, the expression given in equation 5.1.1 leads to a likelihood function (L), which gives the probability of observing the data under the model, given a specific set of parameters, and allowing the model to be fitted by maximum likelihood.[3] The log-likelihood for *OADA* is:

$$\log(L) = \sum_{I=1}^{D}\sum_{i=1}^{N} \log(R_i(t_{I-1}))z_i(t_I)(1 - z_i(t_{I-1})) - \sum_{I=0}^{D} \log\left(\sum_{i=1}^{N} R_i(t_{I-1})\right),$$ 5.1.2

where D is the number of acquisition events observed (where one or more individuals are observed to acquire the trait), $R_i(t)$ is the relative rate of acquisition ($\lambda_i(t)/\lambda_0(t)$), and t_I is the time immediately after the Ith acquisition event (after $z_i(t)$'s are updated). The relative rate of acquisition is used here because the baseline rate function cancels out of the likelihood function. $z_i(t_I)(1 - z_i(t_{I-1}))$ indicates whether i acquired the trait at the Ith acquisition event.

---

[1] We have previously (Hoppitt, Boogert, and Laland 2010) suggested a bounded paramterization for s, where one sets $A = 1 - s$. This means that s can range between 0 (all asocial acquisition) and 1 (all social tranmission). However, we now suggest that this parameterization is more difficult to interpret when individual-level variables are included (box 5.2).

[2] *NBDA* can be seen as a specialized version of survival analysis (or "event-history analysis"), with *OADA* being equivalent to a modified Cox proportional hazards model, and *TADA* being equivalent to a parametric model (Cox and Oakes 1984). In survival analysis, $\lambda_i(t)$ is referred to as the "hazard function" and $\lambda_0(t)$ as the "baseline hazard function." However, in the context of NBDA, we feel "rate of acquisition function" and "baseline rate function" are more intuitive terms.

[3] An optimization algorithm is run to find the set of parameter values that maximizes the likelihood, or equivalently, minimizes the negative log-likelihood.

In the original forms of *TADA* (Franz and Nunn 2009; Hoppitt, Boogert, and Laland 2010), it is assumed that the baseline rate function is constant ($\lambda_0(t) = \lambda_0$). However, Hoppitt, Kandler, et al. (2010) found that *TADA* is susceptible to false positives in the same circumstances as diffusion curve analysis, and suggested fitting a baseline rate function that allowed for systematic changes over time. Hoppitt, Kandler, et al. (2010) suggest a function corresponding to a gamma distribution of times, under asocial conditions, though here we note a Weibull distribution is commonly used in survival analysis, and might also work well for *NBDA*. In principle, any function can be used so long as the user can provide the "hazard function" $\lambda_0(t)$, and "cumulative hazard function" $\Lambda_0(t)$. For many distributions, these functions are readily available in the R statistical environment (R Core Development Team, 2011).

The user may also choose whether to treat time as a continuous variable (continuous *TADA*) (Hoppitt, Boogert, and Laland 2010), or to divide time into a number of discrete steps of equal (Franz and Nunn 2009) or unequal (Hoppitt, Kandler, et al. 2010) length, specifying which step each individual acquired the behavior (discrete *TADA*).

The negative log-likelihood for the continuous *TADA* with a constant baseline rate function is:

$$\log(L) = \sum_{I=1}^{D} \sum_{i=1}^{N} (t_{I-1} - t_I) \lambda_0 R_i(t_{I-1})$$

$$+ \sum_{I=1}^{D} \sum_{i=1}^{N} z_i(t_I)(1 - z_i(t_I))(\log(R_i(t_{I-1}) + z_i(t_{I-1})) + \log(\lambda_0)). \qquad 5.1.3$$

$$+ \sum_{i=1}^{N} R_i(t_D) \lambda_0 (t_D - t_{end})$$

This can be generalized for a nonconstant baseline rate to give:[4]

$$\log(L) = \sum_{I=1}^{D} \sum_{i=1}^{N} R_i(t_{I-1})(\Lambda_0(t_{I-1}) - \Lambda_0(t_I))$$

$$+ \sum_{I=1}^{D} \sum_{i=1}^{N} z_i(t_I)(1 - z_i(t_{I-1}))(\log(R_i(t_{I-1}) + z_i(t_{I-1})) + \log(\lambda_0(t_I))), \qquad 5.1.4$$

$$+ \sum_{i=1}^{N} R_i(t_D)(\Lambda_0(t_D) - \Lambda_0(t_{end}))$$

where t_{end} is the time at which observation ceased (allowing for individuals who did not acquire the trait during the period of observation), and $\Lambda_0(t)$ is the baseline cumulative hazard, in survival analysis terminology, which is related to the cumulative distribution function of the asocial latency distribution, $F_0(t)$, thus:

$$\Lambda_0(t) = -\log(1 - F_0(t)). \qquad 5.1.5$$

Note that if $\lambda_0(t) = \lambda_0$, $\Lambda_0(t) = \lambda_0 t$, equation 5.1.4 reduces to 5.1.3.

For a discrete *TADA*, the data is provided in *P* discrete time steps, for which the time step in which each individual acquired the behavior is known. The log-likelihood is as follows:[5]

---

[4] The middle term here is $\sum_{I=1}^{D} \sum_{i=1}^{N} z_i(t_I)(\log(R_i(t_{I-1})) + \log(\lambda_0(t_I)))$ for any individual who is naïve at time t_{I-1}. Here we modify it such that this component is zero for any individual who is informed at time t_{I-1}, avoiding numerical errors arising from $\log(0) = -\infty$, since $R_i(t_{I-1}) = 0$ for such individuals.

[5] The $z_i(t_{start,p})$ term here ensures that the likelihood is zero for individuals who are informed at the start of period *p*, since $R_i(t_{start,p}) = 0$ for such individuals.

(continued)

Box 5.1   *(continued)*

$$\log(L) = \sum_{p=1}^{P} \sum_{i=1}^{N} (1 - z_i(t_{end,p})) R_i(t_{start,p}) (\Lambda_0(t_{start,p}) - \Lambda_0(t_{end,p}))$$
$$+ \sum_{p=1}^{P} \sum_{i=1}^{N} z_i(t_{end,p}) \log(1 - \exp(R_i(t_{start,p})(\Lambda_0(t_{start,p}) - \Lambda_0(t_{end,p}))) + z_i(t_{start,p}))$$
$$\tag{5.1.6}$$

where $t_{start,p}$ is the start of time period p, and $t_{end,p}$ is the end of time period p. For $\lambda_0(t) = \lambda_0$ this reduces to:

$$\log(L) = \sum_{p=1}^{P} \sum_{i=1}^{N} (1 - z_i(t_{end,p})) R_i(t_{start,p}) (t_{start,p} - t_{end,p})\lambda_0$$
$$+ \sum_{p=1}^{P} \sum_{i=1}^{N} z_i(t_{end,p}) \log(1 - \exp(R_i(t_{start,p})(t_{start,p} - t_{end,p})\lambda_0) + z_i(t_{start,p}))$$
$$\tag{5.1.7}$$

With time steps of equal length, this is equivalent to the model initially proposed by Franz and Nunn (2009), with their parameter τ, the rate of learning per time step, given by $\tau = \lambda_0 sT$, where T is the length of time step. Note that the discrete *TADA* assumes that individuals who acquire the trait in the same time step do not learn from each other, so this may provide a conservative estimate of the rate of social transmission.

---

can check this by calculating the proportion of time units in which individuals are observed performing the trait following the time unit when their performance was initially observed. In figure 5.2 we provide a flowchart to aid the choice of *NBDA* method, between *OADA*, continuous *TADA,* and discrete *TADA*. Box 5.1 provides technical details on the different types of *NBDA*, and how each is fitted to the data.

5.2.1 Inclusion of individual-level variables

A potential problem with *NBDA* is that false positives for social transmission can arise if individuals prefer to associate with others who have a similar asocial rate of acquisition (Hoppitt, Boogert, and Laland 2010). For example, higher-ranking individuals might acquire a trait at a higher rate asocially, and also disproportionately associate with each other, making it appear that the trait is being transmitted among them. As in other statistical models, a researcher can control for the effect of such confounding variables by including them in the model (Hoppitt, Boogert, and Laland 2010; Shipley 1999). There are good reasons for doing this, because even when a variable is not confounded with the social network, statistical power to detect social transmission can be improved by accounting for the variables' effects (Hoppitt, Boogert, and Laland 2010). In addition, it will often be of interest which variables influence the diffusion dynamics (e.g., Boogert et al. 2008).

Hoppitt, Boogert, and Laland (2010) extended *NBDA* to include such "*individual-level variables*" affecting the rate of asocial learning. They recognize two ways in which individual-level variables might be incorporated into the

---

Box 5.2

Inclusion of individual level variables in *NBDA*

In general, we can expand *NBDA* generally to include *V* continuous individual level variables as follows:

$$R_i(t) = (1 - z_i(t))\left(s\exp(\Gamma_i)\sum_{j=1}^{N} a_{i,j}z_j(t) + \exp(B_i)\right)$$

$$B_i = \sum_{k=1}^{V} \beta_k x_{k,i} \hspace{3cm} 5.2.1$$

$$\Gamma_i = \sum_{k=1}^{V} \gamma_k x_{k,i}$$

where $\lambda_i(t) = \lambda_0(t)R_i(t)$, $x_{k,i}$ is the value of the kth variable for individual i; β_k is the coefficient giving the effect of variable k on asocial learning; and γ_k is the coefficient giving the effect of variable k on the rate of social transmission. This general formulation allows the effects of individual level variables on asocial learning and social transmission to differ. In principle, these variables can be fitted in an unconstrained way; alternatively one can fit the additive model defined by Hoppitt, Boogert, and Laland (2010) by constraining $\gamma_k = 0$ for all k, or the multiplicative model by constraining $\gamma_k = \beta_k$ for all k. Categorical variables, or *factors*, with *F* levels can be fitted by defining $F - 1$ indicator variables determining which category each individual lies in, in the same way as for a standard regression analysis (e.g., see Weissberg 2005). The log-likelihood functions given in box 5.1 remain appropriate.

Interpretation of effects requires some explanation. For individual-level variables, β_k gives the additive effect of an increase of one unit of variable k on the log scale, so $\exp(\beta_k)$ gives the multiplicative effect on the rate of acquisition (this is the same as for most standard survival analyses). For example, if we find a coefficient of 1 per cm of body length, this means that, all other things being equal, the model would predict that if an individual A is 1 cm longer than another individual B, and then A would asocially acquire the trait at a rate 2.7 times faster than B.

In contrast, social transmission is modeled as a linear effect, such that it gives the rate of social transmission per unit of connection to informed individuals. In the additive model, this is relative to the baseline level of asocial acquisition (i.e., when $B_i = 0$. We suggest researchers standardize any continuous variables (subtract the mean, then divide by the standard deviation), meaning *s* can be interpreted as the rate of social transmission relative to the average rate of asocial acquisition. If factors are included in the analysis, *s* is relative to the asocial rate for an individual at the baseline level, for each factor. In the multiplicative model, *s* is invariant to the scale of the individual-level variables (see fig. 5.3).

---

model. The additive model assumes that the absolute difference in the rate of acquisition between any two individuals remains constant, for any level of social transmission. The multiplicative model instead assumes that the ratio in the rate of acquisition between any two individuals remains constant for any level of social transmission (see fig. 5.3 and box 5.2).[2]

---

[2] Previously we have suggested that a best fit of the additive model might indicate direct social learning mechanisms, and a best fit of the multiplicative model might indicate indirect mechanisms (see chapter 4; Hoppitt, Boogert, and Laland 2010). However, simulations using algorithms representing either direct or indirect mechanisms (similar to those used in Hoppitt and Laland 2011) did not support this distinction. Researchers should let their data decide whether the additive or multiplicative assumption is most appropriate, or fit a more general model (see box 5.2).

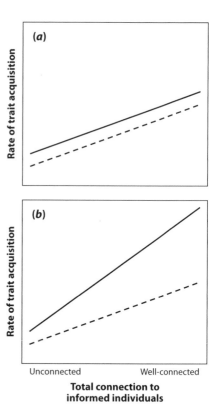

Figure 5.3. A graphical depiction of (*A*) the additive *NBDA* and (*B*) multiplicative NBDA, showing the rate of trait acquisition for two individuals (that differ in their asocial rate of acquisition) as a function of the total connection to informed individuals. At the extreme left of the range, individuals spend no time with any informed individuals, whereas at the extreme right, individuals are extremely well connected to those who have acquired the trait. When unconnected to informed individuals, acquisition is by asocial learning; for all other cases, the rate of acquisition is a combination of social transmission and asocial learning. For both (*A*) and (*B*), the asocial rate of acquisition for individual A (*solid line*) is double that for individual B (*dashed line*). In the additive model, the absolute difference in the rate of acquisition remains constant as the total connection increases, whereas in the multiplicative model, the ratio between the two remains constant. Based on figure 1 in Hoppitt and Laland (2011).

5.2.2 Model selection and inference

To test for social transmission, a researcher must compare a model containing social transmission (henceforth a *social model*) with a model not containing social transmission (an *asocial model*[3]). However, in each case the researcher must decide (*i*) whether to include a constant or nonconstant baseline function (if they wish to employ *TADA*), (*ii*) which individual-level variables to include in each model, and (iii) whether to consider an additive or multiplicative model of social transmission. Model selection for *NBDA* is directly analogous to model selection for a general linear model, and so the same methods can be used. We favor an information theoretic approach (Burnham and Anderson 2002), since in many cases the best asocial model might not be nested in the best model that includes social transmission, meaning that a classical hypothesis test such as a likelihood ratio test (*LRT*) cannot be used. For example, the best asocial model might include a nonconstant baseline (accounting for an acceleratory spread of the trait), whereas the best social model might have a constant baseline function,

---

[3] We note in passing that it is possible for diffusions that are well described by asocial models to reveal evidence for social learning, but not social transmission (Atton et al., 2012). This *prima facie* surprising observation reflects the breadth of the definition of social learning, which allows for forms of social influence on learning that do not qualify as social transmission.

with social transmission providing an alternative explanation for the acceleratory effect. In such cases a hypothesis test cannot be used, and thus information theoretic approaches come into their own. In addition, information theoretic approaches enable a researcher to use a model averaging approach, which allows taking into account model selection uncertainty (see box 5.3).

We suggest fitting models containing every combination of individual-level variables the researcher wishes to consider, with both a constant and nonconstant baseline rate, for an asocial model, additive social model, and multiplicative social model. In each case, the relative support for each individual model can be judged using Akaike's Information Criterion (AIC), or in practice, AIC_c, which is corrected for sample size (see box 5.3). These criteria assess each model based on how well they fit the data, after penalizing for the number of parameters used, with smaller values indicating that a model has greater predictive power. The models can then be ranked according to AIC_c, and the support for each model, or Akaike weight, is calculated from the difference in AIC_c from the best model (this procedure is implemented automatically in the *R* code provided at http://lalandlab.st-andrews.ac.uk/freeware.html).

The evidence for or against social transmission can be assessed by the total Akaike weights for models including social transmission and asocial models. To make this a fair comparison we suggest that a three-way[4] comparison be made between the asocial models, and the additive and multiplicative models of social transmission. The model with the greatest total Akaike weights is the one best supported by the data. The difference in weight between this and the other models indicates the level of support.[5]

Model averaging methods (box 5.3) enable researchers to estimate the strength of social transmission, and to calculate confidence intervals in a way that allows for model selection uncertainty (Burnham and Anderson 2002). Confidence intervals are especially important in cases where there is little support either way for or against social transmission, because they allow a researcher to set an upper plausible limit for the strength of social transmission. This might enable the researcher to make the stronger conclusion that social transmission is unlikely to be important in the acquisition of a trait.

5.2.3 Modeling multiple diffusions

Sometimes a researcher might have access to data from multiple diffusions, either a single trait spreading through multiple groups, or the spread of multiple traits through one or more groups. It might be preferable to include these in a single

---

[4] A two-way comparison between asocial models and models containing social transmission is not a fair comparison if twice as many of the latter are considered. For the same reason, more comparisons should be made, if additional models of social transmission are considered (see section 5.2.4).

[5] If the asocial model has an only slightly lower Akaike weight, this means that there is not strong evidence for social transmission. However, it would not make sense for researchers to "reject" social transmission on grounds of parsimony under such circumstances, since the Akaike weights already factor in model complexity when quantifying the level of support.

statistical model, in order to improve the power to detect social transmission and/or allow comparison between groups. For example, researchers may wish to test whether the rate of social transmission is higher in one context than another. This can be done using *NBDA*, although the exact inference can vary depending on which version of *NBDA* is fitted, and how. In addition, for the model to be meaningful, the social networks for each diffusion must be of the same type in each case (see below).

One option is to fit an *OADA*, assuming a separate baseline rate function, for each diffusion (see box 5.4). Here, minimal assumptions are made, and social

---

Box 5.3

Akaike's information criterion (AIC)

Akaike's Information Criterion (AIC) provides a means to compare the fit of different statistical models that are fitted to the same data. Unlike *p* values, AIC can be used to compare models that are not nested (i.e., when one model is not a constrained version of another). A full description of the theoretical basis for AIC, and a guide to its use are provided by Burnham and Anderson (2002). Here we give a brief summary.

Akaike's Information Criterion is calculated from the log-likelihood for the model (L), where the model parameters have been optimized by maximum likelihood (e.g., see box 5.1):

$$AIC = -2L + 2k$$

where k is the number of parameters in the model. A version of AIC is also available that corrects for sample size:

$$AIC_c = -2L + 2k + \frac{2k(k+1)}{n-k-1}.$$

This should be used unless the sample size is large (in which case there will be little difference).

Models with lower AIC explain the model better after appropriately penalizing for the number of parameters in the model. The degree to which additional parameters are penalized is not arbitrary, since the difference in AIC between any two models fitted to the same data estimates the difference in Kullback-Leibler (K-L) information for the two models. K-L information measures the extent to which the predicted distribution for the dependent variable differs from its true distribution (i.e., the information lost when moving from the true distribution to the model).

Note that AIC only gives a measure of the relative fit of candidate models, not a measure of absolute fit, so it is the absolute *difference* in AIC that determines the relative performance of two models. Given a set of *R* candidate models, a researcher can obtain a relative measure of support for each model *i* by calculating the Akaike weights:

$$w_i = \frac{\exp(-\frac{1}{2}\Delta_i)}{\sum_{r=1}^{R} \exp(-\frac{1}{2}\Delta_r)}$$

where Δ_i gives the difference in AIC between model *i* and the best model in the set; w_i can be thought of as the probability the *i* is the model with best K-L information in the set, accounting for sampling error.

learning is inferred purely from whether the order of acquisition tends to follow the network in each diffusion. Researchers cannot compare whether the rate of acquisition varies between diffusions, and so can only test for whether individual-level variables have an effect if they vary with diffusions. It is possible to test for differences in the relative rate of social transmission (per unit of network strength), but statistical power may be lower than using alternative methods.

Conversely, *TADA* is sensitive to the times at which individuals in each group acquire the trait(s), as well as the order they learn. Therefore, if all individuals

One can obtain measures of support for various features of the model, such as the presence of a variable, by summing Akaike weights over those models that include that feature. For example, imagine we run an *NBDA* with size as an individual-level variable, also considering a number of other individual-level variables, as well as additive, multiplicative, and asocial versions of the model. We get a measure of support for an effect of size by summing the Akaike weights of models that include size as a variable. This gives the probability that size is in the model with the best K-L information of those considered, after accounting for sampling variation. The same process could be used to compare different models of social learning with each other, and with models of asocial learning.

AIC also gives us a method of estimating parameter values that is not subject to the same problems as traditional model selection procedures. The traditional method is to select a "best" model, perhaps based on adjusted R-squared, AIC, or stepwise approaches using *p* values and an arbitrary significance level. Inferences are then based on the best model (i.e., they are *conditional on* that model being true). Such approaches do not take into account the uncertainty in the model selection procedure (i.e., which is *really* the best model). An alternative is to use a *model-averaging* procedure, which uses all the models considered to estimate parameter values, but the contribution of each is weighted by its Akaike weight. One obtains a model-averaged estimate $\hat{\bar{\theta}}$ for a parameter θ as follows:

$$\hat{\bar{\theta}} = \sum_{i=1}^{R} w_i \hat{\theta}_i$$

where $\hat{\theta}_i$ is the maximum likelihood estimator for θ for model i.[1]

Traditional measures of the precision of a parameter estimate, standard errors, and confidence intervals are also conditional on the final model being correct. There are additional methods for adjusting measures of precision such that they take into account model selection uncertainty, yielding unconditional standard errors and confidence intervals.[2] For details, see Burnham and Anderson (2002, 153–167).

---

[1] It may sometimes be desirable to conduct model-averaging across only those models in which the parameter is present, in other cases it may make sense to take the value of the parameter to be zero in such models. See Burnham and Anderson (2002, 150–153) for details.

[2] Though they are still conditional on the set of models considered.

Box 5.4

NBDA **for multiple diffusions**

We can generalize the *NBDA* model for multiple diffusions as follows:

$$\lambda_{il}(t) = \lambda_{0l}(t) R_{il}(t)$$

$$R_{il}(t) = (1 - z_{il}(t)) \left(s_k \exp(\Gamma_{il}) \sum_{j=1}^{N(l)} a_{i,jl} z_{jl}(t) + \exp(B_{il}) \right)'$$

where the subscript *l* refers to the lth diffusion, $N(l)$ is the number of individuals in diffusion *l*, and $\lambda_{0l}(t)$ is the baseline rate function for diffusion *l*. A separate s_l can be fitted for each diffusion, in order to test for differences in the rate of social transmission.

If one fits a *TADA* with $\lambda_{0l}(t) = \lambda_0(t)$, where the shape of $\lambda_0(t)$ is specified, the analysis will be sensitive to differences in times between diffusions. If one fits an *OADA* with a separate unspecified $\lambda_{0l}(t)$ for each diffusion, it will only be sensitive to the order within each diffusion. A compromise is to assume $\lambda_{0l}(t) = \lambda_0(t)$, but to leave the shape of $\lambda_0(t)$ unspecified. This amounts to fitting an *OADA* in which the diffusions are treated as a single diffusion, but with zero connections between pairs of individuals that are in different diffusions.

More generally, one can group diffusions into strata by setting $\lambda_{0l}(t) = \lambda_{0S(l)}(t)$ where $S(l)$ is the stratum for the lth diffusion.

---

associated with a diffusion acquire the trait at a similar time, *TADA* will infer social transmission even if the acquisition within each group does not follow the social network. However, caution must be taken with such findings. It is possible that false positives could arise if there is a different rate of *asocial* acquisition in each diffusion. We therefore advise that a researcher includes in their model selection factors allowing for a difference in rate of acquisition among different groups and different traits, unless there is a good reason to think that such differences could only be the result of social transmission. *TADA* can still infer social transmission on the basis of among-group patterns, if social transmission offers a better explanation than different group coefficients (as judged by AIC_c). For example, if we have data on multiple traits diffusing through multiple groups, it may be that some groups acquire some traits earlier than other groups, and other traits later. This could not be explained by a difference among groups or among tasks in the asocial rate of acquisition, but it might be explained by diffusions "taking off" at different times in each group.[6] Consequently, even in the absence of social network data, *TADA* can provide an "advanced" diffusion curve analysis, which controls for some of the confounding effects inherent in standard diffusion analysis.

A disadvantage of *TADA* is that it requires us to assume a specific form for the baseline rate function that is applicable to all diffusions (or to consider several

---

[6] It is logically possible that we have a group x trait interaction; that is, different groups are faster at acquiring different traits. If this is considered plausible, the interaction can also be considered in model selection.

specified alternatives in model selection). An alternative is to fit a modified $OADA$ in which the researcher assumes that all diffusions have the same baseline rate function, but there is no need to assume any specific form for the baseline rate function. This means that the analysis will be sensitive to the order across groups as well as within groups. This is done by treating all diffusions as a single diffusion, but with zero network connections between each pair of individuals that are in different diffusions. Alternatively, the researcher may wish to "stratify" the analysis, and assume that diffusions with a specified set (or "stratum") have the same baseline rate function, but that those in different strata may differ in baseline rate function. This is done by treating each stratum as a separate diffusion, again with zero network connections between each pair of individuals that are in different strata. In practice, we suggest that a researcher should consider treating different traits as different strata, since the complexity of the behavioral trait is likely to affect the shape of the baseline rate function (Hoppitt, Kandler, et al. 2010).

Differences in the rate of social transmission (per unit of network connection) can be tested for by comparing models with separate social transmission parameters (s) for different diffusions with a model that has a single rate of social transmission for all diffusions. These alternative social transmission models can be included in the model selection procedure described above.[7]

It is important to realize that comparing a network-based model of social transmission with an asocial model, using $TADA$ or the modified $OADA$, does not test whether trait acquisition follows the social network (except in the trivial sense that connections between individuals in different groups are zero). To illustrate: imagine that social transmission occurs equally between all individuals in the group, regardless of the network connection between them. This could occur if all individuals can potentially see anyone else in the group at any given time, or if social transmission occurs via the products left behind by trait performance (see Chapter 4). Unless the network is highly heterogeneous, a network-based model might still provide a better fit to the data than a model in which no social transmission takes place. If a researcher is interested in inferring specifically whether social transmission follows the network, they could either use a standard $OADA$ (with each diffusion treated as its own stratum); compare the network-based model to one in which all within-diffusion network connections are one; or modify the model such that it contains both network-specific and network non-specific components to social learning.

So far we have left aside the issue of whether the acquisition events from multiple diffusions can be considered independent (conditional on the variables in the model). If the researcher is analyzing multiple diffusions on different groups of individuals, we see no reason to doubt this assumption. However, if individuals are present in more than one diffusion, there is good reason to expect that their rate of acquisition might be correlated between diffusions. This can be accounted for by including random effects in the model (Hoppitt et al. 2010a),

---

[7] Note that $OADA$ models with different stratification structures cannot be compared by AIC_c since the models are effectively being fitted to different data.

as one would to account for repeated measures in a linear mixed model (e.g. Pinheiro and Bates 2000).

5.2.4 Choosing a social network

Thus far, we have also left aside the issue of the type of social network that should be used in *NBDA*. There are numerous methods and types of data that can be used to construct social networks (Croft et al. 2008; Whitehead, 2008; Wasserman and Faust 1998). Some result in a binary network, where the connection between each individual is either 0 or 1, whereas others result in weighted networks, where each connection has a strength (as, for example, would be the case if the network represents amount of time spent together). Some networks can be asymmetric, where the connection between two individuals is not necessarily the same in each direction. There is no reason why *NBDA* cannot be used with any of these types of network. When a binary network is used, the model assumes that social transmission occurs between connected individuals and not between unconnected individuals.

As *NBDA* assumes the rate of social transmission between individuals is proportional to the network connection between them, we suggest a network is chosen that reflects the frequency of opportunities for social transmission between individuals. For example, if the trait is a behavior pattern that can be performed at any time, a researcher could use the proportion of time individuals spend within observational distance of one another. However, it is not necessarily the case that social transmission requires observation of another individual. There are a number of cases where it has been shown that social transmission can occur when a naïve individual is exposed to the products of an informed individual's behavior (e.g. Campbell and Heyes 2002; Mitchell et al. 1999; Sherry and Galef 1984; Terkel 1995). In such cases an appropriate social network would seem to be one that reflects how often an individual is likely to encounter those products.

An alternative approach is to use *NBDA* to investigate which type of social network best explains patterns of diffusion, and so make inferences about the manner in which social transmission operates (Franz and Nunn 2009). This can be done using the procedure described above to compare social models based on different networks. For example, a researcher could formulate a number of social networks corresponding to specific theories regarding 'directed social learning' (Coussi-Korbel and Fragaszy 1995). For example, a hypothesis that all social transmission is vertical (from parents to offspring) could correspond to an asymmetric binary social network which has '1' weightings from adults to their offspring, and '0' weightings elsewhere. As noted above, it might also be worth researchers considering a model in which all connections within each diffusion are set to one, in order to assess the evidence that social transmission follows a network at all.[8]

---

[8] This is not worth doing for a standard *OADA* as this will be the same as a model with no social transmission. Social transmission in this case does not affect the order in which individuals acquire the trait.

Another issue that a researcher may wish to consider is the extent to which the network or networks are subject to uncertainty in their structure. For example, estimates of how much time individuals spend together will be subject to sampling error, with the width of confidence intervals associated with each connection dependent on the amount of data collected. *NBDA* in the form given above assumes that there is no statistical error in the network, but it might be desirable to take such uncertainty into account (Lusseau et al. 2008). This can be done by setting *NBDA* in a Bayesian context (Jaynes 1998; Gelman et al. 2004), and instead of providing a number for each network connection, the researcher provides a suitable prior distribution reflecting uncertainty in the connection. Markov Chain Monte Carlo (*MCMC*) techniques would then impute values from these distributions, meaning the posterior samples obtained would account for the uncertainty in the network.

A Bayesian *NBDA* could also be used to infer network structure using time or order of acquisition data. This could be of use in cases where it is known or assumed that a trait or traits are transmitted socially, but relevant association data are hard to collect (Hoppitt, Boogert, and Laland 2010). For example, in humpback whales (*Megaptera novaeangliae*) novel vocalizations are easily recorded, but association data is difficult to obtain in high latitudes (Noad et al. 2000). To infer social network structure, a researcher could simply set an uninformative prior distribution for each network connection, and constrain *NBDA* to have an asocial learning rate of "0."[9] *MCMC* could then be used to generate posterior samples for each network connection.

5.2.5 "Untransmitted" social effects

As in any analysis of observational data, a positive finding in an *NBDA* only indicates a statistical pattern consistent with the causal hypothesis proposed. While a researcher can include any measured individual-level variables that might influence both rate of acquisition and social network structure, it is difficult to exclude the possibility that an unmeasured variable is doing so, but this is a problem common to all observational studies (Shipley 1999).

Hitherto, we have considered known variables that could plausibly result in a false positive for social transmission, and how to control for them statistically. However, in many cases an alternative explanation remains that may be trickier to control. If the trait is constrained to be performed in a particular location, such as the solution to a foraging task, it might be that closely associated individuals are likely to encounter the task at the same time, purely as a result of being together, and so tend to acquire the solution at a similar time (Atton et al., 2012). We call this an "untransmitted" social effect. In this scenario, there is still a causal relationship between the behavior of one individual and the behavior of another, and so this would still be a case of social learning as described under the broad definition in this book. However, it would not constitute a case of social transmission,

---

[9] Each trait diffusion would then have to be started after the first individual acquired the behavior.

since one individual's performance[10] of the trait does not exert a positive causal influence on another individual acquiring the same trait (Hoppitt, Boogert, and Laland 2010). If a researcher observes individual X acquiring the trait at time T, this provides the information that individuals well connected to X are likely to be close to the task at time T. Consequently, these individuals are likely to solve the task sooner after time T than if X had not acquired the trait at time T. Whether or not this distinction is important is a matter for debate, since it seems probable that if individuals move around together this will often result in social learning through local enhancement (see chapter 4). Either way, there is a social influence on learning, triggering behavioral homogeneity in repertoires. Nonetheless, we are investigating methods for distinguishing these two processes.

5.2.6 Related methods

Although *NBDA* was developed independently, similar models had previously been used by social scientists studying the diffusion of innovations in humans. These are described and reviewed by Valente (2005). The approach described by Valente applies a discrete "event history analysis" model similar to the discrete *TADA* described above (Franz and Nunn 2009); however, the relationship between the rate of social transmission and the total network connection to informed individuals takes a different form, by using a logistic regression to model the probability of acquisition within each time unit. Whereas *NBDA* assumes that the rate of social transmission is proportional to network connection, the logistic regression model assumes a threshold relationship, with the rate of social transmission rising suddenly as the total connection to demonstrators reaches the threshold (see box 5.5).

A threshold model has often been considered as a plausible model for the spread of human innovations (Valente 1995) and rumors (Watts 2002). In nonhumans, there is evidence of threshold responses in individuals making decisions based on social information. For example, there is evidence that three-spined sticklebacks use a threshold rule when making movement decisions (A. Ward et al. 2008). However, it is currently less clear whether a threshold model or a linear relationship is more appropriate in the diffusion of innovations through groups of nonhumans; nor is it clear, we would suggest, that a threshold model is appropriate for all human cases. Therefore, it is an empirical question which model is more appropriate for a given case, and we would suggest that further work on humans and nonhumans might compare the fit of models with different mechanistically plausible functional relationships between network connection and rate of transmission. There are a number of other conceivable functions that might be tried, including a dilution effect where a strong connection to one or a few informed individuals

[10] Given that our definition of social transmission places emphasis on the demonstrator's *performance* of the trait, it may seem strange that *NBDA* looks for a relationship between the demonstrator and observers' *acquisition* of the trait. However, in most cases of social transmission there is no conceivable causal path that does not operate through performance, so in practice this is unproblematic.

Box 5.5

Discrete logistic regression model

Here we show the relationship between a discrete logistic regression diffusion analysis, with total connection to informed individuals as a predictor, and *NBDA*. We follow the notation used in boxes 5.1 and 5.2, which allows inclusion of individual level variables. In a logistic regression analysis, time is treated as a number of discrete periods of equal length, *L*. The probability the trait is acquired in a given time period, starting at time t_m, by an individual *i* that is naïve at the start of the period[1], is:

$$p_{acquire}(m) = \frac{\exp\left(\mu + s\sum_{j=1}^{N} a_{i,j} z_j(t_m) + B_i\right)}{1 + \exp\left(\mu + s\sum_{j=1}^{N} a_{i,j} z_j(t_m) + B_i\right)},$$

where μ is the coefficient of the logistic regression. Assuming *i* does not acquire the trait from other individuals who were naïve at the start of period *m*, as is the case in the discrete *TADA*, and as seems to be implied by the model, this means:

$$1 - \exp(-\lambda_i(t_m)L) = \frac{\exp\left(\mu + s\sum_{j=1}^{N} a_{i,j} z_j(t_m) + B_i\right)}{1 + \exp\left(\mu + s\sum_{j=1}^{N} a_{i,j} z_j(t_m) + B_i\right)}$$

$$\exp(-\lambda_i(t_m)L) = \frac{1}{1 + \exp\left(\mu + s\sum_{j=1}^{N} a_{i,j} z_j(t_m) + B_i\right)}.$$

$$\lambda_i(t_m) = \frac{\log\left(1 + \exp\left(\mu + s\sum_{j=1}^{N} a_{i,j} z_j(t_m) + B_i\right)\right)}{L}$$

The resulting functional form is a threshold model, shown in figure 5.4.

[1] The time period is excluded from the analysis for individuals who are informed at t_m, which fulfils the same function as the $(1 - z_i(t))$ term in box 5.1.

has a strong effect on transmission, whereas further connections make little difference. All such functions could be incorporated into the discrete/continuous *TADA* and *OADA* models that we describe above (see box 5.5).

5.2.7 Is NBDA realistic?

The models underlying *NBDA* are likely to be a simplification of the learning processes underlying trait acquisition. In *NBDA*, individuals move directly from a naïve to an informed state, at which point they are assumed to be as effective a demonstrator as all other informed individuals. In reality, it seems likely that both rate of trait performance and competence of performance will increase gradually through asocial learning as individuals are rewarded for trait performance.

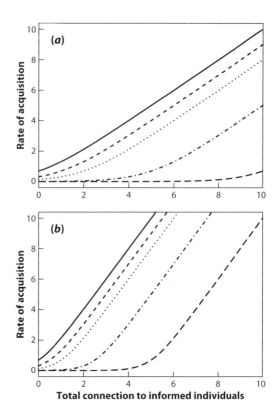

Figure 5.4. The functional relationship between total connection to informed individuals and rate of acquisition assumed by a discrete logistic regression analysis for (*a*) $s = 1$ and (*b*) $s = 2$, and for four different rates of asocial acquisition.

However, this might not be a problem. Previously, we simulated diffusion data from an agent-based model in which individuals' rate of trait performance increased gradually with repeated performance (Hoppitt and Laland 2011). Observation of another's trait performance occurred with a probability proportional to the social network used for analysis, and resulted in a relatively small or transient effect. While the power of *NBDA* was reduced, this reduction was surprisingly small for realistic conditions, suggesting that *NBDA* can still pick up on patterns in the data, even when it is only a crude model of the underlying process.

While initial indications are that *NBDA* provides a sufficiently realistic model of the learning process, it may often be desirable to fit a more complex model when more detailed diffusion data is available. This might allow us to infer more about the way in which social cues influence learning. A simple way to add further structure is to extend *NBDA* to model individuals as moving between multiple states. Such *multistate models* are recognized as a natural extension of survival analysis (Therneau and Grambsch 2000), and so the same applies to *NBDA*. For example, when studying the diffusion of the solution to a foraging task, many researchers collect data on the time of first contact to the task (e.g.. Boogert et al. 2008). This could be included in a multistate *NBDA* by modeling individuals moving between a naïve state, a state of having contacted the task but not solved it, and a state of having solved the task. Alternatively, if there are two different options for solving a task, one could use a competing risks model, in which a naïve individual can move into one of two informed states, one for each option.

5.2.8 Examples

NBDA is a new technique and consequently there are not many published case studies. Boogert et al. (2008) used a randomization technique to test whether the order of diffusion of foraging task solutions in starlings followed association patterns, and Morrell et al. (2008) used a similar technique for guppies. Both failed to find a positive result. We (Hoppitt, Boogert, and Laland 2010) later applied *TADA* (continuous time) and *OADA* to Boogert et al.'s starling data. Using *OADA*, overall we found no evidence that the diffusion followed the patterns of association, but with *TADA* we found reasonable evidence of social transmission. The evidence of social transmission seems to be based in a large part on the within-group homogeneity in the time of solving (see section 5.2.3).

When R. L. Kendal, Custance, et al. (2010) applied Franz and Nunn's (2009) original discrete time *TADA* to diffusion data on wild lemurs, they did not find evidence of social learning based on the *NBDA*. They did, however, find evidence based on the option used to solve the task using the option-bias test (see box 6.1). In such cases, one might obtain a more informative picture by modifying *NBDA* to model different options for solving the task, as we suggested in section 5.2.7. In addition to these examples, we are aware of several other applications of NBDA methodology to nonhuman data. In conclusion, NBDA appears to be a promising technique for using social networks and diffusion data to investigate social transmission in freely interacting groups.

5.3 Spatial Spread of a Behavioral Trait

In a number of cases, the spatial spread of a trait has been used to infer social transmission. Intuitively, if the pattern of trait acquisition appears to spread out from one or a few locations, this suggests a few initial innovators, with other individuals acquiring the trait by social learning. Here we describe the methods that have been used to infer social transmission from spatial data and suggest ways in which these methods might be developed.

5.3.1 Wave of advance models

Wave-of-advance models have been used to infer social transmission in both humans (e.g., Ammerman and Cavalli-Sforza 1984; Aoki 1987) and nonhumans (e.g. Lefebvre 1995). Lefebvre (1995) used a relatively simple version of this analysis to study the spread of the trait of tits (*Parus* spp.) in the U.K. pecking through the tops of milk bottles to obtain the cream inside ; he used data collected from 1921–1947 by Fisher and Hinde (1949). Here we discuss the method he used. This method requires only the times at which a trait was first observed to occur at different points in space. These points need not be specific individuals, but might just be discrete points where an observation was recorded, such as field study sites.

To apply a wave-of-advance model, a researcher must first identify the points at which innovation events are assumed to occur. If one is happy to assume only a single innovation event, then this is simply the point at which the trait was first observed. A researcher must also propose which points belong to which diffusion. In Lefebvre's analysis, there were two clear clusters of points, making it relatively easy to identify the candidate origins of each diffusion and define the area in which each diffusion occurred (see fig. 5.5).

By applying a linear regression, the researcher then tests whether the time at which the trait arose at each point is correlated with the distance from the location of the closest innovation event. The p-value for the regression line gives a hypothesis test against the null hypothesis that the trait was acquired independently at each location, and r-squared gives an indication of how well the model fits the data. Lefebvre (1995) found that a wave-of-advance model, even with two innovation centers did not explain the data well, and did not support the hypothesis that the behavior spread through social transmission (see fig. 5.6). This led him to conclude that the linear wave-of-advance model was not a good model for these data.

In our view, wave-of-advance models have a number of weaknesses. Firstly, they require one to identify putative points of innovation. This is not likely to be a problem if acquisition of the trait by asocial learning is unlikely, but if such events are common, the approach is unlikely to be feasible. In fact, Lefebvre (1995) and Fisher and Hinde (1972) find this explanation plausible for the tit

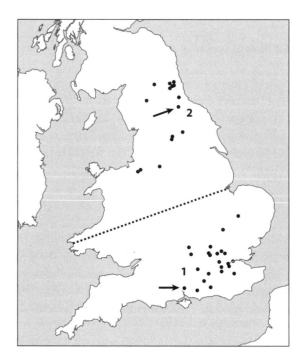

Figure 5.5. U.K. sites where milk bottle-opening by birds had been reported by 1939. The site of the first report in 1921, is indicated by the arrow labeled 1; the site of the inferred second site of origin, in 1926, is indicated by the arrow labeled 2. Based on figure 1 in Lefebvre (1995). Data from Fisher and Hinde (1949).

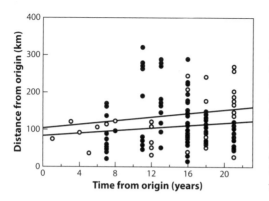

Figure 5.6. Distance of each new bottle-opening site from the presumed source of innovation as a function of years. Distance is from Swaythling, in the south of England, and time is since 1921 for all points southeast of the dividing criterion (see **fig 5.5**); for all points northwest of the dividing criterion, distance is from County Durham and time is since 1926. Based on figure 3 in Lefebvre (1995).

data. A preferable approach would be one that acknowledges that any given acquisition event could have occurred by asocial learning or by social transmission. Secondly, wave-of-advance models treat the spread of a trait in an almost deterministic manner. If a trait, by chance, happens to spread in one direction faster than another at time *t*, then this ought to affect researchers' expectations about acquisitions at subsequent time steps; but the wave of advance model does not take this into account. Thirdly, the wave-of-advance model does not take into account population structure. For example, researchers might expect a trait to spread faster through a heavily populated area than through a sparsely populated area where individuals meet infrequently. A potential solution to these issues is to use *NBDA* to model spatial data.

5.3.2 Other approaches for spatial data

Researchers could use *NBDA* to infer social transmission from spatial data by constructing a social network that reflects how close individuals are to one another. As usual, this should be done in a way that is thought to reflect opportunities for social transmission. For example, one might construct a binary network indicating which individuals have overlapping territories. Alternatively, the researcher might postulate a model in which the probability of social transmission is related to distance, and construct a weighted network based on this. *NBDA* is formulated to fit data where the time of order of acquisition is known for every individual in the group. Usually this will not be the case for large-scale spatial data, where one has only an estimate of the time of first occurrence at a number of locations (e.g., at a number field study sites, from archaeological remains, or historical records). However, *NBDA* could still be used in such cases by treating each site as an individual, with network connections to other sites derived from the distance between them.

Both the wave-of-advance model and spatial *NBDA* detect whether the pattern of spread is consistent with social transmission between neighbors, but there are a number of confounding variables that could result in the same pattern. Perhaps the most obvious concern is that individuals who are close in space have a similar rate of acquisition through asocial learning. This could be the case if a resource promoting acquisition is more abundant in one area than another. For example, in the case of the milk bottle-opening in British tits, milk bottles may be more abundant in some areas (e.g., towns) than others. If factors that influence learning can be identified and quantified, these can be included in a spatial *NBDA* as individual-level variables, and then controlled for statistically. However, other factors might not be so easy to quantify. For example, individuals close to one another might tend to be related, and consequently have correlated acquisition rates.

If the data are not collected at the level of individuals, then movement of individuals might also confound the process. On a short time scale, individuals might acquire the trait asocially in one area, and then move to a nearby location where the trait is also observed, making it appear as though the trait has spread through social transmission. On a longer time scale, problems might arise if the trait has a genetic basis. In such cases, spread of the trait might be caused by the dispersion of descendants of those individuals initially predisposed to display the trait. Of course, such a pattern could also arise if offspring inherit a trait by vertical social transmission, but this would be impossible to distinguish from genetic transmission.

One potential solution to this is to formulate a model in which individuals are allowed to move (perhaps adapting models of the spread of disease), and if the pattern of spread is unlikely to be accounted for by asocial learning and movement alone, then social transmission may be inferred. However, a model including individual movement is likely to be highly complex, making it difficult or even impossible to formulate a likelihood function; this function is necessary for convention techniques, such as maximum likelihood, to be used. A solution to this problem might be to formulate a simulation model, which is then fitted to data by judging how well the simulated data fit the observed data. A technique called approximate Bayesian computation (*ABC*) has arisen which allows posterior distributions for parameters to be estimated for simulation models (Beaumont et al. 2002; Beaumont 2010), and so allow inference in a similar manner to conventional Bayesian techniques such as *MCMC*. The method of *ABC* has now been used quite widely in the study of ecology and evolution (Beaumont 2010; Hartig et al. 2011), but further work is needed to see if it can be successfully employed in this scenario.

5.4 Summary

Diffusion data potentially contains a great deal of information about the rate and pattern of social transmission in groups of animals. Recent theoretical results suggest that the traditional (diffusion curve analysis) method of analyzing such data will often be flawed, and more sophisticated techniques are required. Network-based diffusion analysis infers whether social transmission follows a social network, and can be extended in a number of useful ways. Spatial diffusion data has been used by a number of researchers in the past, but we suggest there is scope for more advanced methods that utilize modern inferential techniques and computing power.

Chapter 6

Repertoire-Based Methods for Detecting and Quantifying Social Transmission

In this chapter, we discuss methods for inferring and quantifying the social transmission of behavior based on a "snapshot" of the behavioral repertoires of individuals or groups. This differs from the approaches described in chapters 5 and 7 in which time-structured data on the diffusion or development of a trait or traits are available. Often repertoire-based methods take the form of a group-contrasts approach (McGrew 1992; Whiten et al. 1999), where the researcher attempts to ascertain whether different groups have different behavioral repertoires, which might be caused by a higher rate of social transmission within groups than between them. Consequently, in section 6.1, we describe methods that can be applied to determine whether group differences in behavior exist.

However, there are other factors that might cause behavioral differences among groups (Laland and Hoppitt 2003; Laland and Janik 2006). Genetic differences between groups might cause differences in behavior. Alternatively, individuals might be acquiring the relevant traits asocially, with their repertoires shaped by the local ecology in a way that causes individuals in the same area to have similar behavior. Here we often adopt the shorthand of referring to these as the "genetic" and "ecological" hypotheses, respectively. In doing so we do not mean to imply that the relevant behavioral traits are exclusively determined by either of these factors in the sense that they are, say, genetically "hardwired" (Bateson and Martin 2000), but rather we mean that group differences in behavior are caused by genetic or ecological differences.

A number of social learning researchers have addressed this issue with an approach known as the "method of exclusion" (Whiten et al. 1999; van Schaik et al. 2003; Krützen et al. 2007). Here a pattern of differences in repertoires, consistent with social transmission, is identified (e.g., by group contrasts), and social transmission is inferred to be the cause of behavioral dissimilarities when the genetic and ecological hypotheses can be ruled out. In section 6.2 we discuss two versions of the method of exclusion, and give our view on the circumstances under which they can be used.

In most circumstances, however, we envisage that it may be better to adopt a model-fitting approach, in which a researcher aims to quantify the effects of genetic differences, ecology, and social transmission. In section 6.3 we discuss a version of this approach that has been advocated by Whitehead (2009). In section 6.4, we then suggest that this approach can be expanded still further to utilize recent developments in causal graphs. We suggest this provides a unified framework for viewing all the methods discussed in the chapter. Here different methods are underpinned by various a priori judgments about which causal pathways are plausible, and what resolution of data is available for fitting models of social transmission.

6.1 The Group-Contrasts Approach

By the "group-contrasts method" we mean any approach that compares the behavioral repertoire of different groups, and uses this to infer social transmission of behavioral traits. The underlying logic is that if behavioral traits occur by innovation and subsequent spread through a group, then the behavioral repertoire of each group will likely differ because different innovations happened to arise and spread. The group-contrasts method has also been called the "geographic method" (van Schaik 2008) in cases where the groups in question are located at distinct locations, such as at specific field study sites. Here we use the former, more general term, which could also be applied in cases where the groups are sympatric (e.g., pods of orcas; Rendell and Whitehead 2001).

The group-contrasts method has been used to infer social transmission in chimpanzees (Whiten et al. 1999, 2001), orangutans (van Schaik et al. 2003), capuchin monkeys (Perry et al. 2003), and bowerbirds (Madden 2008). Here we use Whiten et al.'s (1999) application of this method to chimpanzee data as our primary example, on the grounds that it is the most well-known case, to which many of the methods we describe in section 6.1 have been applied.

The term "method of exclusion" has sometimes been used as a synonymous term to group-contrasts; however, we see it as logically distinct. The key idea associated with the method of exclusion is that behavioral differences can be attributed to social transmission if they cannot be explained by genetic or ecological differences between the units of study. This method has, thus far, been the primary method used on group-contrasts data. However, the method of exclusion can also be applied to explain behavioral differences within a group. For example, Krützen at al. (2005) use this approach to infer social transmission of a tool-using trait; that is, sponging in a single matriline of bottlenose dolphins (*Tursiops* sp.).

We begin by describing the methods used to establish that group-contrasts in behavior exist.[1] The first stage is to identify and define a number of candidate behavioral traits that are thought not to be present in all the groups being studied, or to vary in their frequency across groups. Whiten et al. (1999) used long-term data from seven chimpanzee field sites, five in East Africa and two in West Africa; they also identified 65 candidate traits including tool usage, grooming, and courtship behavior. Instead of determining traits as being merely present or absent in each group, researchers have usually used an ordinate scale based on how many individuals exhibit the trait. For example, Whiten et al. (1999) used the following ordinal scale:

Customary: The trait occurs in all or most able-bodied members of at least one age-sex class (such as adult males).

Habitual: The trait is not customary but has occurred repeatedly in several individuals, consistent with some degree of social transmission.

Present: The trait is neither customary nor habitual but is clearly identified.

Absent: The trait has not been recorded and no ecological explanation is apparent.

Unknown: The trait has not been recorded but there have been inadequate opportunities to observe it.

While this kind of categorical analysis may be all that is feasible, ideally precise quantitative data would be available specifying the frequency of each putative "cultural" trait in each group, since this would yield far greater statistical power to any subsequent analysis. For example, quantitative data would enhance the power of the model-fitting approaches that we advocate (section 6.3) and also allow the researcher to apply other tools, such as the option-bias method (R. L. Kendal, J. R. Kendal, et al. 2009; see box 6.1 for details), as a means of establishing whether learned differences between groups are likely to have been socially transmitted.

When collecting data on group contrasts, researchers usually also consider whether there is an obvious ecological reason why a particular trait is absent. In most of the cases described above, the researcher infers social transmission as being responsible for the group contrast for a trait where there is no obvious ecological reason for the contrast. For example, Whiten et al. (1999) conclude that 39 of their 65 traits are socially transmitted: 7 were ruled out because they were customary or habitual in all groups; 16 were ruled out because they failed to reach habitual status in any groups, and presumably could be isolated innovations that failed to spread; and an additional 3 were ruled out because their absence had an ecological explanation.

---

[1] Here we do not cover the practicalities of the field techniques, such as how much observation time is required to establish an approximately complete behavioural repertoire (van Schaik 2009). Such practicalities, while critical, will differ depending on the species being studied, and will need to be adapted accordingly. For guidance, we suggest the reader start by consulting the examples given in the text.

Box 6.1.

The option-bias method

In many of the cases described above, it is often very clear that individuals in different groups perform unlike variants of a behavioral trait (e.g., ant dipping, see section 7.1.4)). In other cases, a group effect might be subtler, with individuals of the same group being more likely to perform the same variant, but with some stochastic variability both within groups and within individuals. This pattern can be termed an "option bias" (R. L. Kendal, J. R. Kendal, et al. 2009), with each way of performing the trait termed an "option." In such cases it is necessary to ascertain whether the observed option bias could have arisen by chance, or if individuals settle on a preferred option by asocial learning, independently of the choices of others in their group. The null hypothesis is that individuals in the same group are no more likely to choose the same option than those in different groups. A prima facie solution to this would be to apply a Chi-square test to a contingency table of the options performed in each group; however, this is not a valid approach, because the option-choices made by a single individual are unlikely to be independent. A solution is to use a randomization test as follows:

1. Calculate the test statistic T_{DATA} for the contingency table (e.g., Chi-square)
2. Set N = 1, X = 1
3. Reassign individuals to groups, maintaining the same number of individuals in each group
4. Recalculate the test statistic, T_N
5. If $T_N \geq T_{DATA}$, increment X by 1
6. Increment N by 1
7. Repeat steps 3–6 until N is suitably large, say 10,000
8. The *p*-value against the null hypothesis is X/N

This "option-bias test" (R. L. Kendal, J. R. Kendal, et al. 2009) allows for the nonindependence of option choices by the same individual, because the randomization procedure ensures that the option choices made by an individual are always assigned to the same group. The logic underlying this procedure is that if the null hypothesis were true, the assignment of individuals to groups is arbitrary with respect to option choice, so T_{DATA} should be a typical value from the null distribution of T values. The *p*-value quantifies the probably of getting a value of T_{DATA} that is at least as big as that observed under the null hypothesis. For more details on the logic underlying randomization tests, see Manly (2008). This randomization procedure was found to have improved power and more appropriate type 1 error rates than alternative statistical tests, and was also found to typically be more powerful than a Monte Carlo simulation approach (R. L. Kendal, J. R. Kendal, et al. 2009).

   The option-bias test can be modified for groups that were seeded with demonstrators trained to perform a specific option; this modification is accomplished by constraining the demonstrators to remain in their groups during the randomization. The method could also be modified for noncategorical behavioral traits by replacing the test in step 1 with an appropriate test for group differences. For example, for a trait that varies continuously, one could apply an ANOVA with group as a factor in step 1, taking the *F* statistic as the test statistic.

   The option-bias test only allows us to infer whether there is a group-level option bias that needs explaining. It does not, in itself, establish that this difference is a result of social transmission. To do so, a researcher needs to exclude or account for genetic and ecological hypotheses.

6.2 The Method of Exclusion

We define the method of exclusion as the approach whereby a researcher identifies a pattern of variation in the behavioral repertoires of the focal population; the researcher then infers social transmission is at least partly responsible for this pattern on the grounds that they are able to exclude genetic and ecological influences as sufficient explanations for the pattern. We envisage two ways in which this can be achieved, which we term the "basic"[2] and "advanced" methods of exclusion.

6.2.1 Basic and advanced methods of exclusion

The *basic method of exclusion* refers to cases where the researcher simply argues that it is implausible that ecology and/or genetic differences could influence the traits in question, and so social transmission must be responsible. In contrast, the *advanced method of exclusion* refers to instances where the researcher recognizes that ecology and/or genetics influence, or might influence, the trait in question. Here the researcher fits a null model in which ecology and/or genetics influences the traits adopted, and hence shows this to be an insufficient explanation for the observed pattern of repertoires. The strength of evidence against the null model is taken to be strength of evidence for social transmission. Note that a researcher might use one method to exclude the genetic hypothesis and the other to exclude the ecological hypothesis.

The dichotomy we present might be something of a caricature, but we feel it helps to understand and resolve some of the debates that have arisen around the method of exclusion. The most important point to recognize is that methods that have been suggested for addressing the genetic and ecological hypotheses can either be seen as (*a*) potentially refuting an argument for social transmission by the basic method of exclusion, or (*b*) potentially supporting an argument for social transmission by the advanced method of exclusion. For the methods in category *a*, the null hypothesis is "no effect of environment/genetics on the behavioral traits in question," so strong evidence against the null hypothesis is strong evidence *against* the premises used in the basic method of exclusion. For the methods in category *b*, the null hypothesis is "genetics and/or ecology account for the observed pattern of behavioral traits." Strong evidence against this hypothesis in the appropriate direction is strong evidence *in favor* of social transmission. A lack of evidence against type *a* null hypotheses cannot be taken as evidence *for* the basic method of exclusion, and a lack of evidence against type *b* null hypotheses cannot be taken as evidence *against* the advanced method of exclusion, since failure to reject a null hypothesis does not constitute rejection of an alternative.[3]

---

[2] Our use of the term "basic" here should not be seen derisive, as there are cases where we feel this approach is sufficient.

[3] Or, as we prefer: *weak evidence against the null hypothesis does not necessarily constitute strong evidence for an alternative hypothesis*, since "rejection" is based on an arbitrary significance level. Here we are assuming that *p*-values are obtained quantifying the strength of evidence against the null hypothesis,

In the following sections we consider methods for assessing the genetic and ecological hypotheses. For each, we first consider the methods that have been used to refute the basic method of exclusion (type *a* null hypotheses), before moving on to the methods that have been proposed for finding evidence in support of the advanced method of exclusion (type *b* null hypotheses).

6.2.2 Methods for assessing the genetic hypothesis

6.2.2.1 Refuting the basic method of exclusion

Langergraber et al. (2010) assessed the genetic hypothesis for Whiten et al.'s (1999) chimpanzee data by testing whether genetic similarity among the nine groups was associated with behavioral similarity. Their methods could be used to potentially provide evidence against an exclusion of the genetic hypothesis by the basic method, in any case where the appropriate genetic data are available. Langergraber et al. sequenced most of the members of each group for the first hypervariable region of mitochondrial DNA (mtDNA), and calculated genetic dissimilarity using pairwise F_{ST} values (see box 6.2). They then calculated overall behavioral dissimilarity among the groups using both the normalized Hamming distance and Manhattan distance (see box 6.3).

Langergraber et al. (2010) tested whether genetic similarity and behavioral similarity were correlated using both Pearson and Spearman correlation coefficients, and determined the statistical significance of each correlation using matrix permutation methods. They found reasonable evidence for a positive correlation of all combinations of distance measure and correlation coefficient (normalized Hamming distance: Pearson's $r = 0.52$, $p = 0.015$; Spearman's $r_s = 0.37$, $p = 0.031$; and Manhattan distance: Pearson's $r = 0.44$, $p = 0.018$; Spearman's $r_s = 0.36$, $p = 0.029$; see fig. 6.1).

Langergraber et al. (2010) conclude that evidence of a correlation between genetic and overall behavioral dissimilarity indicates that genetic dissimilarity "cannot be excluded as playing a major role in the patterning of behavioural variation among chimpanzee groups" but that it would "not necessarily exclude social learning [transmission]" (411). In our terminology, it is evidence against an argument by the basic method of exclusion (i.e., that genetic differences could not plausibly explain behavioral differences). However, this method of quantifying the relationship between genetic and behavioral differences could be of greater utility in the framework that we discuss in sections 6.3 and 6.4.

These authors repeated their analysis for each behavioral variant individually, finding a range of sample correlations (Spearman's r_s ranging from –0.62 to 0.85). For only 8 of the 38 variants was there reasonable ($p \leq 0.05$) evidence of a positive

---

as has typically been the case. The reader might argue instead that Bayesian or information theoretic approaches might be used to assess the strength of evidence for competing models. Alternatively, a researcher might obtain confidence intervals on the estimated genetic/environmental effects, and thus argue that there is strong evidence against theses effects being large enough to account for the putative pattern of social transmission. We would agree, however, these approaches are more at home in the general frameworks presented in sections 6.3 and 6.4.

> **Box 6.2**
>
> **Genetic similarity/ dissimilarity matrices**
>
> At the individual level, genetic similarity can be given by an estimate of kinship—the proportion of genes shared through common recent ancestry (section 4.2 in Whitehead 2008). Where genealogy is known, this can be estimated directly. In practice, kinship is often inferred by estimates of genetic relatedness using molecular genetic techniques, often with microsatellites (Van de Casteele et al. 2001). A cruder, binary measure is given by whether two individuals share the same mitochondrial DNA (mtDNA) haplotype, which indicates whether individuals are from the same matriline.
>
> At a group level, there are numerous measures of genetic distance among groups (or populations). The most commonly used are pairwise estimates of F_{ST}, which is defined as the proportion of genetic diversity due to allele frequency differences among groups (Holsinger and Weir 2009). A number of related measures are also used (G_{ST}, R_{ST}, Φ_{ST}, and Q_{ST}), depending on the type of genetic data that is available. All of these measures partition genetic diversity into within—and among—population components. For a recent review of these measures and methods of estimation, see Holsinger and Weir (2009). In their review, these authors note that although pairwise estimates of F_{ST}, R_{ST}, and Φ_{ST} provide some insight into the degree to which populations are historically connected, they do not allow us to determine whether that connection is a result of ongoing migration or of recent common ancestry. However, for the purposes described in this chapter, this is unlikely to be an important distinction.

correlation. However, Langergraber et al. (2010) note that there is very low power in such analyses to reject the null hypothesis that there is no relationship between genetic and behavioral variation. Consequently, they only use the analysis to explore general patterns as to which behavioral variants are more or less associated with patterns of genetic dissimilarity. They found no obvious pattern regarding which types of behavior (e.g., fishing action, exploitation of leaf properties) had a distribution that was more or less strongly predicted by genetic dissimilarity. However, we would add that the low statistical power noted by Langergraber et al. (2010) will inevitably be associated with a lack of precision in the estimates of the correlation for each variant (at the population level), which might act to obscure any such pattern.

In general, we suspect there will be few cases where a researcher can, a priori, really rule out any influences of genetic differences on the behavioral traits in question. To illustrate, Whiten et al. (1999) sampled across two chimpanzee subspecies—*verus* at the western sites and *schweinfurthii* in the east. A third of the chimpanzee "cultural" variants are observed in one subspecies alone. Some chimpanzee subpopulations have been genetically isolated for hundreds of thousands of years, and *verus* and *schweinfurtii* occupy distinct branches of a neighbor-joining tree based on genetic distances for mtDNA haplotypes (Gagneux et al. 2001). Thus, it is not implausible that some behavioral differences between chimpanzees have a genetic origin. Likewise, van Schaik et al. (2003) sampled two orangutan species in Borneo (*P. pygmaeus*) and Sumatra (*P. abelii*), and half of the orangutan variants are only seen in one of the two species, rendering a genetic hypothesis tenable here too. More generally, we would expect many geographically distributed groups of animals to differ genetically.

In our view, a large part of the value of Langergraber et al.'s (2010) study is that it reinforces the view that the genetic hypothesis is an alternative explanation to social transmission that needs to be taken seriously, and cannot be ruled out a priori. But how can this be done? Below we attempt to evaluate both the success of attempts to assess the genetic hypothesis using the chimpanzee dataset of Whiten et al. (1999), and the extent to which such attempts offer a general solution for excluding the genetic hypothesis applicable to other species.

6.2.2.2 Advanced method of exclusion

Langergraber et al. (2010) undertook an additional analysis that could potentially exclude the genetic hypothesis. The authors determined whether each

---

Box 6.3

Calculating behavioral similarity/ dissimilarity matrices

The first point to note is that some methods require similarity matrices, whereas others require dissimilarity (or "distance") matrices. An easy way to convert from a similarity to dissimilarity matrix, or vice versa, is to subtract every value from the maximum value in the original matrix. For the matrix regression techniques described in box 6.6, if researchers transform the dependent matrix in this way, it merely has the effect of reversing the sign of the standardized regression coefficients. Therefore, if researchers are consistent in using similarity or dissimilarity matrices for all variables, they will get exactly the same results whichever they choose to use.

The first step is to decide whether to calculate behavioral similarity for a single behavioral trait, or for a collection of traits. The collection could perhaps consist of all known traits suspected of being socially transmitted (e.g., Langergraber 2011) or of a particular part of the behavioral repertoire such as foraging (e.g., Matthews 2009). A sensible approach might be to apply the relevant techniques at a repertoire level to test for an effect of social transmission in general, and then break the analysis down by trait to examine which ones are specifically influenced (Matthews 2009).

The form a behavioral similarity matrix takes depends on the trait that is of interest. Researchers might be interested in a single behavioral trait that has a number of different variants, such as chimpanzee ant dipping (Whiten et al. 1999), where they could use a binary matrix with a 1 for pairs of individuals who perform the same variant, and a 0 for those that do not (Whitehead 2009). Alternatively, behavioral similarity could be continuous; for example, researchers could measure vocal dialect similarity by the proportion of call types shared (Whitehead 2009). For a more complete discussion of the ways in which different types of data can be treated, see Kaufman and Rousseeuw (2005, ch. 1). Here we restrict ourselves to describing the methods used in the examples discussed in the main text (Matthews 2009; Langergraber et al. 2011), which include a useful general algorithm for computing behavioral similarities.

Langergraber et al. (2011) calculated behavioral dissimilarities based on the categories assigned by Whiten et al. (1999) for each trait. They assigned the following values: 3 = customary, 2 = habitual, 1 = present, and 0 = absent. If a trait was "unknown" for either group in a pair, it was not used to calculate the dissimilarity for that pair. Langergraber et al. used two measures: (1) the normalized Hamming distance, which is the number of traits with a different code, divided by the number of traits for which the pair of groups was compared; and (2) the Manhattan distance, the sum of the absolute values of the differences between trait values.

behavioral variant differed between groups that were not genetically differentiated, as determined by two criteria: (*a*) nonsignificance (p > 0.05) of genetic differences in a permutation test, and (*b*) groups occupying the same block of forest. Langergraber et al. (2010) note that that criterion *a* should be treated with caution, since it depends on the statistical power of each comparison, but note that *b* is a conservative criterion by which "gene flow is almost certainly insufficient to generate differences in behaviour" (411). The logic here is that a genetic explanation can be excluded for variants that vary among genetically similar groups. Using criterion *a*, the genetic hypothesis could be excluded for approximately half of the variants, and using criterion *b*, it could be excluded for 5 of the 38 variants.

The Manhattan distance is designed for linear continuous measures, but is often used for ordinal data of this kind (Kaufman and Rousseeuw 2005). The latter approach has the advantage that it acknowledges that a pair of groups with a 2 and a 3 is more similar than a pair with a 0 and a 3. It has the potential disadvantage that it assumes a linear scale (e.g., that the difference between presence and absence is the same as the difference between customary and habitual). The approach of repeating the analysis using both measures seems a sensible compromise for this kind of data.

More generally, researchers might be interested in a range of traits forming some part of the repertoire, say foraging behavior (e.g., Matthews 2009), some of which vary on a continuous scale, and some of which are discrete variants. A solution to this disparity is to use the general dissimilarity coefficient of Gower (1971; cited by Kaufman and Rousseeuw 2005), which can handle any combination of continuous, nominal, ordinal, and binary data. A further advantage of Gower's dissimilarity coefficient is that it can allow for asymmetrical binary variables, meaning that when both individuals share the value 0, it is considered to be uninformative. Matthews (2009) suggests an asymmetrical variable be used for the presence/absence of a particular trait, since not possessing it is presumably the default condition, and so provides no information.

Gower's general dissimilarity coefficient between two individuals *i* and *j* is given by:

$$d(i,j) = \frac{\sum\limits_{f=1}^{p} \delta_{ij,f} d_{ij,f}}{\sum\limits_{f=1}^{p} \delta_{ij,f}}.$$

This is a weighted mean of the contribution of each of *p* variables, with a variable *f* carrying weight $\delta_{ij,f}$. This weight is zero if the relevant data are missing, or if *f* is an asymmetrical variable and its values for *i* and *j* are both zero, and $\delta_{ij,f} = 1$ otherwise. The value $d_{ij,f}$ is the unweighted distance between *i* and *j* with respect to variable *f*. In the case of nominal or binary variables, $d_{ij,f}$ is set to 0 if *i* and *j* are equal and 1 otherwise. In the case of ordinal or continuous variables, $d_{ij,f}$ is the absolute difference between *i* and *j* divided by the total range of the variable (see Kaufman and Rousseeuw 2005 for more details). At the time of writing, Gower's general dissimilarity coefficient can be calculated in the R statistical environment (R Core Development Team 2011) using the "daisy" function in the "cluster" package (Maechler et al. 2005).

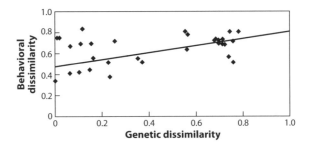

Figure 6.1. The relationship between behavioral and genetic dissimilarity in 36 chimpanzee groups. Based on figure 1 from Langergraber et al. (2011).

We believe this approach could be used to exclude the genetic hypothesis, provided it is applied using a conservative criterion like b. The problem with a is essentially the same as with the other methods—a failure to find genetic differentiation between groups is not evidence that the groups are undifferentiated. The advantage of b is that additional biological knowledge can be brought to bear; in this case, this additional information specifies that chimpanzee groups inhabiting the same block of forest are unlikely to be sufficiently different genetically to support behavioral differences. However, it is unclear how often a researcher would be able to find enough genetically nondifferentiated groups to be able to take this approach. In addition, populations may coexist sympatrically in some species, but without interbreeding, thus allowing genetic differences to become established. Moreover, we note that the method could be conservative in ruling out the genetic hypothesis. First, the same behavioral variants might happen to arise as independent innovations in genetically nondifferentiated groups; second, there may be cultural transfer between groups, resulting in the same behavioral variants being present in each one.

An alternative framework for testing, and potentially ruling out, the genetic hypothesis is the use of phylogenetic analysis. One possibility would be to test whether the pattern of behavioral traits across groups is likely to be have been generated by biological evolution, given a phylogeny of the genetic relationships between those groups. This can be done using the Kishino-Hasegawa (K-H) test (Kishino and Hasegawa 1989; cited in Lycett et al. 2009). The K-H test can be used to test hypothesized processes that might account for the structure of a cladogram; this is done by comparing the maximum parsimony (MP) cladogram with cladograms reflecting those hypothesized processes. A researcher could obtain a phylogeny for their groups, derived from genetic data, and use this as the hypothesized tree against which the MP cladogram for the behavioral traits is tested. The p-value yielded by the K-H test then tells us the extent to which the behavioral data are incompatible with the genetic hypothesis. This has the advantage that the genetic hypothesis is treated as the null hypothesis, which seems to us to be critical for the method of exclusion.

Lycett et al. (2009) applied this approach to the chimpanzee data collected by Whiten et al. (1999), and found that there was not strong evidence against the vertical transmission of traits ($p = 0.77$). However, Lycett argues (pers. comm.) that this is not a good test of the genetic hypothesis, because even if a trait is culturally transmitted, genetic differences will still be correlated with behavioral differences due to a shared evolutionary history between groups. Indeed, Lycett et al. were using the method to determine whether traits they believe to be cultural are primarily vertically transmitted between groups; they argue that these traits are *less* subject to horizontal transfer than genes (see below). Nonetheless, if a researcher found strong evidence against vertical transmission of a behavioral trait in animal populations using the K-H test, it would offer a compelling refutation of the genetic hypothesis.

Lycett et al. (2007) argue that a more discriminating approach, at least in the case of the chimpanzee data, is to test whether aspects of the structure of the phylogeny is reflected in the structure of a tree constructed from the behavioral data. The authors note that chimpanzees living in East Africa (*P. t. schweinfurthii*) and West Africa (*P. t. verus*) are genetically well-differentiated subspecies, whereas those living in East Africa are genetically similar (i.e., all members of the same subspecies). They reason that if the genetic hypothesis were correct, cladograms using the behavioral traits should exhibit the same pattern. The authors performed one cladistic analysis of the data from the five East African groups, and another using all seven groups across the continent (see fig. 6.2). In each case, they calculated the Retention Index (RI), which is a measure of the number of homoplastic changes, correcting for the length (total number of changes) in the cladogram. Lycett et al. reasoned that if the genetic hypothesis is correct, the RI yielded by the continental analysis should be higher than the RI yielded by the East African analysis. In contrast, they found that RI is lower in the continental analysis (0.44) than in the East African analysis (0.53), and conclude that the genetic hypothesis is refuted (i.e., genetic differences among groups cannot fully explain differences in behavior).

Lycett et al. (2007) argue that the observed mismatch between the genetic and behavioral data can be explained if the traits are culturally transmitted. It

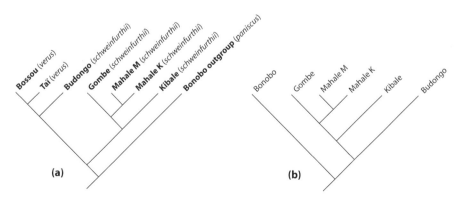

Figure 6.2. Maximum parsimony tree produced by cladistic analysis of chimpanzee behavioral data, for (*a*) East and West African populations and (*b*) East African populations alone. Based on figure 1 from Lycett et al. (2007).

is likely that dispersing females are responsible for both genetic and cultural transmission among groups, since males remain in their natal group and do not engage in prosocial interactions with males from other groups. Lycett et al. (2007) reason that the observed genetic/behavioral mismatch is likely to be a consequence of females transmitting genes among groups at a greater rate than they transmit culture. They suggest this could be because females adopt the behavior of the group they move into and/or tend not to be copied by members of their new group.

However, a number of researchers have expressed concerns that Lycett et al.'s conclusion might not follow. First, Lycett et al. (2007) provide no statistical test to show how unlikely the observed pattern of RI is under the genetic hypothesis (box 6.4; see also Langergraber et al. 2011; Laland et al. 2009). This could be done using simulations of trait evolution under the genetic hypothesis. However, simulations along these lines (Nunn et al. 2010) suggest that the pattern of RIs

---

Box 6.4

Statistical tests and the RI method

Lycett (2010) has argued that in the chimpanzee case, a statistical test is not necessary for excluding the genetic hypothesis on the basis of tree structure:

> Note that it is not necessary to determine whether these RI values are statistically different in order to be in contradiction with the genetic hypothesis. This is because the genetic hypothesis predicts a marked decrease in phylogenetic structure when using the reduced data set as opposed to the all-regions data set. Thus, even an equal RI value (i.e., nonsignificantly different) in both parsimony analyses would be inconsistent with the predictions of the genetic hypothesis. Hence, a statistical difference between the RI values is not essential to reject the genetic hypothesis in the face of an increased RI value when moving from the extended data set to the reduced data sets used in these analyses. (257)

While this argument has prima facie appeal, in general it does not follow that if the results go in the opposite direction to that predicted by a hypothesis, then there is strong evidence against that hypothesis. This is best illustrated with an analogous example. Imagine we have a coin that, when flipped, we suspect of being biased toward tails (the "tails hypothesis"). We flip the coin a single time, and it lands heads up. In parallel with Lycett (2010), we can argue that to show the coin is biased toward tails, we need to show that it lands tails up significantly more often than heads up. The observed result goes in the opposite direction to that predicted by the tails hypothesis, and so the hypothesis is not supported. However, it is clear in this case that we do not have strong evidence *against* the tails hypothesis. In fact, the Wilson's 95% confidence interval for the probability of tails is 0–0.949, showing that it is still entirely possible that the coin is biased toward tails. Alternatively, using a Bayesian approach, the posterior probability that the coin is biased toward tails is still 0.25 (if we start with a uniform prior U[0,1] then the posterior distribution is Beta[1,2]).

Again the problem here is that the genetic hypothesis is not being treated as the null hypothesis. If a researcher's interest is in showing the genetic hypothesis alone is unlikely (the advanced method of exclusion), they must show how unlikely the observed pattern of RIs is under the genetic hypothesis.

observed by Lycett at el. (2007) might be consistent with the genetic hypotheses, under certain circumstances. Nunn et al. argue that if behavioral variations were genetically inherited and not learned, then a sufficiently high evolutionary rate could cause a reduction of RI in a tree with deeper phylogenetic separation. They conclude that the RI-based test conducted by Lycett et al. (2007) is insufficient to rule out genetic inheritance of the behavioral variations.[4] Lycett et al. (2011) have responded to Nunn et al.'s criticism by pointing out that their conclusion is not drawn solely from the pattern of RIs across the behavioral tree, but instead from the fact that this pattern is different from the genetic data.

While Lycett et al.'s (2007) finding is inconsistent with the hypothesis that the putative cultural variation is determined *primarily* by genetic differences among populations, whether it provides strong support for the culture hypothesis is a moot point (though we do not know how strong the evidence is for the inconsistency; see box 6.4). The method is applied to repertoire patterns across all behavioral traits in the data set, and does not preclude the possibility that differences in a subset of the traits are underpinned by genetic differences.[5] Moreover, their findings are such that it remains tenable that considerable genetic and ecological variation account for differences in chimpanzee behavior. The low RIs for both maximum parsimony trees indicate considerable homoplasy in the data, an observation consistent with parallel independent evolution, or independent learning, in response to local ecology (or alternatively, diverse cultural lineages for different behavior patterns), a point we return to in the next section.

Lycett et al.'s (2007) method could probably be adapted to address many of the concerns above, but it seems unlikely that it could provide a general solution for detecting social transmission in animals. This is because researchers require variation in enough behavioral traits to construct a "cultural" cladogram, so the methods could not be used when they have only a few candidate traits. Furthermore, researchers require a genetic phylogeny that suggests a particular pattern in structure if the genetic hypothesis were true, such as the differentiation between East and West African chimpanzees, which might not exist in all cases. Nonetheless, we do not doubt that, in general, phylogenetic analysis of culture is a valuable exercise, which might tell us a great deal about routes of transmission among groups, and the adaptive value of culturally transmitted traits (see section 9.2.8; Lycett 2010; Nunn et al. 2010).

One further complication is worthy of mention here: researchers may need to think carefully about the kind of genetic data they collect in order to assess genetic similarity. For example, the use of genetic distance in mitochondrial DNA markers as a proxy for the genetic distance between the largely nuclear genes that

---

[4] Lycett et al. (2009) have gone on to repeat their analysis using alternative measures of phylogenetic structure, the phylogenetic bootstrap and the permutation tail probability (PTP) test, both of which use randomization methods to quantify relative phylogenetic signal. In both cases they find more structure when only the East African groups are included, which they take to refute the genetic hypothesis. Unfortunately, these alternative measures were not considered by Nunn et al. (2010) in their simulations but, to our understanding, it is possible that they are affected by rates of evolution in the same way as RI.

[5] This limitation also applies to the methods we discuss below, when numerous traits are analyzed.

actually underlie behavior is likely to be low resolution for male dispersing species.[6] The dispersal of males that carry genotypes expressed in behavior from one population to another will not show up as genetic similarity among populations, since they do not pass on the mitochondrial DNA; this leads to an apparent mismatch between genes and behavior. To avoid or minimize the chances of this bias operating, researchers might be well advised to utilize Y chromosomes, or microsatellites, in addition to mtDNA, or to choose the type of genetic data carefully.

Further insight into these issues can be gained from Krützen et al.'s (2005) analysis of sponging in bottlenose dolphins. Sponging refers to when a "dolphin breaks a marine sponge off the seafloor and wears it over its closed rostrum to apparently probe into the substrate for fish" (Krützen et al. 2005, 8939). Krützen et al.'s study is not based on a group-contrasts approach, but instead on the fact that the sponging behavior is almost exclusive to a single matriline (12 of the 13 sponging dolphins sampled had the same mtDNA haplotype). Krützen et al. hypothesized that this was a result of social transmission of sponging from mother to (mainly) female offspring. As anticipated by both this social transmission and the genetic hypothesis, relatedness between spongers, as estimated using nuclear DNA markers, was higher than would be expected by chance. However, Krützen et al. argue that the genetic hypothesis can be ruled out because the distribution of sponging does not fit with any of 10 single locus models of genetic inheritance (see Krützen et al. 2005, table 3 for details). Some models predict no sex bias in sponging, whereas of the 15 spongers that have been recorded, only one is a male.[7] Similarly, the authors claim they can rule out other genetic models that predict sponging in only females or males because sponging has been reported in both sexes. Krützen et al. (2005) also rule out the hypothesis that sponging is inherited though mtDNA, not only due to the observed sex bias, but because not all individuals with the focal haplotype are spongers.

A criticism of Krützen et al.'s (2005) exclusion of the genetic hypothesis is that it relies heavily on the observation of a single male sponger, which rules out four genetic models that predict only female spongers. If we allow for the possibility that any "sponging" locus merely affects the probability that sponging will develop, these genetic models might still be plausible. We illustrate this using simple statistical models in box 6.5. Nonetheless, when adapted in this manner, we believe that testing the predictions of specific models of genetic transmission is one of the more promising methods for excluding the genetic hypothesis.

There remains the concern, however, that in many, and perhaps in most, cases social transmission is likely to result in a pattern of inheritance that is also consistent with at least one genetic hypothesis, and so will result in a false negative for social transmission under the method of exclusion. This reflects a more general

---

[6] The problems associated with the use of mtDNA for male-dispersal species extend beyond low resolution—under some circumstances the genome and mitochondria in these species can actually be discordant in their real evolutionary histories (Matthews 2012).

[7] Krützen et al. conclude that these models are not possible (see table 3 in Krützen et al 2005). Strictly speaking, it is *possible* to obtain the observed distribution if there was no sex bias: it would, however, be very unlikely.

issue, recognized by all researchers working on this problem, that genetic differences will often be correlated with behavioral differences, even if the latter are caused by social transmission.[8]

6.2.3 Methods for assessing the ecological hypothesis

6.2.3.1 Refuting the basic method of exclusion

On the face of it, the ecological hypothesis seems easier to exclude than the genetic hypothesis, because ecological factors are easier to observe or measure. Researchers using the basic method of exclusion have frequently attempted to identify when the absence of a trait can be explained by local ecology, usually because the resources required to perform the trait are absent. For example, Whiten et al. (1999) rule out algae fishing as a "cultural" trait (since its absence can be fully explained by the rarity of algae), and ground nesting at night (the absence of which can be explained by high predator risk). The logic is that if a trait often does not occur in locations where the appropriate resources are available, then in places where it does occur, it is likely to be an innovation that has spread by social transmission (van Schaik 2009). This is what we term the basic method of exclusion: the ecological hypothesis is excluded because there is argued to be no ecological influence on differences in the traits in question. Using this logic, van Schaik et al. (2006) conclude that 33/43 geographically variable behavior patterns in orangutans could be considered as socially transmitted innovations, and the remaining 10 as probable innovations.

Van Schaik (2009) suggests that this reasoning can be supplemented with experiments, in which the researcher provides the materials necessary for the behavioral trait to be performed in areas where they are missing. Here the null hypothesis is that the subjects will not perform the trait, which corresponds to the assumptions of the basic method of exclusion. A positive result refutes the argument for exclusion of the ecological hypothesis by the basic method. However, a negative result does not constitute rejection of the ecological hypothesis. As van Schaik notes, a negative result might reflect flaws in the experimental design, such as the materials not being provided for long enough or at the right stage of development.

In general, there may be reasons (aside from the genetic hypothesis) why a trait might not be performed when the necessary resources are available. Prey selection models in optimal foraging theory suggest that use of a prey item is independent of its abundance and is instead dependent on the availability of prey items with a higher profitability (Stephens and J. R. Krebs 1986; J. R. Krebs and Davies 1993, 61). The same logic might be applied in the case of some foraging traits. A foraging trait might not be performed, even when the necessary resources are available, because more profitable means of gaining food are freely available. Moreover, a priori there is no reason to expect the more profitable

---

[8] If we employ a model-fitting approach (section 6.3), this would manifest itself as uncertainty in the parameters representing the strength of the effects of genetic and social transmission.

Using likelihood-based models to test genetic and ecological hypotheses

Often researchers judge the genetic and ecological hypotheses to be ruled out by the data on informal grounds. If researchers are committed to using the method of exclusion, we advocate instead that they, where possible, use likelihood-based models. These can be used to formally quantify the support for, or evidence against, the genetic and ecological hypotheses relative to the alternative social transmission model. We illustrate this with analyses applied to the data on sponging in bottlenose dolphins collected by Krützen et al. (we use their dataset 1; see table 2 in Krützen et al. 2005). Krützen et al.'s exclusion of both genetic and ecological hypotheses has been questioned as part of an ongoing debate about the usefulness of the method of exclusion (Laland and Janik 2006; Krützen et al. 2007; Laland and Janik 2007). The specific models we apply here could almost certainly be improved upon, but our aim is to illustrate how such formal quantification of the evidence can be achieved in principle.

As discussed in the main text, Krützen et al. (2005) rule out 10 single-locus models of inheritance of sponging, because they make predictions about the distribution of the trait that are incompatible with the data. Four such models are ruled out because they predict that only females would inherit the trait, which is ruled out by a single male sponger. However, we note that these models cannot be immediately ruled out as impossible if one considers that the hypothetical sponging locus merely acts to make sponging more likely to develop. Krützen et al.'s (2005) favored hypothesis is that of "almost exclusive vertical social transmission within a single matriline from mother to female offspring" (8939). Both hypotheses predict females with haplotype H to be more likely than other dolphins to develop sponging. This can be modeled with a probability of sponging for females with haplotype H. However, these probabilities are estimated to be the same under each model, and so we can ignore this data when assessing the relative fit of the two.

Under the modified genetic hypothesis, males with any haplotype and females without haplotype H are equally likely to develop sponging; this hypothesis is fitted with a single parameter, with a maximum likelihood estimator (MLE) = 2/110. This gives a log-likelihood of:

$$\log\left(\left(\frac{2}{110}\right)^{2}\times\left(\frac{108}{110}\right)^{108}\right) = 2\log(2) + 108\log(108) - 110\log(110) = -10.0.$$

We can then compare this with a social transmission model in which males with haplotype H develop sponging with a probability with MLE = 1/10, (one male out of 10 learned), and all individuals without haplotype H have a probability with MLE = 1/100 (one female out of 100 females and males without haplotype H), giving a log-likelihood of:

$$\log\left(\left(\frac{1}{10}\right)\times\left(\frac{9}{10}\right)^{9}\times\left(\frac{1}{100}\right)\times\left(\frac{99}{100}\right)^{99}\right)$$
$$= 9\log(9) + 99\log(99) - 10\log(10) - 100\log(100) = -8.85.$$

From this we can calculate the difference in AIC (see box 5.3), which is $-2\times$ the difference in log-likelihood, plus $2\times$ the difference in degrees of freedom (1) $= -2 \times (-8.85 + 10.0) + 2 = -0.29$. This corresponds to only $\exp(-0.5 \times -0.29) = 1.16\times$ more support for the social transmission hypothesis (Akaike weight = 53.6%). Alternatively, since these are nested models, we could use a likelihood ratio test (LRT): LR $= 2 \times (-8.85 + 10.0) = 2.3$, which we look up in the upper tail of the cdf of a Chi-square distribution with 1 d.f. to give $p = 0.13$. This is the strength of evidence against the simpler model (i.e., the genetic model). These models are only a simple attempt to address the question, but are sufficient to show that the genetic hypothesis is not so easy to rule out if we view the hypothetical sponging locus as merely making sponging more likely.

We can use a similar approach for specified ecological hypotheses. Krützen et al.'s (2005) study on bottlenose dolphin attempts to rule out ecological explanations using the range overlap method; they note that sponging is mainly confined to deep-water channels, but that not all individuals that use this habitat are "spongers." In particular, males are regularly seen foraging in the channels, and as we noted earlier, sponging has only been observed in single male. One might argue that females are predisposed to sponging, and that those that forage in the channels tend to develop sponging. Krützen at al. counter this claim by pointing out that there are at least four females without haplotype H that forage in the channels and are not spongers. Laland and Janik (2006) suggest that this is not a large enough sample to reject the hypothesis that all females exposed to deep-channel foraging are equally likely to develop sponging.

Here, we show how a simple likelihood-based analysis can be used to evaluate such equivocal cases. We assume all 20 haplotype H females used the channels, of which 11 developed sponging; and that there were 5 females without haplotype H, of which 1 developed sponging. Under the null (ecological) hypothesis, we assume all females that forage in the channels have an equal probability of developing sponging, with MLE= 12/25. This gives a log-likelihood of:

$$13 \log(13/25) + 12 \log(12/25) = -17.31.$$

Under the alternative hypothesis, haplotype H females are more likely to develop sponging (MLE = 11/20) than females without haplotype H (MLE = 1/5), giving a log-likelihood of:

$$\log(1/5) + 4 \log(4/5) + 11 \log(11/20) + 9 \log(9/20) = -16.26.$$

This corresponds to an AIC difference of only 0.09 in favor of the alternative hypothesis ($= 2 \times (17.31 - 16.26) - 2$), indicating approximately equal support for each (Akaike weight = 51.1%), or using an LRT: LR = 2.09, d.f. = 1, $p = 0.15$. So this analysis indicates that Krützen et al.'s (2005) data are approximately equally consistent with hypotheses in which dolphins learn sponging asocially while foraging in deep channels (ecological), and females are more likely to develop sponging than males (genetic). We emphasize the fact that the numbers used here are chosen solely because they are given in Krützen et al. (2005), and the above analyses do not afford any general conclusions. Our main purpose is twofold: to show how a simple likelihood-based approach can be used to assess the evidence in such cases, and to illustrate the fact that both genetic and ecological hypotheses cannot be ruled out with confidence in this case. If researchers really want to claim that their data supports social transmission hypotheses better than genetic hypotheses, they would be advised to evaluate each using likelihood-based methods, rather than "ruling out" completely deterministic genetic models on the basis of a small number of observations.

Note: These models can all be fitted in R (R Core Development Team 2011) using the glm function. The two models used to test the genetic hypothesis can be fitted (including females with haplotype H) using the following code:

```
#Enter number of spongers and non-spongers for females with
haplotype H, males with haplotype H and individuals without
haplotype H
```

(continued)

Box 6.5   *(continued)*

```
spongers<-c(11,1,1)
nonspongers<-c(9,9,99)
#Bind into a matrix
y<-cbind(spongers,nonspongers)
#Enter indicator variables showing which individuals are
haplotype H females, and which are haplotype H males
femaleH<-c(1,0,0)
maleH<-c(0,1,0)

#Fit the genetic model
geneticModel<- glm(y~femaleH, family=binomial)
#Fit the social transmission model
socialModel<- glm(y~femaleH+maleH, family=binomial)

#Get AIC difference
AIC(socialModel)-AIC(geneticModel)
#Get relative support for social transmission model
exp(-0.5*(AIC(socialModel)-AIC(geneticModel)))
#Get Akaike weight for social transmission model
1/(1+exp(0.5*(AIC(socialModel)-AIC(geneticModel))))

#Do LRT to compare the two models
anova(geneticModel, socialModel,test='Chi')
```

Here is the R code for the two models used to test the ecological hypothesis:

```
#Enter number of spongers and non-spongers for females with
haplotype H; and females without haplotype H, out of those
that use the channels
spongers<-c(11,1)
nonspongers<-c(9,4)
#Bind into a matrix
y<-cbind(spongers,nonspongers)
#Enter an indicator variable showing which individuals are
haplotype H females
femaleH<-c(1,0)

#Fit the ecological model
ecologicalModel<- glm(y~1, family=binomial)
#Fit the social transmission model
socialModel<- glm(y~femaleH, family=binomial)

#Get AIC difference
AIC(socialModel)-AIC(ecologicalModel)
#Get relative support for social transmission model
exp(-0.5*(AIC(socialModel)-AIC(ecologicalModel)))
```

```
#Get Akaike weight for social transmission model
1/(1+exp(0.5*(AIC(socialModel)-AIC(ecologicalModel))))

#Do LRT to compare the two models
anova(ecologicalModel, socialModel,test='Chi')
```

Note that the absolute log-likelihoods and AICs will not match with those calculated manually, since in this code we are not treating each individual as a binary data point. However, the differences in AIC and log-likelihoods are exactly the same. The data could be entered as binary data points, yielding the same AICs and log-likelihoods as we calculated manually, but this method is more "fiddly."

resource, or resources, to necessarily resemble the unexploited one. This means that it may be very difficult for researchers to identify the vital ecological variable, or variables, that are causing a behavioral difference, leading to a danger of false positive claims for culture.

There is a relevant cautionary tale here that illustrates some inherent dangers in applying the basic method of exclusion to discount the ecological hypothesis. Whiten et al. (1999) suggested that chimpanzees "ant dipping" is a particularly compelling case of a socially transmitted trait. This is because differences between communities are not related solely to whether the behavior is present or absent; rather, two different methods of ant dipping are employed. In the "pull through" method, used in Gombe National Park in Tanzania, a long wand is held in one hand and a ball of ants is wiped off with the other. In the "direct mouthing" method, customary at the Taï National Park, Côte d'Ivoire, a short stick is held in one hand and used to collect a smaller number of ants, which are transferred directly to the mouth. Whiten et al. were satisfied that ecological differences between the Gombe and Taï sites could not explain this variation.

However, Humle and Matsuzawa (2002) investigated ant dipping among chimpanzees of Bossou, in southeastern Guinea, a site at which both forms of ant dipping are observed. They found that both the length of the tools and the technique employed were strongly influenced by the nature of the prey. Several species of ants were preyed on, and these differed in their density, aggressiveness, and in whether they were migratory or at nest sites. In situations in which ants were abundant, aggressive, or had severe bites, the first method, with its long wand, was employed; it was probably associated with fewer bites than the second method, while in other circumstances the direct mouthing method was used. Skeptics would find it easy to envisage that chimpanzees could individually be shaped by aggressive insects to use the strategy that resulted in the fewest bites. Humle and Matsuzawa's findings suggest that an ecological explanation for ant dipping cannot be ruled out.

Nonetheless, there are some cases where there is seemingly no conceivable way in which the trait in question *could* be shaped by local ecological conditions to account for the observed behavioral differences (van Schaik 2009). These include variants in traits used for social interaction, such as "kiss-squeaks" in orangutans (van Schaik 2009) and numerous communication variants in chimpanzees (Whiten et al. 2001). Nonetheless, this does not solve the problem for those traits that are not used for social interactions. Is there any way in which unknown ecological variables can be ruled out using the basic method? Perhaps the most compelling application of this approach is to groups of animals with overlapping ranges, which researchers consequently know to be exposed to the same ecological conditions. For example, orcas (*Orcinus orca*) show a number of pod-specific behavior patterns, including foraging specializations, migration patterns, and vocal dialects. Rendell and Whitehead (2001) argue that the ecological hypothesis can be ruled out, since sympatric pods develop different pod-specific behaviors.

6.2.3.2 Advanced method of exclusion

Application of the advanced method of exclusion here would attempt to identify ecological variables that might influence the development of a trait; if these variables (and the genetic hypothesis) cannot fully account for the observed pattern (e.g., a pattern of group contrasts), then a role is inferred for social transmission. Krützen et al.'s (2005) study of sponging in bottlenose dolphins attempted to rule out ecological explanations in this way; they argued that sponging is mainly confined to deep-water channels, but that not all individuals that use this habitat are "spongers." In particular, males are regularly seen foraging in the channels, and as we noted earlier, sponging has only been observed in single male. A skeptic might argue that females are predisposed to sponging, and those that forage in the channels tend to develop sponging behavior. Krützen at al. counter by pointing out that there are at least four females without the focal haplotype that forage in the channels and are not spongers. Laland and Janik (2006) suggest that this is not a large enough sample to reject the hypothesis that all females exposed to deep-channel foraging are equally likely to develop sponging. In box 6.5 we show how simple likelihood models can be used to quantify the evidence in such cases.

In general, provided the genetic hypothesis can also be ruled out, in cases of behaviorally diverse populations with relevant range overlap there are reasonable grounds for thinking that the behavioral differences might be caused by social transmission. It is possible that social interaction within groups causes individuals in different groups to be exposed to different parts of the environment (e.g., if deep-channel foraging were socially transmitted between mother and offspring in bottlenose dolphins), which in turn shapes their behavior through otherwise asocial learning. We see no reason why this should not be classified as social transmission. Sterelny (2009, 2012) even argues this process of "social niche construction" is likely to be one of the most important social learning processes in development. Nonetheless, it is possible that a researcher wishes to rule out this explanation, perhaps if they wish to know whether a trait is acquired directly by observation. In section 6.4 we argue that this is logically not possible using repertoire-based data, and the best a researcher can hope for is to eliminate explanations involving exposure to specified ecological variables.

The fundamental problem with applying the method of exclusion to ecological data is that there are a virtually infinite set of ecological parameters to choose from, and the researchers may simply not have chosen the ecological variables that account for behavioral variation. While advocates might counter, on grounds of pragmatism, that it is not unreasonable to focus on those ecological variables that are the most likely candidates to account for behavioral variation, there are additional problems associated with the method of exclusion that render it vulnerable to false positive (and false negative) claims of culture. We address these in the following section.

6.2.4 Further problems with the method of exclusion

To summarize the above conclusions about the method of exclusion, we do not think it will ever be possible to exclude the genetic hypothesis using the basic method of exclusion. Researchers committed to the method of exclusion should focus on the advanced version, in which models of genetic transmission are assessed with relevant data—and indeed this is the way the field is going. It would seem that the ecological hypothesis can only be entirely ruled out in some rather specific cases, where the researcher is able to dismiss the possibility that there might be an unmeasured ecological variable of importance. If we combine this with the limited cases in which the genetic hypothesis can be excluded, we are left with a very small set of circumstances in which both hypotheses can be excluded. Moreover, there are good reasons to expect the distribution of many socially transmitted traits to covary with both genes and ecology. As a consequence, when *rigorously applied*, the method of exclusion is likely to grossly underestimate the importance of social transmission in generating differences in behavioral repertoires. We emphasize "rigorously applied" because we are not convinced that this method has always been made use of as meticulously as it might have been. Frequently, genetic variation has been casually dismissed as a source of behavioral variation, very few ecological variables have explicitly been considered, and these have not been listed, quantified, or incorporated in statistical analyses (Laland and Janik 2006; Laland et al. 2009). We suspect that part of the problem here is that diligent application of the method of exclusion would lead to a large number of false negatives; that is, rigorous application of the method of exclusion would fail to detect the social transmission that researchers intuitively believe to be present.

However, the method of exclusion is problematic in another important respect. It is fundamentally flawed because there is no quantification of cultural variation; rather culture is inferred as what is left over when other sources fail to account for the observed pattern of variation. This leaves claims of culture potentially inflated by a multitude of other unconsidered factors, including measurement error, low statistical power, and the failure to use appropriate predictor variables. It is not hard to come up with candidates for factors that could be inflating the estimate of cultural variation. We have already described how collection of the wrong kind of genetic data, and failure to incorporate the right ecological variables, can undermine the reliability of this approach. In addition, there is inevitably error in the behavioral measures. For example, traits are likely to be present in some populations but registered as absent because they are not yet unrecorded by researchers, while estimates of the frequency of other characters will be inflated by sampling.

While we reach a negative conclusion about the method of exclusion, we acknowledge it has been responsible for some landmark studies in the history of social transmission research, as well as being instrumental in the collation of some highly valuable sets of repertoire-based data. Furthermore, we do not wish to imply that there is anything inherently problematic about either the group-contrasts approach, or attempts to partition variance among sources, as a means to study social transmission. Rather, we call for a shift in emphasis (following Whitehead 2009), where the focus is not on asking "is it social transmission?" or "is it culture?" by excluding genetic and ecological explanations, but rather on quantifying the effect of genetic differences, ecological differences and social transmission on differences in behavioral repertoires. Many of the approaches described above could be incorporated into such a framework.

6.3 A Model-Fitting Approach

Social learning researchers require a broader framework that allows for cases of social transmission where genetic and ecological differences cannot be ruled out using the methods described above. The model-fitting approach that we advocate assumes that there are, or might be, genetic, ecological, and cultural (social transmission) influences on differences in repertoires. An attempt is made to quantify the effects of each, by specifying a model for the way in which each affects the repertoire. The model is fitted to the data to estimate the relative importance of each component. The strength of evidence for social transmission is the strength of evidence against a model in which the effects of social transmission are constrained to be zero.

There are numerous ways in which such model fitting could be achieved, depending on our assumptions about the transmission process and the data available. For example, Otto et al. (1995) use a path analysis approach to quantify the effects of cultural, environmental, and genetic influences on IQ in humans. We focus on a particular version of the method of quantification based on that described by Whitehead (2009),[9] but we suggest some extensions, many based on the work of Matthews (2009).

6.3.1 A matrix regression approach

This method takes, as data, matrices giving the genetic, ecological, social, and behavioral differences (or, equivalently, similarities) among individuals or groups.

[9] We note Madden et al. (2004) used a very similar approach in their study of bowerbirds (see section 6.3.2).

Matrix regressions are used to tease apart and quantify the relationship among these variables.

We discussed methods for calculating genetic similarity and behavioral similarity among individuals in boxes 6.2 and 6.3, respectively. The social similarity matrix is essentially a social network (see chapter 5) designed to quantify opportunities for social transmission. As with *NBDA*, the social network should be chosen such that it reflects the hypothesized process of social transmission (see Whitehead 2009 for additional discussion). The only difference is that in *NBDA* the network could be asymmetrical, allowing for the fact that the trait might pass more rapidly in one direction between two individuals than in the other direction. This does not make sense for repertoire-based data, because the model does not specify who is learning from whom. Asymmetrical networks could be converted to symmetrical ones by replacing each a_{ij} and a_{ji} with their mean value.

Ecological similarity is perhaps the trickiest matrix to construct. Whitehead (2009) recommends that for small-scale studies researchers use range overlap from geographic information systems, and for large scales they employ ecological similarity measures (C. R. Krebs 1989, 293–309). Alternatively, a researcher could break down ecological similarity into a number of more specific variables that might plausibly influence behavioral similarity.

Whitehead (2009) reasoned that genetic, ecological, and social similarity were all likely to play some role in determining behavioral similarity, and our task as social learning researchers is to quantify the relative effects of each. He suggested that a researcher use a multiple regression with behavioral similarity as the dependent variable, and genetic, ecological, and social similarity as the independent variables, to estimate standard partial regression coefficients and then use permutation tests (box 6.6) to test the significance of the partial correlations.

The main difference between this approach and the advanced method of exclusion, aside from the emphasis on quantification, is that a specific component of social transmission is quantified as the effects of the social similarity matrix. Nonetheless, the advanced method of exclusion, when applied to group contrasts, can be seen as a subset of this approach where the social similarity matrix merely gives information about which individuals are in the same group (see section 6.3.2(b)).

6.3.2 Examples

There are a couple of studies that have used a similar technique to that advocated by Whitehead (2009). One is on spotted bowerbirds (*Chlamydera maculata*) (Madden et al. 2004) and another is on sperm whales (*Physeter macrocephalus*) (Rendell, Mesnick, et al. 2011).

Male bowerbirds attract females by constructing bowers, which they decorate with items and use as a staging area for their display to females. Madden et al. (2004) studied spotted bowerbirds in Taunton National Park, Queensland, Australia, during 1998–2000. They noted that males whose bowers were in close

---

Box 6.6

Matrix correlations and regressions

When testing for the statistical significance of correlations or regression coefficients using similarity matrices, one cannot use the standard distributions, since the values are not independent. Take, for example, a social network based on associations; if individual A is strongly connected to B, and B is strongly connected to C, it is probable that A is also strongly connected to C. Using permutation tests can solve this problem.

The most commonly used method is the Mantel test. The standardized Mantel test statistic, r_M, is calculated for matrices **X** and **Y**, both of size $n \times n$ as:

$$r_M = \frac{1}{d-1} \sum_{i=1}^{n-1} \sum_{j=i+1}^{n} \left(\frac{x_{ij} - \bar{x}}{s_x} \right) \left(\frac{y_{ij} - \bar{y}}{s_y} \right),$$

where $d = [n(n-1)/2]$ is the number of entries in the upper triangle of each matrix; x_{ij} is the entry in the ith row and j^{th} column of **X**; and $\bar{x}$ and s_x are the mean and standard deviation of the upper triangle of **X**. This statistic acts like a correlation coefficient varying between −1 and 1, with 0 indicating no linear relationship. If the matrices have a monotonic but nonlinear relationship, one can replace r_M with the Spearman correlation coefficient (Dietz 1983, in Legendre and Legendre 1998).

Permutation is achieved by swapping around the identity of each individual and reordering one of the matrices accordingly, before recalculating the test statistic. This is done numerous times, say 10,000, to generate the null distribution for the test statistic, which is then used to find a p-value in the usual way.

The partial Mantel test extends the Mantel test to calculate and test partial correlations between similarity matrices (Smouse et al. 1986, in Legendre and Legendre 1998). The numerical procedure for three matrices is described in Legendre and Legendre (1998). At the time of writing, the "mantel" function in the "ecodist" package (Goslee and Urban 2007) of the R statistical environment (R Core Development Team 2011) can be used to perform partial Mantel tests with multiple matrices.

Legendre and Legendre (1998) note that "model matrices" representing a specific hypothesis can also be included in a Mantel test procedure. This justifies using a social similarity matrix hypothesized to reflect the degree of social transmission among individuals, such as a group delineation matrix, with 1's for individuals from the same group and 0's for those from different groups (Whitehead 2009).

---

proximity tended to use similar bower decorations, leading to the hypothesis that preferences for particular decorations are socially transmitted between neighbors. Madden et al. measured behavioral dissimilarity by adapting genetic distance measures, considering each decoration type as a "locus" and the presence/absence of the decoration as "alleles." They used a similar measure to estimate ecological dissimilarity, based on the presence/absence of particular bush species in the canopy. They then estimated relatedness using genetic techniques, and social dissimilarity (in Whitehead's terminology) was the distance between bowers.

Madden et al. (2004) found no strong evidence of a relationship of canopy composition (ES) with either decoration usage (BS) (Mantel tests, 1998: 18 bowers, $r_M = 0.075$, $P = 0.32$; 1999: 18 bowers, $r_M = 0.15$, $P = 0.10$), or with bower

Multiple regression can be performed on matrices by taking the upper diagonal of each matrix, and using multiple regression to estimate coefficients (Legendre and Legendre 1998). To obtain standard partial regression coefficients, the researcher first standardizes each matrix (i.e., subtracts the mean and divides by the standard deviation). This is the method suggested by Whitehead (2009), and provides a means to estimate the path coefficients in a causal model (see section 6.4). The statistical significance of each standard partial regression coefficient can be tested by permuting the response matrix in the same manner as for a Mantel test. Unfortunately, bootstrap methods cannot be used to calculate standard errors and confidence intervals for matrix regressions (see Whitehead 2008, 32); Whitehead (2008, 2009) suggests using the jackknife standard error in such situations. To do this, a researcher omits each unit from the data in turn, and recalculates the parameter estimate, with $\hat{s}(-i)$, referring to estimate from the data with unit i *omitted*. The researcher then calculates pseudovalues:

$$\varphi_i = n\hat{s} - (n-1)\hat{s}(-i),$$

where n is the number of units and $\hat{s}$ is the original estimate. The jackknife standard error is then the standard error of the sample of pseudovalues. Whitehead (2008, 33) notes that jackknife standard errors tend to be imprecise and overestimate standard errors, but the jackknife currently appears to be the best option for estimating precision in these cases.

At this stage, it is worth giving a note of caution. Though widely used, matrix permutation procedures that involve permutation of the raw data, such as the Mantel test, are susceptible to greatly increased type I error rates when certain restrictive assumptions are not met (Dekker et al. 2007). Dekker et al. suggest that researchers instead use the Freedman–Lane semi-partialling (FLSP) approach, or their double semi-partialling (DSP) approach, both of which involve permutation of the residuals. We suggest researchers consult Dekker et al. (2007) to check whether the relevant assumptions are reasonable for their data before using Mantel tests, and if they are not, DSP can be implemented in the R package "sna" (Butts 2010) using the netlm function.[1]

[1] Many thanks to Luke Matthews for bringing this issue to our attention.

proximity (SS) (Mantel tests, 1998: 18 bowers, $r_M = 0.161$, $P = 0.09$; 1999: 18 bowers, $r_M = 0.163$, $P = 0.08$). They concluded that decoration usage does not appear to covary with canopy type.

They then went on to test whether decoration usage (BS) was related to bower proximity (SS) after controlling for genetic similarity using partial Mantel tests. Overall, there was reasonable evidence of a positive association (partial Mantel tests, 1998: 13 bowers, $r_M = 0.25$, $P = 0.03$; 1999: 12 bowers, $r_M = 0.33$, $P = 0.02$; 2000: 11 bowers, $r_M = 0.24$, $P = 0.08$). They then tested for relations between genetic and decoration matrices when controlling for geographical associations. They found reasonable evidence in only one year (partial Mantel tests, 1998:13 bowers, $r_M = 0.012$, $P = 0.46$, 1999: 12 bowers, $r_M = 0.27$, $P = 0.04$; 2000:

11 bowers, $r_M = 0.003$, $P = 0.48$). However, it is possible that 1999 is a chance result, because when Madden at al. combined p-values using Fisher's C test (see section 6.4), there was little evidence against the hypothesis of no relationship between GS and BS ($c_2^6 = 9.5$, $p = 0.30$[10]). These findings led Madden at el. (2004, 759) to suggest that these similarities in bower decoration "are products of local tradition, either culturally transmitted by neighbouring males who regularly inspect neighbours' bowers, or as localized responses to variable individual female preferences."

Rendell, Mesnick, et al. (2011) used similar techniques to investigate vocal dialect variation in sperm whales. Sperm whale social groups can be assigned to vocal clans based on their production of codas, which are short stereotyped patterns of clicks. The authors investigated mitochondrial DNA (mtDNA) variation among sympatric vocal clans in the Pacific Ocean. A total of 194 individuals were sampled from 30 social groups, belonging to one of three vocal clans.

Correlations among genetic, vocal, and geographic distance matrices using Mantel and partial Mantel tests were sought by the authors. They found strong evidence that genetic similarity[11] was positively correlated with vocal similarity ($r = 0.17$; $p = 0.003$), but no evidence for a relationship with geographic distance ($r = -0.07$, $p = 0.220$). In the partial Mantel tests the results were little affected by the addition of extra factors (genetic vs. vocal controlling for geographic distance: $r = 0.17$, $p = 0.009$). This is a striking result, considering the sampling took place over thousands of kilometers.

However, Rendell, Mesnick, et al. (2011) concluded that genetic differences are unlikely to be the cause of vocal differences. They note that the two most common haplotypes were present in significant quantities in each clan, suggesting that genetic variation cannot account for behavioral variation between clans. They instead suggest that the causal pathways parallel the situation in humans, where vertical (parent to offspring) transmission of language has resulted in correlations between variation in language and neutral genes.

6.3.3 A return to group comparisons

Group comparisons can be brought into the matrix regression framework described in section 6.3.1 by constructing a group delineation matrix, indicating which individuals are in the same group, and using it as the social similarity matrix (Whitehead 2009).[12] A further way one might want to expand this analysis is to investigate whether group membership is sufficient to explain differences in behavior, or whether the fine-grained social structure, among and within

---

[10] Fisher's C test (see section 6.4) for the relationship between BS and SS, controlling for GS, gave: ($c_2^6 = 19.9$, $p = 0.006$).

[11] An estimate of the probability that two individuals chosen randomly, one from each group, will share the same haplotype. Other measures of genetic distance were also considered.

[12] This method could be used as an alternative to the option-bias method (box 6.1) if option-choice is used to construct the behavioral similarity matrix. This approach has the advantage that it is easier to incorporate into the causal modeling framework proposed here.

groups, also has an effect. Matthews (2009) developed a method to address such questions.[13] He defined two separate hypotheses of social learning. The first is the *social clique* hypothesis, in which individuals within a group acquire traits from each other and so are behaviorally similar, but the patterns of association within each group and among groups do not affect social transmission. The second is the *diffuse network* hypothesis, in which individuals tend to acquire traits from other individuals in proportion to the network connection between them.

Matthews (2009) notes that these hypotheses can be distinguished by testing for correlations between social similarity and behavioral similarity in three different subsets of the data: (*a*) using all the data (i.e., a matrix containing an entry for all pairs of individuals), (*b*) using only pairwise relations within groups, and (*c*) using pairwise relations between groups. These correlations can all be tested using the matrix permutation methods given in box 6.6. If the diffuse network hypothesis were correct, then one would expect all three correlations to be significant. In contrast, if the social clique hypothesis is correct, we would expect only correlation *a* to be significant.[14] Matthews suggests a further comparison to test whether behavioral similarity is greater within groups than between groups; this could be accomplished by testing for a matrix correlation between behavioral similarity and a group delineation matrix (showing which individuals are in the same group).[15]

Matthews (2009) applied this method to data from a group of capuchin monkeys (*Cebus albifrons*), based on two foraging behavior patterns: palm tree breaking and Sloanea fruit processing. The former was either present or absent in an individual's repertoire, whereas the latter was always present but performed using two variants. He used Gower's dissimilarity coefficient (box 6.3) to quantify behavioral similarity between individuals across the two traits. He collected proximity data to construct a social network, and used cluster analysis techniques (box 6.7) to identify cliques within the network. Matthews found that behavioral and social similarity were correlated across all the data (correlation *a*), but not within (*b*), or between (*c*) clusters. In addition, individuals within a cluster were more behaviorally similar than those in different clusters, supporting the social clique hypothesis.

6.4 A Causal Modeling Framework

In this section we suggest an extension of the matrix regression approach described in section 6.3.1 using causal graphs and the associated methods for inferring causal relationships from observed statistical patterns between variables

---

[13] Matthews applies his method to clusters of individuals identified using the methods given in box 6.7, but the method could also be applied where groups are identified a priori; consequently we describe the method in terms of "groups" rather than "proximity clusters."

[14] There is a concern that correlations *b* and *c* are tested with less statistical power than *a*, since they use less of the data, and thus the social clique hypothesis is favored. Matthews (2009) describes a jackknife technique to assess whether this is likely to have been a problem for a specific dataset.

[15] Matthews (2009) uses a different permutation procedure, but later realized that this does not fully account for nonindependence in the matrices. On applying an improved analysis he found that the statistical significance of his own results was not affected (pers. comm.).

Box 6.7

Identifying groups using cluster analysis

A number of repertoire-based methods described here require that individuals be assigned to groups. In some cases, such groups might be obvious (e.g., Whiten et al. 1999), but in other cases they might not be. Researchers might be studying a population in which all individuals could potentially interact, but in which they suspect there are distinct cliques. Matthews (2009) suggests that cluster analysis can be used to identify groups in such cases.

Cluster analysis techniques take a dissimilarity matrix for a group of individuals, and assign them to groups by detecting clusters of individuals that are close together. Some methods assign individuals definitively to groups, whereas other "fuzzy" cluster algorithms assign membership coefficients (summing to 1) to each individual that determine the extent to which it can be assigned to each cluster. In the case of the algorithm described below, these can be interpreted either as the probability of discrete membership to each cluster (Rousseeuw 1995) or as the degree of membership to each cluster (Kandel et al. 1995). Matthews (2009) recommends that fuzzy cluster algorithms be used, hypothesizing that social learning cliques are unlikely to be completely discrete phenomena.

Fuzzy clustering can be accomplished using the function "fanny" of the "cluster" package (Maechler et al. 2005) in the R statistical environment (R Core Development Team 2011). This finds clusters by minimizing the distance between pairs within each cluster. This is done by choosing the set of membership coefficients, $u(i,v)$, denoting the membership of i to cluster v, which minimizes the following objective function:

$$F = \sum_{v=1}^{k} \left[\frac{\sum_{i,j=1}^{n} u_{iv}^r u_{jv}^r d(i,j)}{2 \sum_{j=1}^{n} u_{jv}^r} \right],$$

where n is the number of individuals, k is the number of clusters, and d_{ij} is the dissimilarity between observations i and j. The parameter r gives the crispness of the clustering, with $r \to 1$ alloting increasingly crisper clusters, and $r \to \infty$ giving increasing fuzziness. By default this parameter is set to 2, and the user is advised if a smaller value is required. The "fanny" algorithm has the potential disadvantage that the user must specify the number of clusters a priori. Alternative algorithms are available that choose the number of clusters automatically (Milligan and Cooper 1985). The R package "fpc" (Hennig 2010) includes procedures for doing this.

Most of the group comparison methods described in this book require an individual to be assigned to a specific cluster, in which case the researcher can assign each individual to the cluster for which it has the highest coefficient (Matthews 2009). However, there is scope for further development of methods that take more advantage of the fuzzy cluster coefficients. One possibility is to apply cluster analysis to the behavioral similarity matrix as well, and test whether the cluster coefficients for behavior and proximity are correlated (Matthews 2009).

(Pearl 2009; Shipley 1999).[16] With this approach, causal graphs are postulated, representing hypotheses about the causal relationships among ecological similarity, genetic similarity, social similarity, and behavioral similarity. The researcher ascertains which set of graphs are consistent with the data, and estimates the strength of causal influences conditional on each graph in the set. All of the

---

[16] Other versions of the method of quantification could also be extended to employ causal graphs.

approaches discussed so far in the chapter can be seen as a subset of this approach, where different a priori assumptions are made about the causal structure linking the variables of interest. We return to this point in sections 6.4.1 and 6.4.4.

Early on in their statistical training, most scientists are taught the phrase "correlation does not imply causation," but this is inaccurate. Instead, as Shipley (2000, 3) notes, "a simple correlation implies an unresolved causal structure." If variables X and Y are correlated, then there could either be a causal pathway leading from X to Y (X → Y), from Y to X (X ← Y), or causal pathways leading to both X and Y from other variables (e.g., X ← Z → Y). This problem has led most scientists to believe that randomized experiments are the only way to reliably infer the presence and direction of a causal link between variables. However, many causal hypotheses involve variables that cannot be directly manipulated, so such hypotheses cannot be tested using a randomized experiment. This is true in most cases in which we wish to infer that social transmission is responsible for similarities in behavioral repertoires.

Fortunately, methods now exist that allow us to translate a causal hypothesis into a corresponding statistical model, and thus distinguish between competing causal hypotheses using observational data. These methods are described in an accessible manner by Shipley (2000), and are given a more technical and comprehensive treatment by Pearl (2009). Here we offer a brief overview of causal modeling, which will be sufficient for readers to understand the methods that we advocate.

The first stage is to express a causal hypothesis as a directed graph (Pearl 2009; Spirtes et al. 2001), in which variables (often called "nodes") are connected by "edges" (which essentially are causal arrows) representing direct causal effects (see fig. 6.3).

In figure 6.3, A is a direct cause of C, and this means that changes in A will result in changes in C irrespective of the behavior of any other variables in the graph. In contrast, A is an indirect cause of D, because changes in A will only induce changes in D by causing changes in C. Causes are only direct or indirect relative to the variables included in the graph; if D were excluded from the graph in figure 6.3, C would become a direct cause of F. A and B are called the *causal parents* of C, and E and D are its *causal children*. More generally, A, B, and C are the *causal ancestors* of D, and E, D, and F are the *causal descendents* of C. Applying this to the problem at hand, figure 6.4 proposes two alternative causal graphs for the relationship between genetic, ecological, social, and behavioral similarity. (Note, these are not the only two possibilities and, as we describe below, there are many other possible causal graphs for these variables.) In both graphs in figure 6.4, genetic similarity causes social similarity (perhaps individuals prefer to associate with relatives), and social similarity causes ecological similarity (perhaps

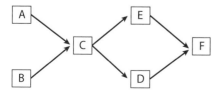

Figure 6.3. Example of a causal graph with 6 variables. This is also a directed acyclic graph (DAG) since there are no feedback loops. Based on figure 2.5 in Shipley (2000).

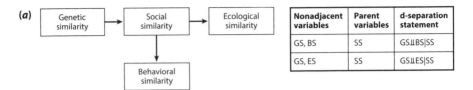

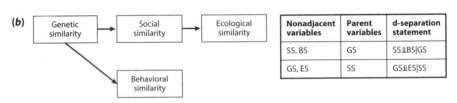

Figure 6.4. Examples of causal graphs showing two hypotheses for the relationships between genetic, social, ecological, and behavioral similarity. A set of d-separation statements sufficient for testing each hypothesis is given in each case.

attraction to another individual causes them to utilize the same parts of the environment). However, in *a* it is social similarity that causes behavioral similarity (i.e., if individuals learn from close associates), whereas in *b* it is genetic similarity that causes behavioral similarity (i.e., relatives have similar behavior).

We can translate a causal graph into a statistical model using the property of d-separation (see box 6.8). This is a criterion for deciding, from a given causal graph, whether a variable *X* is independent of another variable *Y*, given a set of variables *Z*. The idea is to associate "dependence" with "connectedness" (i.e., the existence of a connecting path), and "independence" with "unconnectedness" or "separation." Every d-separation event that exists in the causal graph will be reflected by an equivalent statistical independence in the data, if the causal graph is correct (Shipley 2000). For example, in figure 6.4, applying the d-separation criterion results in a different set of statements about statistical independence for each hypothesis. For example, if *a* is correct, then once we have statistically controlled for the effects of social similarity on behavioral similarity, then behavioral similarity will be independent of genetic similarity. Conversely, if *b* is correct, then controlling for social similarity would not affect the relationship between genetic similarity and behavioral similarity. We can use this to test whether our data are consistent with a specific causal hypothesis, or to determine which of several hypotheses best fit the data. Most frequently this is done using the methods of structural equations modeling (SEM; see box 6.9). However, these methods are not applicable to matrices, since they do not account for the nonindependence of the elements in a matrix (see box 6.6).

A solution is provided by the "d-sep test" pioneered by Shipley (2000). The above phrase, "once we have statistically controlled for the effects of social similarity on behavioral similarity then behavioral similarity will be independent of genetic similarity," is an example of a conditional independence statement. More generally, such statements take the form "once we have statistically controlled

Box 6.8
Property of d-separation

Here we briefly explain the property of d-separation; this is a summary of the explanation given by Shipley (2000). We first need to introduce some additional terminology. For a given graph, an *undirected path* exists between variables X and Y if it is possible to traverse between them along the edges (arrows), ignoring the direction of those edges. This path is also a *directed path* between X and Y if the route in either direction follows the direction of the edges. A *collider* on a specified path is a variable with arrows pointing into it in both directions. For example, consider the graph given in figure 6.3. The variables A and B are connected by an undirected path, and A and F by a directed path. Variable F is a collider on the undirected path between D and E.

The variables X and Y are unconditionally d-separated if and only if

1. there is no undirected path between X and Y; or
2. there is a collider on every undirected path between them.

The variables X and Y are d-separated conditional on a set of other variables **Q**, if and only if

1. there is no undirected path between X and Y; or
2. every undirected path between X and Y either

   a. includes at least one non-collider that is in set **Q**, or
   b. includes at least one collider that is not in set **Q**, and does not have a descendent in set **Q**.

We can express this d-separation event using the notation $X \perp\!\!\!\perp Y | Q$, meaning X and Y are d-separated given **Q**. For example, in figure 6.3, A and B are unconditionally d-separated since there is a collider at C. This is expressed as $A \perp\!\!\!\perp B$. The variables A and B are not d-separated if the conditioning set includes any of the variables C, D, E, or F, since C is a collider on the undirected path between A and B, and D, E, and F are all descendents of C. The variable C is d-separated from F given E and D ($C \perp\!\!\!\perp F | ED$). Note the both E and D have to be in the conditioning set, otherwise there is an undirected path between A and B that does not meet criteria 2a or 2b.

Pearl 1988, cited in Shipley 2000, has shown that for any causal graph without feedback loops (a directed acyclic graph or DAG), every d-separation statement implies a corresponding independence relation in the joint distribution of the random variables in the graph. For example, $C \perp\!\!\!\perp F | ED$ implies that C and F will be probabilistically independent conditional on E and D. We can use this fact to test whether our data are consistent with the causal hypothesis specified by G. For a DAG, this relationship does not depend on any assumptions about error distributions, or the functional form of the causal relationship between variables. For a cyclical causal graph (with feedback loops) the relationship holds as long as the relationships between variables are linear (Spirtes 1995, cited in Shipley 2000), but unfortunately the relationship does not hold for cyclical graphs in general.

It is worth noting that the statistical models or procedures used to test conditional independence statements may make assumptions about distributions and functional form. The permutation tests advocated in this chapter do not require a specific error distribution, and use of the Spearman correlation coefficient requires only a monotonic relationship between variables (see box 6.6). However, the method presented here assumes that each direct causal influence exerts an additive effect on the variable in question. If the researcher wishes to test a hypothesis in which there are interactions between variables, the d-separation property is still valid, but the statistical tests need to be adjusted accordingly.

Box 6.9

Path analysis and structural equations modeling (SEM)

Path analysis is a statistical technique invented by Sewall Wright (1921) that can now be seen as a part of modern causal modeling. Wright saw path analysis as a means to estimate direct causal effects, given an assumed causal graph. Path analysis is an extension of multiple linear regression, and so makes the same assumptions (i.e., the error for each variable is normally distributed with a constant variance, and the causal influences are linear). Each direct causal link is assigned a path coefficient, which determines the strength of that direct causal influence. The path coefficients for the direct causes of a variable X can be estimated using a multiple regression with X as a dependent variable, and including all the causal parents of X as dependent variables.

For example, consider the causal graph in figure 6.5.

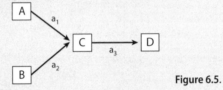

Figure 6.5.

The path coefficients can be estimated using two regressions: a_1 and a_2 are estimated with C as a dependent variable, and A and B as independent variables; a_3 is estimated with D as a dependent variable; and C as the independent variable. Confidence intervals and statistical tests for the path coefficients use the standard method for regression. If a researcher is using matrix regression (box 6.6), the path coefficients can be estimated in the same way, but statistical tests and confidence intervals cannot (Legendre and Legendre 1998; see box 6.6).

The overall structure of a causal graph cannot be tested using the original methods of path analysis, and for Wright, the structure of the causal graph had to be established independently through theory or experimentation. Structural equations modeling (SEM), developed in the social sciences, extends path analysis to include unmeasured or latent variables, and includes methods for testing a specific structural equations model (see Shipley 2000). However, these tests (1) require large sample sizes, (2) assume linear relationships between variables, and (3) do not account for the nonindependence in similarity matrices. Consequently they cannot be used in the causal modeling framework described here, but instead we can use Shipley's (2000) d-sep test (see main text).

for the effects of variable Z on variable X, then variable X will be independent of variable Y." To conduct the d-sep test, the researcher tests each conditional independence statement using a statistical test appropriate for the data, and combines them into a single *p*-value reflecting the strength of evidence against the causal hypothesis in question. In this case, one can use methods for testing for matrix correlations and partial correlations to obtain *p*-values for each statement (see box 6.6). To use the d-sep test, the researcher must first generate a list of independent d-separation statements that specifies the causal graph. This can be done by listing in one column each pair of nonadjacent variables (i.e., those not connected by an edge) and listing in a second column the causal parents of each pair. The pairs of variables in the first column are d-separated by the set of variables in the second, and provide a sufficient list of statistically independent variables to specify the causal graph (Shipley 2000; e.g., fig. 6.4). The researcher then tests each statement using the permutation methods given in box 6.6 to generate a p-value for each. The overall p-value is then calculated using Fisher's C test (retrospectively named by Shipley 2000) for combining p-values. The test statistic is:

$$C = -2\sum_{i=1}^{k} \ln(p_i)$$

where p_i is the p-value for the ith statistical independence statement. If all k statements are true, C will follow a chi-square distribution with $2k$ degrees of freedom. This can be used to generate a p-value that quantifies the strength of the evidence against the null hypothesis; if the p-value is not small, the data do not provide grounds to reject it as a causal hypothesis. As Shipley (2000) notes, the situation is somewhat different to a standard case of hypothesis testing, in that the hypothesis being tested can be considered a null hypothesis. However, it is directly comparable to the inference made when using of a goodness of fit test, such as a chi-squared test. Ideally researchers would, in preference, obtain the relative strength of evidence for different causal hypotheses using Bayesian or information theoretic criteria (compare to *NBDA* methods in chapter 5). However, if inference on the relationships between matrices is based on permutation rather than likelihood, as it is here, it is not clear how this could be done.[17] Nonetheless, this approach allows a researcher to rule out some causal hypotheses as improbable, which in turn increases their knowledge about which hypotheses are likely.

It is worth noting at this stage that some causal graphs provide equivalent sets of d-separation statements, and thus cannot be distinguished using observational data. Consequently, the test described above is really testing a specific set of causal hypotheses. This set can be found using the method given in box 6.10. In some cases, researchers might be able to rule out some graphs that are part of a set that is consistent with the data on logical grounds. For example, researchers know that if two individuals have a similar behavioral repertoire, this does not cause them to be more or less similar genetically.

Once the researcher has established which candidate causal graphs are consistent with the data, he or she can then estimate the relative strength of each causal pathway. As in path analysis (box 6.9), the strength of the direct causal influences on a variable Y can be estimated by using a multiple regression in which all the direct causal influences are included as independent variables. For example, to

---

[17] Likelihood-based models of social networks are available (e.g., see Leenders [2002]; de Nooy [2010]; O'Malley and Christakis [2011]). Such approaches might ultimately provide a better means of comparing alternative causal models. The researcher may have specific candidate mechanisms for the way in which genetic, ecological, and/or social factors influence the behavioral trait(s) in question. For example, Krützen et al. (2005) consider specific hypothetical mechanisms as possible causes of sponging in bottlenose dolphins (section 6.2.2.2). In box 6.4 we suggest ways in which these hypotheses could be implemented as likelihood-based models. In principle, such models could be incorporated into a causal modeling framework.

Box 6.10
Equivalent causal graphs

As mentioned in the main text, different causal processes can result in the same pattern of statistical dependencies in the data, in which case they are said to be observationally equivalent (and imply the same d-separation statements). This is an important concept since it may be impossible to distinguish some or all of the causal hypotheses being considered using observational data. Furthermore, we suggest that if a researcher finds there is not strong evidence against a candidate causal model, then they should ascertain which set of equivalent causal graphs are also consistent with the data.

An equivalent set of graphs (directed acyclic graphs, or DAGs) can be economically expressed using a partially oriented inducing path graph or POIPG (Spirtes et al. 2001; see chapter 8 of Shipley 2000). In a POIPG, the o symbol is used where it is not known whether there is an arrowhead. For example, if we have Xo→Y, this could be X→Y or X←→Y, but not X←Y. Further constraints on the causal structure are expressed using the underline notation, which shows the absence of an unshielded collider (denoted a "definite noncollider"). That is, if we have Xo -oZo - oY, this means we do *not* have Xo→Y←oZ.

We can find the POIPG equivalent to a specified DAG by applying the following algorithm (Verma and Pearl 1991; details in Shipley 2000, 265):

1. Change all arrows in the graph to lines.
2. Draw the o symbol at either end of each line.
3. Redraw each unshielded collider in the original causal graph (e.g., if there was an unshielded collider X→Z←Y in the original graph, replace Xo - oZo - oY with Xo→Z←oY).
4. Add an underline to every noncollider triplet (e.g., if there was X→Z→Y, X←Z←Y or X←Z→Y in the original graph, replace Xo - oZo - oY with Xo - <u>oZo</u> - oY).

An example of this process is shown in figure 6.6.

Any graph that can be created with the same "skeleton," without removing any unshielded colliders or adding any new unshielded colliders, is equivalent to the original causal graph. There

estimate the direct causal effects on behavioral similarity given the causal graph shown in fig. 6.4a, we include only social similarity as an independent variable. The resulting standardized partial regression coefficients give the direct causal influence (or "path coefficient") of each variable. In this case, we use the matrix regression techniques described in box 6.6. Whitehead (2009) suggests that we can then use the jackknife method to estimate the precision of the path coefficients (see box 6.6).

In summary, the method we recommend for analyzing repertoire-based data is the following:

1. Obtain matrices quantifying behavioral, social, genetic, and ecological similarities between individuals (boxes 6.1 and 6.2).
2. Express competing causal hypotheses as causal graphs.
3. For each graph, obtain the sufficient set of d-separation statements, and thus the corresponding set of conditional independent variables for each graph.

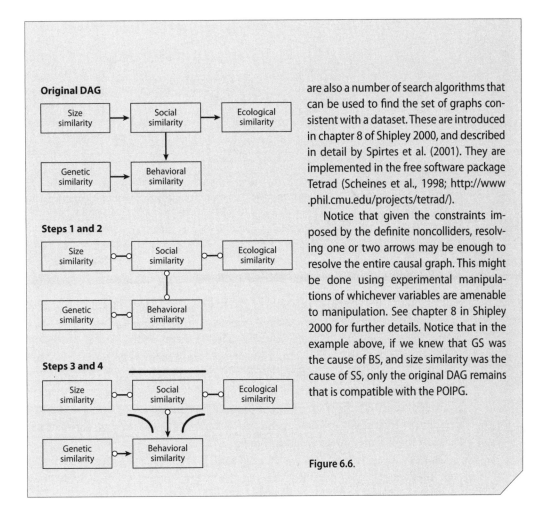

Original DAG

Size similarity → Social similarity → Ecological similarity

Social similarity → Behavioral similarity

Genetic similarity → Behavioral similarity

Steps 1 and 2

Size similarity — Social similarity — Ecological similarity

Social similarity — Behavioral similarity

Genetic similarity — Behavioral similarity

Steps 3 and 4

Size similarity — Social similarity — Ecological similarity

Social similarity → Behavioral similarity

Genetic similarity → Behavioral similarity

are also a number of search algorithms that can be used to find the set of graphs consistent with a dataset. These are introduced in chapter 8 of Shipley 2000, and described in detail by Spirtes et al. (2001). They are implemented in the free software package Tetrad (Scheines et al., 1998; http://www.phil.cmu.edu/projects/tetrad/).

Notice that given the constraints imposed by the definite noncolliders, resolving one or two arrows may be enough to resolve the entire causal graph. This might be done using experimental manipulations of whichever variables are amenable to manipulation. See chapter 8 in Shipley 2000 for further details. Notice that in the example above, if we knew that GS was the cause of BS, and size similarity was the cause of SS, only the original DAG remains that is compatible with the POIPG.

Figure 6.6.

4. Use scatterplots to determine whether matrices are linearly related.[18]
5. Use the appropriate correlation statistic and a permutation test to get a p-value for the strength of evidence against each conditional independence statement obtained in step 3.
6. Combine the relevant p-values obtained in step 5 using Fisher's C test, to obtain an overall p-value quantifying the strength of evidence against each candidate causal graph.
7. For each candidate causal graph that is consistent with the data from step 6, obtain the set of equivalent causal graphs (box 6.10).
8. For any of the causal graphs from step 7 that are mechanistically plausible, obtain the path coefficients and standard errors (box 6.6).

---

[18] This determines whether to use Spearman's correlation coefficient to test the conditional independence statements (box 6.8). In addition, if the relationships are not linear then (a) the researcher will not be able to test cyclical causal graphs (see box 6.8), and (b) the data should be transformed to obtain path coefficients in step 8, which will in turn affect the way the path coefficients are interpreted.

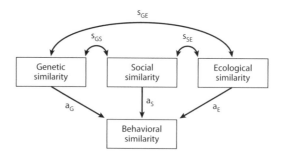

Figure 6.7. The matrix regression method (Whitehead 2009) as a causal graph. There are unspecified correlations between GS, SS, and BS, with the path coefficients, a_G, a_S and a_E to be estimated, and tested against zero in the analysis. This is the relationship between standard multiple regression and path analysis in general (Shipley 2000, 126–130).

6.4.1 Relationship to the matrix regression approach

At this stage we will pause to see how the matrix regression method described in section 6.3.1 relates to the causal modeling framework described above. The matrix regression approach (Whitehead 2009) relates to the casual modeling framework in the same way that multiple regression relates to path analysis (Shipley 2000). The original justification for multiple regression was to consider the independent variables as potential direct causes of the dependent variable. This is equivalent to a path diagram in which there are direct causal influences from the independent variables to the dependent variable, with unspecified correlations between the independent variables (see fig. 6.7).[19]

One can then test whether each standard partial regression coefficient (or path coefficient) is significantly different from zero. However, one cannot test whether the causal graph itself is consistent with the data, since there are no implied d-separation events (Shipley 2000). Consequently, a significant partial regression coefficient does not necessarily mean that the direct causal pathway is supported by the data; instead, the estimate of the pathway is conditional on the causal model proposed in figure 6.7. The prior application of Shipley's d-sep test increases the power of the approach considerably, for two reasons. First, it potentially allows researchers to rule out some causal hypotheses. Second, it yields simpler and better-specified causal graphs on which to base estimates of direct causal effects. If the data do not allow researchers to distinguish between competing causal graphs, then nothing has been lost in the attempt.

6.4.2 Ruling out homophily

It is tempting to argue that if researchers are only interested in direct causal pathways leading to behavioral similarity, then they can be content with figure

---

[19] A double-headed arrow denotes unspecified correlations. In SEM this means that the errors are correlated, though this is equivalent, in terms of d-separation, to a model in which there is a latent variable that is a common cause of both (Shipley 2000, 265).

6.7 as the underlying causal graph. However, there are two additional, highly plausible types of causal influence that are statistically compatible with this model, and might need to be ruled out using d-separation. The first is known as "homophily": the idea that being behaviorally similar will cause individuals to spend more time together. For example, monkeys able to crack nuts (e.g., Viselberghi et al. 2007) might spend more time in the vicinity of trees bearing nuts, causing them to spend more time together. This is represented in the causal graph by an arrow from behavioral similarity to social similarity (e.g., fig. 6.8b). Whitehead (2009, 140) argues: "If it [behavioural similarity] is not [caused by] social learning, then it must be genes, environment, and/or something else (such as gender or ontogeny). If these factors are captured by the other independent factors . . . in the regression, then the standard partial regression coefficients will be small, and social learning will not be supported, even though social and behavioural similarities are well correlated." Unfortunately, this is not the case; if there is a direct causal link between social and behavioral similarity, then they will not be d-separated conditional on any other set of variables in the graph, regardless of the direction of the causal link. Furthermore, d-separation does give us another means to distinguish homophily from social transmission. We suggest the key is to find a variable that is a cause of social similarity; for example, individuals might prefer to associate with others of the same sex or a similar size. If this variable is included in the causal graph, one can see that the d-separation statements can be different for social transmission and homophily (see fig. 6.8).

In figure 6.8, size similarity in *a* (social transmission) will be unconditionally correlated with behavioral similarity, because there is no collider on the path between them (see box 6.8). Furthermore, size similarity will be d-separated from behavioral similarity by social similarity. In contrast, in *b* (homophily), size similarity will not be unconditionally correlated with behavioral similarity, since there is a collider (social similarity) on the only pathway between them. However, they are no longer d-separated conditional on social similarity. Consequently, these two causal graphs could be distinguished statistically.[20] To illustrate this, we present an example based on simulated data, simulated to follow the causal model in figure 6.8a. The set of sufficient d-separation statements for 6.8a is shown in table 6.1. We tested each of these using partial Mantel tests to obtain p-values for each d-separation event.

We then combined these to get a test statistic of $C = -2(\log(0.791) + \log(0.696) + \log(0.762) + \log(0.861) + \log(0.310)) = 4.38$. Looking this up in a chi-squared distribution with $2 \times 5 = 10$ d.f., we get $p = 0.929$, and there is no evidence against the causal model. In contrast, if we apply the same procedure to the homophily model in figure 6.8b (see table 6.2), we get a test statistic of $C = -2(\log(0.791) + \log(0.001) + \log(0.762) + \log(0.006) + \log(0.310)) = 27.40$.

---

[20] In some cases, a researcher might be able to rule out homophily and unknown ecological variables without the addition of an extra variable. For example, in figure 6.8 the d-separation of GS for SS also differs between *a* and *b*. However, the inclusion of an additional variable increases the resolution between the two hypotheses.

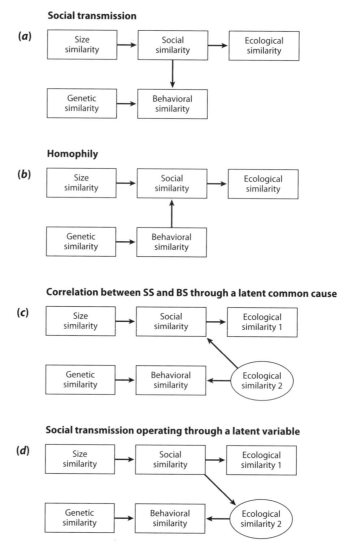

Figure 6.8. Different causal hypotheses relating GS, SS, ES, BS, and the differences in size between individuals. In this hypothetical example, size is known to be a cause of social similarity, with individuals of a comparable size tending to associate. Variables shown in circles are latent (unmeasured) variables. See text for an explanation.

Looking this up in a chi-squared distribution with $2 \times 5 = 10$ d.f., we get $p = 0.002$, and strong evidence against the causal model in figure 6.8b.

An alternative or complementary approach to ruling out homophily is suggested by Matthews' (2009) work on capuchin monkeys. He found that when behavioral similarity was based on the foraging behavior patterns that were present in each individual's repertoire, behavioral similarity was correlated with social similarity. In contrast, when behavioral similarity was based on the relative frequency with which those different behavior patterns were used, there was little

Table 6.1

Nonadjacent variables	Parent variables	d-separation statement	*p*-value from partial Mantel test
Size, GS	None	GS ⊥⊥ BS	0.791
Size, BS	SS,GS	Size ⊥⊥ BS|SS,GS	0.696
Size, ES	SS	Size ⊥⊥ ES|SS	0.762
SS, GS	Size	SS ⊥⊥ GS|Size	0.861
GS, ES	SS	GS ⊥⊥ ES|SS	0.310

Table 6.2

Nonadjacent variables	Parent variables	d-separation statement	*p*-value from partial Mantel test
Size, GS	None	GS ⊥⊥ BS	0.791
Size, BS	GS	Size ⊥⊥ BS|GS	0.001
Size, ES	SS	Size ⊥⊥ ES|SS	0.762
SS, GS	Size, BS	SS ⊥⊥ GS|Size, BS	0.006
GS, ES	SS	GS ⊥⊥ ES|SS	0.310

evidence of such a correlation. If behavioral similarity causes social similarity, one would expect a stronger correlation between social similarity and the frequency with which different behavior patterns are performed, than with the presence of behavior patterns in the repertoire.

6.4.3 Ruling out unknown ecological variables

The second type of causal influence that might be ruled out using d-separation is the problem that we identified earlier in the chapter: even if researchers quantify and include ecological variables in their analysis, there might be an unmeasured ecological variable that is driving behavioral differences. This is only a problem if similarity between individuals in the unmeasured ecological variable also causes social similarity (see fig. 6.8c), because this would cause an unconditional correlation between social similarity and behavioral similarity. This explanation might also be ruled out using d-separation; in figure 6.8c, size similarity is, again, unconditionally d-separated from behavioral similarity, but it is not d-separated from behavioral similarity given social similarity.

Another possibility might present itself to the reader at this stage; social similarity might cause similarity in an unmeasured ecological variable (fig. 6.8d), which in turn causes behavioral similarity. This graph is entirely compatible with figure 6.8a, since causes are only "direct" relative to the variables included in the graph. Whether or not this is a concern depends on the specific interests of the researcher. Even if a causal pathway leads through a *measured* ecological variable, social similarity is still a cause of behavioral similarity in the sense in which "causes" are defined in the axioms of causal modeling (Shipley 2000).[21] If the cause is direct, it merely means that there is a causal pathway that does not operate through any other variables in the graph. However, whether this issue is a problem or not, it is also inherent when detecting causal pathways by experimental manipulation of variables. Any causal effect on variable Z, detected by manipulation of another variable A, could, and presumably does, operate through any number of variables, B, C, D, and so forth, that are not held constant in the experimental procedure. Shipley (2000) discusses the issue of direct and indirect causation in more detail.

6.4.4 Relationship to the method of exclusion

The causal modeling framework described here has a straightforward relationship to the basic and advanced methods of exclusion described in section 6.2.1, at least when applied to group-contrasts data. The basic method of exclusion rules out the causal pathways GS→BS and/or ES→BS a priori (see fig. 6.9a). The effect SS→BS is thus inferred to exist from evidence against unconditional independence between SS and BS; in the group-contrasts case, this evidence consists of group specific differences in behavior. In section 6.2.2.1 we argued that a researcher will rarely, if ever, be able to rule out the GS→BS pathway a priori, whereas in section 6.2.3.1 we concede there are cases where the ES→BS pathway can be ruled out. In such cases we get the causal graph shown in figure 6.9b.

In the advanced method of exclusion, a researcher acknowledges the possible existence of pathways GS→BS and ES→BS, and judges whether these causal pathways alone can account for a putative pattern of social transmission. This corresponds to a rejection of the graph shown in figure 6.9c in favor of the graph shown in figure 6.9d.

In general, a causal graph encodes assumptions or knowledge about causal structure. Critically, the *missing* arrows represent a commitment to assuming that a causal effect *does not exist*, since an arrow that is present can be fitted with a path coefficient of zero (Pearl 2009). Therefore, the methods of exclusion are justified to the extent that the missing arrows in figure 6.9 can be justified.

6.5 Conclusions

We hope to have persuaded the reader that the time is ripe to move away from the method of exclusion (section 6.2), commonly used in social learning field studies, and to move to a broader framework. A model-fitting approach is the obvious next step, and in section 6.3 we describe some specific methods that have

---

[21] If a causal graph including the pathway SS → ES → BS is supported, we suggest the researcher break down the ecological similarity matrix into more specific components, because it is possible that social similarity causes similarity in a different aspect of ecology than that which causes behavioral similarity. This exercise would also elucidate further the exact causal pathway between SS and BS, if it exists.

Basic method of exclusion

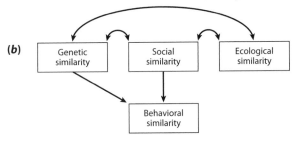

Basic method of exclusion (ES only)

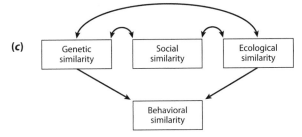

Advanced method of exclusion: null hypothesis

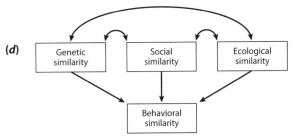

Advanced method of exclusion: alternative hypothesis

Figure 6.9. The method of exclusion as causal graphs. *a*) The basic method of exclusion for both the ecological and genetic hypothesis; *b*) the basic method for the ecological hypothesis only; *c*) the null hypothesis for the advanced method of exclusion; *d*) the alternative hypothesis for the advanced method of exclusion.

been proposed to accomplish this. We advocate going a step further and moving to a causal modeling framework, and suggest ways in which this might be done in section 6.4.

We are not suggesting that these alternative frameworks provide a "magic bullet" that will instantly resolve debates about animal culture. Rather, we feel they are a better way to quantify our knowledge about the role of social transmission in generating differences in behavioral repertoires. For example, with the model-fitting approach, when genetic, ecological, and social similarities are highly correlated, it will be difficult to distinguish their effects. Consequently, without a great deal of data researchers will have a lot of uncertainty about the effects of each; however, this would be an accurate representation of our state of knowledge. This is an improvement over "rejecting" social transmission as unimportant in such circumstances, when using the method of exclusion.

Likewise, a researcher using casual modeling methods might find that two or more competing causal hypotheses cannot be distinguished by the data, that is, there is not strong evidence against either. There are two reasons this can occur, and it is important that a researcher recognizes the difference between them. First, the competing hypotheses might correspond to different statistical models, but the data do not provide strong evidence against either. In this case, collecting enough data, if possible, should resolve the difference. Second, the competing hypotheses might correspond to equivalent graphs (box 6.10), and consequently to the same statistical model. No amount of observational data will resolve this difference, and a suitable experimental manipulation will be required to distinguish them (Shipley 2000; Pearl 2009). This should not be taken as a weakness of the causal modeling approach. It is instead a strength, in that it tells us what the logical limitations are on inferences from our observational data.

We conclude this chapter with a general cautionary note on the interpretation of repertoire-based methods for inferring social transmission. As we noted at the outset, these methods are concerned with the extent to which differences in behavior are the result of social transmission. Given that much interest in animal social learning is concerned with its relationship to human culture, this concern with the role of social transmission in producing differences in behavior is not inappropriate. Nonetheless, it is worth noting that just because social learning does not cause differences in a particular trait, it does not necessarily mean that social learning is not important in the development of that trait. This mirrors exactly the limitations of genetic heritability, which quantifies the amount of variance in a trait that can be assigned to genetic, rather than environmental, differences. Just because a trait has low heritability does not mean genetics do not influence its development. Bateson and Martin (2000) illustrate this with the heritability of "walking on two legs" in humans. People walk on fewer than two legs only as the result of environmental factors, such as accidents, disease, war injuries, and exposure to teratogenic toxins before birth; this has a heritability of zero. But it would be absurd to say that genetics does not influence walking on two legs, since genetics clearly underlies this difference between humans and other species.

We suggest social learning researchers familiarize themselves with critiques of heritability (e.g., Bateson and Martin 2000) to avoid falling into the same traps and engaging in the stale debates that have already occurred in the field of genetics. Such critiques advocate studying the role of both genes and environment during the developmental process. Correspondingly, we argue that this limitation of repertoire-based methods means that social learning researchers require additional methods that study the role of social learning in shaping behavior over time (chapters 5 and 7) if they are to acquire a full understanding of the process, rather than focusing purely on repertoire data.

Chapter 7

Developmental Methods
for Studying Social Learning

We ended the previous chapter by drawing attention to the limitations of studying social learning based solely on differences in repertoires, and suggested that developmental approaches are also required. We are not alone in seeing virtues to a developmental perspective on social learning. For example, Dorothy Fragaszy (2012a) writes: "Comparing behavior across groups cannot tell us about the developmental origins of a behavior, and thus such comparisons provide an inadequate basis to determine if a behavior is a tradition." Fragaszy maintains: "The social setting in which young animals develop shapes the traditions they will acquire." She further suggests: "To determine how social partners contribute to skill learning, we should relate individual differences in skill development with individual histories of watching others." Other researchers have taken a similar line (Fragaszy and Perry 2003; Biro et al. 2003; Lonsdorf 2006; Perry 2009; Sargeant and Mann 2009; Reader and Biro 2010). Here we provide examples of some of the developmental methods that have been used to study social learning, and suggest some ways in which the methodologies might be developed further.

We take a broad view of what constitutes a developmental approach to include any approach that aims to elucidate the role of social influences in the development of a behavioral trait.[1] Developmental approaches can be broadly divided into two types. The first type comprises approaches that involve collecting observational data on the development of a trait and the opportunities that arise for social learning, as well as attempting to infer the role of social learning. We discuss

---

[1] We do not aim to cover the methods used specifically by psychologists studying human development (see section 2.5), but rather focus on those that can be used on a broad range of species. That is not to say the methods described here could not be used fruitfully on humans.

such approaches in section 7.1. These approaches have much in common with the methods discussed in chapter 5, especially where the data are analyzed using time-structured models. Indeed there is no reason why the methods of chapter 5 could not be adapted for developmental data, and the methods presented in section 7.1 for diffusion data, since the only real difference is one of time scale.

In section 7.2 we go on to discuss developmental methods that involve experimental manipulations. Strictly speaking, most of the laboratory experiments described in chapters 3 and 4 would constitute developmental methods. However, here we restrict ourselves to approaches that can be applied to animals in their natural environment, or in captivity in more naturalistic settings, such as a freely interacting group within an enclosure.

7.1 Observational Data

We begin by concentrating on some of the methods that have been applied to observational data on the development of traits in order to elucidate the social influences on development. For the cases we discuss, the researchers' aim was often to build up a picture of all factors influencing the development of a trait, rather than solely to investigate the social influences on learning throughout development. However, we will focus on those aspects of each case that are pertinent to investigation of social influences on the transmission of knowledge. As in previous chapters, we will dwell on the methods used and their potential utility as general tools for social learning researchers.

7.1.1 Describing the developmental process

A critical role for developmental studies is to document the typical process by which a behavioral trait is acquired, through detailed longitudinal observations on a number of individuals. Such studies often seek to identify certain stages in the development of a character, and the opportunities for social learning that occur at each stage. In doing so, the researcher sets out to identify at which stages social learning plausibly operates, and what type of social experience might be necessary for development in each case.

These advantages are illustrated by the work of Dora Biro and colleagues (2003, 2006) on nut cracking and leaf folding[2] in chimpanzees at Bossou in the Republic of Guinea. Nut cracking involves placing the nut on a stone, or "anvil," and using a second stone as a "hammer" to crack the nut open. Leaf folding involves folding leaves inside the mouth into an "accordion-like" shape; these are then dipped into water and retrieved to obtain water for drinking. Aside from being instances of nonhuman tool use, both of these traits are also among those claimed as aspects of chimpanzee culture, because some groups display these traits whereas others

---

[2] Biro et al. (2006) is in part a review of a number of studies on these traits at Bossou. The work on leaf folding was later expanded by Sousa at el. (2009).

do not (Whiten et al. 1999; see chapter 6). Consequently, there is considerable interest in obtaining developmental evidence that these traits are socially transmitted, and if so, how this occurs.

The research was conducted at an "outdoor laboratory" at Bossou, established by Tetsuro Matsuzawa in 1988 with the aim of increasing encounter rates with the chimpanzees of the Bossou group, and providing opportunities for researchers to observe tool-using behavior. The laboratory is opened for one to two months per year during the dry season, and during this time it is closely monitored by researchers from seven a.m. to six p.m. every day. The outdoor laboratory includes a tree with an enlarged natural hollow, allowing leaf folding and subsequent leaf tool use to be performed. In addition, researchers stock the lab with oil-palm nuts, and stones of known size and weight allowing chimpanzees to perform nut cracking. All chimpanzees in the group frequently visit the site, allowing Biro and her colleagues to analyze the development of each trait in individuals over time.

Biro et al. 2003 and Biro et al. 2006 report that no nut cracking was found before the age of 3 to 3.5 years, and that if an individual did not acquire nut cracking by the age of 7 years, they were never seen to perform the trait later in life. This led to the proposition that there is a critical period during which nut cracking can develop (Matsuzawa 1994). Further analysis of the observations on individuals of 0.5 to 3.5 years of age has enabled Biro and her colleagues to determine the stages of learning involved in nut cracking. First chimpanzees engage in single object manipulations of nuts and stones, followed by multiple actions on multiple objects in sequences of increasing complexity. By 1.5 years of age, infants had performed the component actions of the trait, but it typically took another 2 years until these were combined in the correct sequence, enabling infants to crack a nut by themselves.

Drinking water using folded leaves consists of two phases: tool making and tool using. At around 2 years of age chimpanzees start using folded leaves to drink water, but use tools discarded by older individuals. Just before the age of 3.5 years, infants were observed to start constructing their own tools, but initially these were discarded in favor of tools made by older chimpanzees. Sometime after the age of 3.5, infants would start to use their own tools for drinking.

The data collected at Bossou also allowed Biro et al. (2003) and Biro et al. (2006) to determine the extent to which there were opportunities for social learning. They note that although both tool-using traits are solitary traits, they tend to be performed synchronously with others. Furthermore, infants always travel with their mother, and so are exposed to her tool-using behavior and that of the chimpanzees with whom she associates. Consequently, there are opportunities for social interaction to influence development. Biro and her colleagues note two distinct types of possible influence: those that operate through direct observation of other individuals' tool-use behavior, and those that operate through exposure to the products of tool use. For nut cracking, the latter consists of scrounging cracked nuts from their mothers, which is the infants' only tangible source of reinforcement in the nut-cracking context. It seems plausible that such experience would cause an increase in the frequency of interaction with nuts and stones in the future. In addition, infants are also allowed to interact with stones during their mother's nut-cracking bout (e.g., touching the anvil stone and holding the mother's hammering arm while she is delivering blows). Biro et al. (2006) note that juveniles at Bossou are not granted the same liberties to scrounge as infants, and usually are chased away if they approach an adult engaged in nut cracking. Biro et al. (2006)suggest that this might contribute to the end of the critical period for learning.

Similar social opportunities arise for leaf folding. While they are being held by a mother engaged in leaf-tool use, infants are able to reach into the tree hole and obtain water with their hands. Though this is not interacting with the products of behavior, Biro and her colleagues consider it a parallel to the scrounging of palm nut kernels, since infants may be "getting a taste" of the rewards of successful leaf-tool use. As noted above, infants are also able to engage in leaf-tool use before they are able to construct their own tools. In our terminology (chapter 4), this constitutes a case of opportunity providing, since the products of the "demonstrator's" behavior provide the infant with an easier version of the task (requiring only tool use, rather than tool making and then tool use). It seems plausible that acquiring the tool-using component of the trait might be a necessary prerequisite for obtaining the tool-making component, which would make access to discarded tools critical in the acquisition of the full trait.

Biro and her colleagues (2006) also recorded frequent opportunities for the observation of conspecifics using tools, and went on to analyze the pattern of observation between individuals of different age classes (adults, juveniles, and infants). They defined an episode of observation of one chimpanzee by another as occurring when "the latter approached the former to within about 1 meter and remained with gaze fixed upon the target individual's face or hands for at least 3s [3 seconds]" (Biro et al. 2006, 495). The authors then calculated the rates at which each age class was observed by each other age class, dividing by the total time during which observation could have occurred. This was done by calculating the time each individual was present in the company of at least one other individual who was engaged in the relevant activity (nut cracking/handling or drinking water with the aid of leaf tools) (Biro, pers. comm.). The time was then summed across individuals in the observing age class.[3] Biro et al. (2006) concluded that (*a*) adults were more likely to be observed performing the relevant activity than juveniles or infants, (*b*) juveniles and infants were more likely to observe than adults, and (*c*) chimpanzees almost exclusively observed conspecifics in the same age group or older than themselves (see fig. 7.1). Biro et al. (2006) note that this pattern of observation has implications for any model of social transmission. The pattern fits with the hypothesis that individuals tend to observe others who are proficient in the target trait.

This measure does not completely separate out the effects of chimpanzees choosing to watch individuals of a particular age class from differences in the

---

[3] Biro et al. (2003) also calculated rates of observation for each observer individual; this allowed them to run statistical models revealing strong evidence of the effects reported.

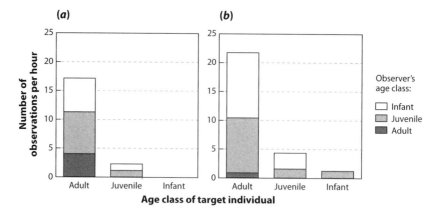

Figure 7.1. Chimpanzees in different age classes as targets of observation during (*a*) nut cracking, and (*b*) using leaves to drink water. See main text for an explanation of how rates of observation were calculated. Based on figure 11 in Biro et al. (2006).

number of opportunities to watch individuals of different age classes. For example, the latter effect could have arisen if there were more adults then juveniles/infants engaged in nut handling activities at any given time. In this case, Biro (pers. comm.) reasons that it is more likely to be the former effect since, in most instances, the groups present were comprised of roughly equal numbers of young and adults. Nonetheless, in general, a researcher could isolate the former effect by calculating the rate of observation of class B by class A (R_{AB}) as follows:

1. Calculate O_{AB}, the total number of observation episodes in which the observer was of class A and the demonstrator of class B.
2. For each pair of individuals in the group, *i* and *j*, calculate T_{ij}, the total time during which *i* was present and *j* was engaged in the target activity.
3. Sum T_{ij} across all pairs of individuals for which *i* is of the target observer class A, and *j* is of the target demonstrator class B, in order to obtain T_{AB}.
4. Calculate $R_{AB} = O_{AB} / T_{AB}$.[4]

This measure would allow for the number of potential demonstrators of class B that are available at any given time as well as the number of potential observers of class A. As such, it should give an indication of differences in a tendency to choose to observe individuals of specific classes.

The research conducted at Bossou illustrates the merits of a developmental approach to understanding the role of social learning in naturally occurring traits. By obtaining longitudinal data, a social learning researcher can identify whether there is a critical or sensitive period in development during which the trait is usually learned. The researcher can also identify the stages in learning, and assess the extent to which opportunities for social learning arise at each of these stages.

---

[4] In addition, a researcher could analyze data at the individual observer level, by summing only across *j*, to assess the strength of evidence for differences in rates of observation between classes of individuals.

It is of note that these aims were achieved without requiring a complete history of each individual's attempts at performing each trait, nor were observations obtained throughout the year. Instead, a one to two month intensive period of observation each year enabled Biro and colleagues to infer what stage each individual had reached in the acquisition of each trait. Similarly, a complete history of observation opportunities was not obtained—instead a sample was obtained for a specific location and time window. The study thus gives researchers some idea of the social learning opportunities for the chimpanzees at Bossou in general. Although the artificial provisioning of resources might have caused chimpanzees to aggregate more frequently than under fully natural conditions, it still provides researchers with an idea of the extent to which such aggregations, when they do occur naturally, afford opportunities for social learning.

However, identifying opportunities for social learning does not provide evidence that this learning occurs as a result of receiving such opportunities; nor does it estimate the importance of such opportunities relative to other factors influencing development. Over recent years a number of approaches have been made to attempt to do this, to which we now turn.

7.1.2 Modeling probability of acquisition

For traits that are not acquired by all individuals in the population, researchers can investigate which factors influence the probability that the trait will be acquired. To ascertain whether social learning is involved, variables can be included in the analysis that represent an individual's opportunities for social learning. This approach allows a researcher to include, and statistically control for, other variables that might cause changes in the probability of acquisition.

Sargeant and Mann (2009) used this approach to investigate the acquisition of seven different foraging tactics in bottlenose dolphins, including sponging, a strategy discussed in chapter 6. The work was conducted at Shark Bay in Western Australia, which has been the site of a longitudinal research project since 1984. As part of the project, dolphin mothers and calves were observed from 1990 to 2004 during a mother-calf study, using a focal animal follow protocol. Among the activities recorded by researchers was the use of the seven different foraging tactics.

Sargeant and Mann used logistic regression to investigate the factors influencing the probability that a dolphin would acquire each foraging tactic (trait). They reasoned that there could be vertical social transmission, from the mother, so included whether the mother exhibited the foraging tactic as a factor in their analysis. Alternatively, there could be oblique or horizontal social transmission from other dolphins with which the subject associated. They defined an associate as a dolphin that had been seen in a group with the subject, and determined how many of these had been observed to use the foraging tactic in question. The number of associates known to use the tactic was then included in the analysis.

Sargeant and Mann (2009) also wished to include the effect of habitat use by each calf. They quantified this by the average water depth, which they recorded every five minutes during focal follows. They note that though this is a rough measure of habitat use, it reflects the different zones of the study area, which could be categorized as shallow sea grass banks and deeper areas with little sea grass. They had also found depth to be an important predictor of the occurrence of several foraging tactics in previous analyses (Sargeant et al. 2007).[5] Finally, Sergeant and Mann included the number of hours each individual had been observed to statistically correct for the fact that a tactic is more likely to be observed in an individual that is observed more (but see box 7.1).[6]

The fitted models then allowed Sergeant and Mann to estimate the effect each variable has on the probability of acquisition of each trait. In order to interpret the effects of a logistic regression, it is important to realize that the model assumes that each variable has a linear effect on the natural logarithm of the odds ratio (the log-odds scale), where the odds are the probability of acquisition divided by the probability of nonacquisition.[7] By exponentially back-transforming these coefficients and their associated confidence intervals, a researcher yields an estimate of the multiplicative effect on the odds ratio. For example, for "bottom grubbing," Sergeant and Mann (2009) estimate an effect of 0.17 per meter (95% confidence interval = 0.02– 0.52), which means that for an increase of one meter in average water depth, the log odds of bottom grubbing are multiplied by 0.17. Effects are interpreted in a similar manner for categorical variables. For example, the effect of having a mother who was a sponger is estimated at 143.97, which means that the odds of sponging were estimated at 143.97 times higher than for calves with a nonsponging mother. A 95% confidence interval is calculated as 5.59 to 1,3900,000, showing that that the actual effect on the odds ratio is very unlikely to be less than 5 times.

Sergeant and Mann (2009) found strong evidence that three foraging tactics were affected by maternal use of that tactic: sponge carrying, mill-foraging, and rooster tail foraging[8] (see fig. 7.3). As noted above, there was evidence bottom-grubbing was associated with average depth; this supported the hypothesis that

---

[5] Sergeant and Mann (2009) also tried an alternative measure of habitat use for sponging, with very similar results.

[6] Sergeant and Mann (2009) used a backward stepwise elimination procedure, with "hours observed" forced to remain in the model. We would advocate instead using a model averaging procedure using AIC_c (Burnham and Anderson 1998), which allows for model selection uncertainty and overcomes the problems of using an arbitrary significance level (see box 5.3).

[7] There is no functional form for factors (categorical variables), since a separate coefficient is fitted for each level of the factor. However, the effects of a factor are assumed to combine with the effect of other variables additively on the log-odds scale. Consequently, if there are other variables in the model, it is often only possible to interpret effects on the odds scale rather than directly in terms of probability. Sergeant and Mann (2009) do provide estimates of the effect of some variables in terms of probability (see fig. 7.2), but these are for cases where there is no other important variable in the model.

[8] These traits are defined by Sergeant and Mann (2009) as follows:

Sponge carrying: Dolphin carries a sponge on its rostrum during stereotyped tail-out dive/ peduncle dive foraging.

Mill foraging: Dolphin surfaces irregularly and changes directions with each surface while foraging, often with the surfacing being rapid.

Bottom grubbing: Dolphin orientates toward and pokes rostrum into sea grass or the sea floor, with body positioned vertically.

Rooster-tail foraging: Dolphin swims rapidly near the surface so that a sheet of water trails off the dorsal fin.

habitat use drives acquisition of this trait. In all of these cases, there were no other variables remaining in the model (aside from "hours observed," which was forced into the model and for which there was not strong evidence of an effect). For "leap and porpoise feeding" there was weak evidence of an effect of maternal foraging, associate foraging, and water depth. For the remaining two tactics, there was little if any evidence (not at the 15% significance level) that they were affected by the variables considered.

At this stage it is worth considering what the causal interpretation of such models can be. Evidence of a positive effect on acquisition of having a mother that exhibits that trait seems to constitute evidence of a causal effect from mother to offspring. The reverse causation is ruled out, since as far as we are aware, no mother acquired a foraging tactic after a calf did. It is logically possible that a third variable might be causing both mother and calf to acquire the traits, but the most likely candidate for such a third variable would be habitat usage, which is ruled out, at least when using the proxy of water depth. The analysis in itself does not rule out genetic inheritance of these traits from mothers to offspring, though Sargeant and Mann (2009) consider it unlikely in this case based on other lines of evidence (see section 6.2.2.2). If so, then the social effect on these traits qualifies as strong cases of social transmission of a trait.

In light of the causal modeling techniques discussed in section 6.4, it is worth noting that if social transmission operated through a variable that is included in the analysis, it would not be detected by these methods. For example, it might be that a mother acquiring the bottom-grubbing trait causes her to spend more time in shallow water. This in turn is likely to cause her calves to spend more time in shallow water, which might also cause them to be more likely to acquire the trait of bottom grubbing. This would also qualify as social transmission of a trait by our definition (chapter 1). However, since maternal foraging is d-separated (see box 6.8) from probability of acquisition by water depth, it would not be included in the final model.

7.1.3 Modeling time of acquisition

This is a conceptually similar process to modeling the probability of acquisition. However, it can also be used when all individuals in the population reliably acquire the trait, but there is variation in the time at which they do so (i.e., they acquire the trait at different rates; see chapters 1 and 5). Both types of information (who acquired the trait and when they did so) could be combined by using a survival analysis in which those who do not acquire the trait are "censored" and thus taken into account when estimating the effect each variable has on rate of acquisition.[9] However, the cases we are aware of do not use this approach because all individuals in the population acquire the trait.

---

[9] This would make the analysis similar in form to network-based diffusion analysis (*NBDA*) discussed in chapter 5.

Box 7.1.

Controlling for observation time when modeling probability of acquisition

If a researcher is studying the factors influencing the probability of acquisition of a trait, it might be necessary to control for observation effort. In other words, given an individual has the trait, we are more likely to know about it in individuals we have observed more. This is especially important if observation time is correlated with any of the variables of interest. Sargeant and Mann (2009) statistically controlled for observation effort by including the number of hours of observation as a variable in the linear predictor of the logistic regression. Though this is a reasonable approach, we feel it can be improved upon. The problem with including observation effort in the linear predictor is that the model assumes that the log odds of observing the trait increase indefinitely with observation effort (tending to a probability of 1). So if an individual has a set of predictors that in reality give it a very low probability of acquiring the trait, the model unrealistically assumes that this can be offset by observation time. If such individuals are observed frequently, the effect of observation time will be estimated as being low, meaning that it will fail to correct fully for those individuals that are observed very little.

A preferable approach is to model the probability of observation of the trait in individual i, conditional on i having that trait as a function of observation time t. If the trait is performed at a rate of λ, then conditional probabilities are as follows:

$$p(O_i = 1 \mid X_i = 1) = 1 - \exp(-\lambda t)$$
$$p(O_i = 0 \mid X_i = 1) = \exp(-\lambda t)$$
$$p(O_i = 1 \mid X_i = 0) = 0$$
$$p(O_i = 0 \mid X_i = 0) = 1$$

Here X_i denotes whether i has the trait (1) or not (0), and O_i denotes whether i has been observed performing the trait (1) or not (0) by time t. The marginal probabilities are

$$p(O_i = 1) = (1 - \exp(-\lambda t))p(X_i = 1)$$
$$p(O_i = 0) = \exp(-\lambda t)p(X_i = 1) + p(X_i = 0)$$

If we combine this with a logistic model, we get:

$$p(O_i = 1) = (1 - \exp(-\lambda t))\exp(LP_i)/(1 + \exp(LP_i))$$
$$p(O_i = 0) = \exp(-\lambda t)xp(LP_i)/(1 + \exp(LP_i)) + 1/(1 + \exp(LP_i))'$$

where LP_i is the linear predictor for the individual i. This means the log-likelihood if i is observed to perform the trait $O_i = 1$ is:

$$\log(p(O_i = 1)) = LP_i + \log(1 - \exp(-\lambda t_i)) - \log(1 + \exp(LP_i)),$$

and if i is not observed to perform the trait, then

$$\log(p(O_i = 0)) = \log[\exp(-\lambda t_i)\exp(LP_i)/(1 + \exp(LP_i)) + 1/(1 + \exp(LP_i))].$$

Here t_i is the observation time for individual i. These expressions can be used to calculate a log-likelihood for a given set of parameters (both those in LP_i and λ, which quantifies the rate at which the trait is observed in informed individuals):

$$\log Lik = \sum_{i=1}^{N} O_i[LP_i + \log(1 - \exp(-\lambda t_i)) - \log(1 + \exp(LP_i))]$$

$$+ \sum_{i=1}^{N} (1 - O_i)\log[\exp(-\lambda t_i)\exp(LP_i)/(1 + \exp(LP_i)) + 1/(1 + \exp(LP_i))].$$

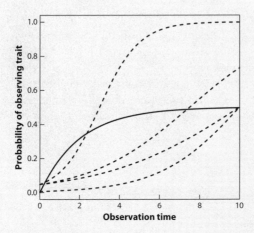

Figure 7.2. The solid line shows the probability of observing a trait in a specific individual as a function of observation time. Here there is a 0.4 probability that individual has the trait, and if it has, the trait is observed being performed at a constant rate. It is impossible to find a logistic function that models this relationship well, as illustrated by the examples (dashed lines).

An optimization algorithm, such as the optim or nlminb functions in R, can then be used to find the maximum likelihood estimators and standard errors (B.J.T. Morgan 2010), and subsequently calculate AIC$_c$ or conduct likelihood ratio tests. However, if random effects are required, it might be easier to use Markov Chain Monte Carlo methods using, for example, WinBUGS (Spiegelhalter et al. 2003).

The function given in figure 7.2 enables the probability of observation to tend to a specific value for a given value of the linear predictor as observation time increases (unbroken line). If observation time is included in the linear predictor, this is not the case, because the function is constrained to tend to one. Consequently, it is impossible to find a combination of parameters that fits the data well for individuals that are unlikely to have the trait (e.g., broken lines).

The above analysis assumes that the trait is always acquired before observation time starts; in practice, a period of observation time t' might occur during which developmental period when the trait can be acquired (possibly all of it; i.e., $t = t'$). In this case, given the trait has been acquired after A observation time units,

$$p(O_i = 1 \mid X_i = 1, A) = 1 - \exp(-\lambda(t - A)).$$

Now, assuming the trait is equally likely to have been acquired at any time during t' (probability density function: $f(A) = 1/t'$), and using the law of total probability we get:

$$p(O_i = 1 \mid X_i = 1) = \int_1^{t'} (1 - \exp(-\lambda(t - A)))(1/t') \, dA$$
$$= (1/t')\left[A - \tfrac{1}{\lambda}\exp(-\lambda(t - A)) \right]_0^{t'}$$
$$= 1 + (1/\lambda t')\left[\exp(-\lambda t) - \exp(-\lambda(t - t')) \right]$$

for $0 < t' \leq t$. Therefore,

$$p(O_i = 0 \mid X_i = 1) = (1/\lambda t')\left[\exp(-\lambda(t - t')) - \exp(-\lambda t) \right],$$

which gives us the following expressions:

(continued)

Box 7.1   *(continued)*

$$\log(p(O_i = 1)) = LP_i + \log(1 + (1/\lambda t')[\exp(-\lambda t) - \exp(-\lambda(t - t'))]) - \log(1 + \exp(LP_i))$$
$$\log(p(O_i = 0)) = \log[(1/\lambda t')[\exp(-\lambda(t - t')) - \exp(-\lambda t)]\exp(LP_i)/(1 + \exp(LP_i)) + 1/1 + \exp(LP_i)]^{\cdot}$$

These can be used to calculate the total log-likelihood as above.

A further advantage of modeling observation effort in this way is that observation time does not appear in the linear predictor, making it easier to interpret the effects on probability of acquisition, which are interpreted in the standard way for a logistic regression described in section 7.1.2.

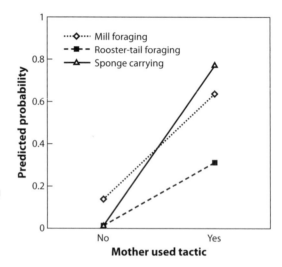

Figure 7.3. Predicted probability a dolphin would use three different foraging tactics, as a function of whether their mother used the tactic. Observation time is set to 34.8 hours. Based on figure 1 in Sargeant and Mann (2009).

Lonsdorf (2006) used this approach to study the development of termite fishing in chimpanzees at Gombe National Park in Tanzania. She defined the age of acquisition to be the age at which an individual had been observed successfully performing the trait three times. Lonsdorf collected data allowing her to calculate a number of measures of the mother's frequency of performing the skill, and her level of sociality while doing so. As chimpanzees remain with their mother until the age of around 7 years, the former measure is likely to reflect a chimpanzee's opportunities to practice termite fishing and observe their mother termite fishing, with the latter reflecting opportunities to watch and interact with other conspecifics performing the trait. These considerations yielded the following predictor variables:

1. Mother's overall percent time spent fishing.
2. Mother's overall percent time spent "alone" (with dependent offspring only).
3. Mother's overall percent time spent "alone" or with maternal family.
4. Mother's percent time alone while fishing.
5. Mother's percent time alone or with family while fishing.

6. Mother's percent time with a small group (fewer than 3 other maternally related individuals) while fishing.
7. Mother's proficiency (see below).
8. Offspring gender.

A measure of the mother's proficiency (see section 7.1.5) was also included by Lonsdorf (2006), along with the sex of the subject as a factor. The data were analyzed using the linear model framework,[10] and forward selection starting from a null model. There was not strong evidence of an effect of any of the measures (averaged over years) on age of acquisition, but the sample size was small with only six individuals monitored.

Such small sample sizes are likely to be common for intensive longitudinal studies, where individuals must be observed from a very young age. Ideally, more powerful methods are required that take advantage of detailed longitudinal data, that is, by including information about the age at which individuals reach different stages on the way to acquiring the trait. Lonsdorf (2006) noted that the termite fishing trait is composed of five "critical elements": (1) identification of a hole on the mound, (2) making a tool, (3) manipulating a tool, (4) inserting the tool into the hole, and (5) extracting termites. She recorded the number of critical elements (CEs) that had been acquired by each individual at ages 1.5, 2.5, and 3.5 years. These represented the response variable in linear models with the same predictors as for the age of acquisition model, except each was entered into the model separately for each year prior to and during the year in which an offspring skill level was measured. For example, a model of the number of CEs by age 1.5 might include mother's time alone at offspring age 0.5 and mother's time alone at offspring age 1.5.

Lonsdorf did not find strong evidence of any effects influencing the number of CEs acquired by age 1.5. However, the final model for the number of CEs acquired by age 2.5 included the mother's time spent fishing in a small group at offspring age 1.5, and the mother's time spent fishing alone at offspring age 2.5. Both variables were positively correlated with the number of CEs. For offspring in age class 3.5, the final model included mother's time alone or with family at offspring age 2.5, and mother's time alone or with family at offspring age 3.5. Both variables were, again, positively correlated with number of CEs. Lonsdorf (2009) notes that up to 1.5 years, infants spend most of their time watching their mother, whereas by 2.5 years, they spend most of time manipulating mounds and tools. She suggests that it is beneficial to have others to watch early in life, but detrimental later on when trying to experiment.

In general, the strategy of tracking the time at which individuals attain different stages of development, such as the number of critical elements acquired, seems to be a promising route forward for developmental data. However, the approach is limited by the use of separate statistical models for the number of CEs at each age. Moreover, the approach does not take into account the fact that

---

[10] In Lonsdorf (2006) these are described as GLMs (general linear models), but we prefer the terminology of linear models (LMs) to distinguish them from *generalized* linear models.

elements acquired by individual X at age Y will also be acquired at all future ages, or the fact that some CEs might be necessary precursors to others. In addition, the rate of acquisition of some CEs might be a function of social experience, whereas others might not. A useful further development would be a multistate model in which the transition rate between developmental states, by individuals, is modeled as a function of social experience.

7.1.4 Modeling proficiency of trait performance

It seems reasonable to suppose that social influences might act on development in a way that affects the proficiency with which the trait is performed, in addition to, or instead of, how soon the trait is acquired, or how often it is performed. Consequently, some researchers have obtained measures of the proficiency with which individuals perform the trait. For example, in her aforementioned study of termite fishing in chimpanzees, Lonsdorf (2006) quantified two measures of proficiency, the mean number of termites per dip (which she terms "proficiency") and the mean number of termites per minute (which she terms "efficiency"). She did not find any strong evidence of an effect of any of the variables quantifying opportunities for social learning (see section 7.1.3). Previous work (Lonsdorf et al. 2004) established that proficiency increased with age, consistent with the effects of practice, and that females tended to be more proficient than males. Lonsdorf notes that it is premature to rule out a correlation between a mother's proficiency and her offspring's, because this might be obscured by the sex effect.

Humle et al. (2009) used a developmental approach, similar to Lonsdorf's, to study the maternal influences on the acquisition of ant dipping in chimpanzees at Bossou. Ant dipping is another tool-using behavior that has been claimed as a component of chimpanzee culture (Whiten et al. 1999). Ant dipping involves the use of a stick or stalk of vegetation as a tool to gather army ants (*Dorylus* spp.). A chimpanzee holds the tool between its index and middle finger and performs a back and forth movement of the tool, to stimulate the ants to attack the tool. The chimpanzee then ingests any ants that climb the tool.

Humle et al. investigated whether social learning affects the proficiency of ant dipping. Using human ant-dipping experiments, they found that longer dips tended to increase the number of ants gathered, so they investigated the variables influencing dip duration in chimpanzee infants. Humle et al. observed 89 ant-dipping sessions, and this allowed them to calculate the mean percentage observation time each mother chimpanzee spent ant dipping across seasons. They used this to classify offspring (0–10 years) into two categories—high or low learning opportunity—depending on how much time their mother spent ant dipping (threshold = 0.2%).[11] High-opportunity offspring tended to dip for longer at trail[12] than low-opportunity young, and dip duration was correlated between

---

[11] In general, such variables might also be considered in a continuous form.

[12] Ant dipping is performed at ant trails (low risk) and at ant nests (high risk due to increased aggression). There was not strong evidence of an effect of prior opportunity at nests. Humle et al. analyzed these

mother and offspring. The authors suggest that since the time mothers and their offspring spent ant dipping was also positively correlated, the similarity in their proficiency might be the result of experience. They conclude that these results are consistent with the hypothesis that the opportunity for observation and practice, especially provided by the mother, are determinants of proficiency during development.

The approach of studying social influences on proficiency raises the question of what type of social transmission is operating (if observing others performing trait X causes an individual to become more proficient at trait X). It is also possible that social transmission might not be occurring, for example, if observers are learning what not to do, as a result of observing the mistakes of others (e.g., by observational *R-S* learning; see chapter 4). Alternatively, the social interaction in question could provide the subject with favorable opportunities to practice the trait (opportunity providing or opportunity teaching; see chapter 4 and section 7.2.1).

However, if there is evidence that observation of proficient individuals causes increased proficiency in the observer, it seems more likely that something is being socially transmitted. Should researchers regard this as another type of social transmission: the social transmission of trait proficiency? We argue against this. It seems likely that refinement of trait performance will involve learning something about the appropriate contexts in which an action is performed, or about some changes in the form of the action. For example, Humle et al. (2009) found that proficiency of ant dipping was related to the duration of dip. We see this as social transmission of a trait; the trait X' just happens to be a modification of a trait X that is already in the repertoire. So long as other individuals acquiring X' exert a positive causal influence on the transition $X \rightarrow X'$, then this is social transmission of trait X'. In practice the exact differences in trait performance between proficient and nonproficient individuals might be difficult to pin down, and more easily quantified by measures of proficiency.

7.1.5 Modeling option choice

Where a trait can be performed using two or more variants, or "options," a researcher can investigate the factors influencing option choice. For example, does a stable option preference develop over time, and if so, is this socially influenced? If option preferences fluctuate over time, do they do so in a synchronized manner within a group, and if so, is this caused by social learning?

Perry (2009) used this approach as part of her study of foraging techniques in white-faced capuchin monkeys (*Cebus capucinus*). The traits in question were two alternative techniques for releasing seeds from *Leuhea candida* fruits: pounding and scrubbing. Perry defines pounding as the repeated striking of the fruit against a substrate, and scrubbing as repeatedly moving the fruit back and forth while it is firmly in contact with a rough surface. The study was conducted at

---

data separately since there were differences in dip duration between the two contexts. These results are for offspring more than 5 years old.

Lomas Barbudal Biological Reserve, in Costa Rica. Perry has studied this population since 1990, and consequently kinship relations are known for virtually all members of the study groups. In the study population, capuchins live in multi-male, multifemale groups ranging in size from 5 to 38 individuals. Perry describes the forest occupied by each group as being "essentially identical ecologically" and notes that monkeys within a group forage as a fairly cohesive unit, visiting the same *Luehea* trees at roughly the same time. However, she also states that groups have broadly overlapping home ranges, such that individuals of adjacent groups frequently visited the same *Luehea* trees, though never at the same time.

To collect data, Perry followed a group and recorded the processing techniques that were employed. At any one time she watched a focal animal, but switched focal animals every time it finished processing a fruit. She also recorded the identities of all other monkeys foraging on *Luehea* within ten body lengths (400 cm) of the focal animal, all instances in which the focal monkey's gaze was oriented toward another monkey foraging within 400 cm of the focal, and the identity of the observed monkey.

The main aim of the analysis was to ascertain whether individuals' relative rate of performance of each variant (option choice) was influenced by the choices they observed other individuals making. To this end, for each monkey i in group j for each year k, Perry calculated an exposure score:

$$e_{ijk} = \frac{\sum_{l=1}^{Gj} n_{ijkl} p_{ljk}}{N_{ijk}},$$

where n_{ijkl} is the number of samples during which i was in proximity to another individual l in year k (where, proximity to self was $n_{ijki} = 0$); $p_{ljk} = y(1)_{lik}/(y(1)_{lik} + y(2)_{lik})$, where $y(1)_{lik}$ and $y(2)_{lik}$ are the number of fruits observed to be processed by l in year k, that were pounded and scrubbed respectively; and Gj is then number of monkeys in group j, and N_{ijk} is the number of recorded foraging events that i could have seen in year k. Therefore, e_{ijk} is an estimate of i's relative exposure to pounding versus scrubbing in year k.

Perry was also interested in the effects of "developmental year" on option choice (i.e., the number of *Luehea* seasons that the monkey has experienced). In our notation, k is "actual" year of the study. The developmental year can then be denoted by d_{ijk}, which, for example, takes the value 1 if year k was individual i's first *Luehea* season. Perry modeled the proportion of pounding, p_{ijk}, as a function of e_{ijk}, with sex and developmental year as factors; she also included two-way interactions in the model. This was conducted on a sample including 79 cases from 48 subjects (21 females and 27 males total, with 24 of those subjects contributing data to just 1 developmental year, 17 to 2 developmental years, and 7 to 3 developmental years). Perry used a Poisson regression with standard errors corrected for repeated measures on individuals.[13]

---

[13] Rather than a Poisson regression, we would advise instead using a logistic model (i.e., a GLMM with a binomial error structure, logit link function, and individual as a random effect) for such data; the differences between these two approaches are described in box 7.2.

Box 7.2

Using Poisson models for modeling proportions

The logistic model, with a binomial error structure, and logit link function is most commonly used to model proportions (another link function, the probit link function, is sometimes used). In contrast, a standard Poisson model (with Poisson error structure and log link function) is used to model counts, rather than proportions out of a given total. However, there are two ways in which a Poisson model can be modified to model proportions.

The first is used by Perry (2009) in her study on capuchin monkeys. The general form of a Poisson model is:

$$\log(E[Y_i \mid LP_i]) = LP_i,$$

where LP_i is the linear predictor for data point i, and $E[Y_i \mid LP_i]$ is the expected number of counts for the dependent variable. The model then calculates the likelihood, given the observed count from a Poisson distribution with mean $E[Y_i \mid LP_i]$, which enables the model to be fitted by maximum likelihood. The Poisson model can then be modified to model rates (i.e., the expected number of counts per unit). For example, we might want to model the rate at which an animal performs a trait, given counts that were made over differing periods of time. The length of each period would then be called the exposure, e_i.[1] The model is modified as follows:

$$\log\left(\frac{E[Y_i \mid LP_i]}{e_i}\right) = LP_i$$
$$\log(E[Y_i \mid LP_i]) = \log(e_i) + LP_i$$

which amounts to putting $\log(e_i)$ in the model as an offset (this means putting $\log(e_i)$ in as a variable with a coefficient fixed to 1). Perry (2009) used the number of instances of fruit pounds as the dependent variable, and included the total number of instances of pounding plus scrubbing for the individual per year. This means that the estimated effects can be interpreted on the log probability scale rather than the log-odds scale, meaning that when they are exponentially back transformed, they give the estimated multiplicative effect of a single unit change in the variable.

However, we prefer that a logistic model be used to model such data, because unlike rates, proportions are constrained to be ≤ 1. The logistic model imposes this constraint, whereas the Poisson model with exposure does not. In addition, the Poisson model assumes that the variance is equal to the mean, so that the variance is higher for proportions closer to 1. This is unrealistic, because if a "success" has a probability of 1, the variance has to be 0, as is assumed by a logistic model. This means that the results of the model will be sensitive to which option is chosen as the response variable, whereas in a logistic model one would obtain an equivalent model. This is not to say that the results of Poisson models used in this way are meaningless, since the model will probably fit coefficients, such that the expected proportion is in the range 0 to 1 for the values of the variables that were observed.

Another way in which the Poisson model can be applied to model proportions is to use what is called a log linear model, usually used for contingency tables. A description of this technique is beyond the scope of this book, but for further details see a book on GLMs, such as Faraway (2006). Essentially, the model can be used for proportions if the linear predictor is constrained to contain certain variables and interaction terms. If the response variable has two categories, the results obtained will be identical to an equivalent logistic model. Log linear models have the advantage that they can be used to model a response variable with more than two categories. However, one can also do this using a multinomial logistic model, which is an extension of the binomial logistic model.

[1] Not to be confused with the "exposure score" e_{ijk} calculated in Perry's (2009) study, which quantifies the social exposure to one option relative to the other.

Perry found that for females, there was at least reasonable (5% significance level) evidence for an effect of social impact, and that this effect was weaker for males. In addition, she found evidence that the effect of exposure decreased over time, with a smaller effect in the developmental years 4 and 5 than in years 1 to 3.

Use of the Poisson or logistic regression models implies a specific relationship between exposure and option choice; they are often used for statistical convenience, rather than because they provide a plausible model of this relationship. An alternative method would be to fit alternative models in which the functional form of the relationship corresponds to specific social learning strategies. McElreath et al. (2009) provide a method for doing this, which we describe in chapter 8 (box 8.2).

7.1.6 Limitations of observational data

One concern with observational data on behavior development is that there might be compensatory processes operating that mask a causal effect of social experience. For example, an individual might seek out more observational experience if they are failing to acquire a trait. However, the main limitation of observational developmental data is the complication associated with establishing the direction of causality. If there is a positive association between, say, a measure of observation of trait performance and one of the measures of the rate of trait acquisition, then this is clearly consistent with social transmission of the trait. However, there will be other causal hypotheses that result in the same statistical pattern between the two variables. For example, acquiring a trait early might cause an individual to observe the trait being performed more, perhaps because it is more likely to visit locations at which the trait is performed. This explanation could be ruled out if the model is suitably time structured. More difficult to resolve, is the possible presence of an unmeasured variable that exerts a positive causal influence on both the measure of observation and rate of acquisition. It may be that such alternatives can be ruled out informally as being implausible. However, it is important that such alternative explanations, and their plausibility, are considered.

Future work might focus on using a causal modeling approach, similar to that described in section 6.4, to distinguish alternative causal hypotheses. As we noted earlier (section 7.1.2), this approach might also detect pathways of social transmission that would, prima facie, appear to be ruled out by a standard model selection approach. Of particular interest is the extent to which genetic and environmental hypotheses, that could cause a "spurious" correlation, can be ruled out by inclusion in the model (see chapter 6). The issue of establishing the direction of causation is most easily dealt with (in a conceptual sense, though it might be very difficult in practice) by experimental manipulation, which we now discuss.

7.2 Experimental Manipulations

Strictly speaking, traditional demonstrator-observer experiments constitute a developmental methodology. The experimenter manipulates the social cues available to the observer, and estimates what effect these have on the development of the acquisition of a trait. Such experiments can be integrated into a study of the development of a trait. For example, Kenward et al. (2005) studied the development of tool use in New Caledonian crows (*Corvus moneduloides*), a species renown for its tool use in the wild. These researchers documented the response of captive-bred crows when presented with twigs over a seven-week period early in their lives. Furthermore, the four crows were divided into a tutored group, which observed tool-using behavior performed by a human demonstrator, and a nontutored group. They found that the two birds in the tutored group spent more time carrying twigs and inserting them into crevices.

Unfortunately, demonstrator-observer experiments have the disadvantage that they do not estimate the importance of social cues in the development of a behavioral trait in natural circumstances, because the researcher does not know how often the social cues are experienced. Here we briefly discuss some of the experimental manipulations that can be performed in field conditions that compensate for this disadvantage to some extent. We do not provide a complete review of field experimental techniques (see instead Reader and Biro 2010) but instead focus on the relationship of field experiments to the methods we have hitherto discussed for observational data.

7.2.1 Diffusion experiments

A relatively simple form of field experiment is an "artificial" or "controlled" diffusion in which a task, usually a foraging task or some other opportunity to exploit a novel resource, is made available to a natural group of animals, and the pattern of solving is observed. In a seeded diffusion, a trained demonstrator is also introduced into the population, whereas in an unseeded diffusion, all individuals are naïve to begin with (Lefebvre 1995). Each approach has its advantages. For example, over a number of replicates of an unseeded diffusion, a researcher can also make inferences about what variables influence who is the innovator, that is, the first individual to solve the task (Laland and Reader 1999). This approach is flexible enough to used in a laboratory setting (e.g., Boogert et al. 2006; Boogert et al. 2008); in captive facilities, such as zoos (e.g., Day et al. 2003); and in the field (e.g., Lefebvre 1986; Biro et al. 2003). Conversely, when using seeded diffusions a researcher can, in principle, infer social transmission more easily by seeding different groups with demonstrators trained to solve the task in different ways. If individuals tend to prefer the same option as the demonstrator in their group, this can be taken as evidence of social transmission. This could be tested using the option-bias test (box 6.1). The mechanism of social learning might also be inferred from such experiments, based on how the options differ. For example, if the options are directed to the same location but use different actions, then a positive result suggests a mechanism of social transmission that is action specific (see chapter 4).

A researcher could still test for such an effect in an unseeded diffusion, since a pattern of within-group homogeneity would be consistent with (option-specific) social transmission. However, it is harder to rule out an alternative hypothesis with an unknown variable causing individuals in each group to converge on the same option by asocial learning. This unknown confounding variable is less problematic for an artificial diffusion than for naturally occurring differences in repertoires (chapter 6), because in the former the researcher can ensure that an essentially identical task is presented to each group. However, it is logically possible that differing prior experience or genetics of the two groups predisposes them to different options (though a case may be made that this is unlikely where the options are arbitrarily different, such as the same action in different directions).

This problem is alleviated, in principle, in seeded diffusions, because the option seeded in each group is controlled or chosen at random by the experimenter, and so cannot be caused by any variables that influence option choice in the naïve individuals of the group. However, there still remains the practical problem of statistical inference. If groups are composed at random from a pool of individuals, then they can be treated as independent data points in the analysis, since their choices will be independent under the null hypothesis (asocial learning). However, this is not the case if the groups in question are genetically homogeneous and have more similar prior experience within groups than between groups, as will usually be the case for field experiments. In this case, the experimental manipulation is at the level of the group, and if researchers wish to rule out alternative explanations experimentally, the analysis should ideally be performed at this level (or it should incorporate a group-level effect into the analysis). For example, if a researcher were to seed two groups of individuals with a demonstrator performing option A, and a further two groups with a demonstrator performing option B, and all four groups prefer the option performed by their demonstrator, then a one-sided Fisher exact test gives a p-value of 0.167. One could make an argument in such a case that the effect is too strong at the individual level for a prior disposition to be a plausible explanation for the effect. A researcher could formalize and test this argument by presenting the task separately to control individuals from a number of other groups;[14] this would quantify the magnitude of any underlying group-level bias and statistically control for it.

In some cases it might be possible to attain experimental control, equivalent to seeding demonstrators, by manipulating the *apparent* demonstrator behavior

---

[14] The data could be analyzed at the individual level, with the response variable being the number of choices of each option. For a two-option task, data could then be modeled using a GLMM with a binomial error structure and logit link function, with group as a random effect. Then, one could code a dummy continuous variable representing the effects of a demonstrator, with 1/-1 for a demonstrator trained to options A and B respectively, and 0 for the controls. The coefficient would subsequently estimate the strength of the effect of the demonstrator relative to controls on the log odds scale (see section 7.1.2). If groups tended to have a strong disposition to one option, the random effect would be estimated to have a large variance, resulting in a high correlation between individuals within experimental groups, meaning the result would tend to the group-level analysis. If on the other hand, there was little or no group-level predisposition, the random effect would be estimated to have a small variance, and the analysis would tend toward one in which individuals are independent. A more direct approach would be to remove individuals from the experimental groups themselves, such that they are not involved in the diffusion; this would quantify each group's bias, if it existed, more directly. However, this would also disrupt the social dynamics of the group for the diffusion.

(Reader and Biro 2010). For example, Van de Waal et al. (2010) conducted diffusion experiments on wild groups of vervet monkeys (*Chlorocebus pygerythrus*) without training demonstrators. These researchers took advantage of the fact that a single dominant individual would dominate the artificial foraging task, and thereby act as a demonstrator. The task was a box with two doors that could be opened to obtain food. When the task was presented to the demonstrator, one door was blocked, forcing the demonstrator to open the other door to obtain food. When observers of this were tested later, observers of female demonstrators were found to prefer her "choice" of door, but this effect was not found for male demonstrators. Reader and Biro (2010) discuss a number of other cases where researchers have manipulated the apparent demonstrator behavior.

Further evidence of social transmission, and its mode of operation, might be gleaned by analyzing the patterns within each diffusion in finer detail. This can be done using the methods presented in chapters 5 and section 7.1 for observational data. It is worth noting that though the data would be derived from an experiment, they would still be observational data, in that the variables quantifying social experience are not under the direct control of the experimenter. A good example of this approach is provided by Biro et al. (2003), who extended their study of nut cracking in chimpanzees (see section 7.1.1) by presenting the Bossou group with types of nut they were unfamiliar with. They then used much the same methods to analyze the pattern of spread as they did to investigate the development in Bossou infants of cracking native nuts.

7.2.2 Manipulation of social experience

The most direct way to obtain evidence of a causal link from a type of social experience to learning is to experimentally manipulate the social experience in question. Any evidence of an association between the manipulation and learning (perhaps using one of the measures described in section 7.1) can be interpreted as evidence of a causal effect. Ideally, one would remove all naturally occurring social experience of the kind proposed to have an effect, before artificially replacing it at the required level. However, this is unlikely to be feasible or ethical in many cases. One alternative is to supplement natural levels of social cues, and look for evidence of a causal link between level of supplementation and learning. This procedure might fail to find evidence of a causal link if subjects are already "saturated" with the target social experience, or if variation in natural levels drowns out the manipulation. Nonetheless, a positive result can be taken as strong evidence of a causal effect. Thornton and McAuliffe (2006) used this approach in their research on meerkats, described below (see also Raihani and Ridley 2008). Another approach is to provide varying levels of social experience to those who would not usually receive it, perhaps to individuals from a population in which the trait is not present.

The approach in general would seem to depend on the possibility of providing an artificial cue that is a close substitute for the social experience in question. This is probably easiest for testing hypotheses of social transmission that operate through the products of a demonstrator's behavior.[15] Thornton and McAuliffe (2006) used this approach to illustrate strong evidence of teaching (see chapter 4) in meerkats (*Suricata suricatta*). This involved showing that the behavior in question satisfied all of Caro and Hauser's (1992) criteria for teaching (see chapter 4), but here we will focus on how they established that the behavior causes learning in the pupil.

The work was conducted at the Kalahari Meerkat Project in South Africa, which was established in 1993 by Tim Clutton Brock. The project involves the close monitoring of 11 meerkat groups, which are habituated to human presence, and therefore amenable to the kind of experimental manipulations performed by Thornton and McAuliffe. Meerkats are cooperative breeders that live in groups of 5 to 20 animals, with a single breeding pair and a number of adult "helpers" who aid in bringing up the pups. One of the helpers' roles is to provide the pups with food, which includes scorpions and other dangerous prey. As is the case in a number of the Carnivora (Caro and Hauser 1992), the pups are often provided with live prey, which allows them to practice their hunting skills. This has been hypothesized to be a case of nonhuman teaching, with the function of teaching young how to catch prey (Caro and Hauser 1992). In the case of meerkats, the prey is presented to pups in a state suitable for their level of experience. For example, very young pups tend to be presented with dead scorpions; older individuals with disabled scorpions, with stinger and/or claws bitten off; and older individuals with intact scorpions.

Thornton and McAuliffe (2006) found that pups' skills of handling scorpions increased over time, but noted that this could be an effect of age rather than experience. To investigate whether exposure to live prey benefited pup handling skills, they presented each of three littermates each day with either (i) four dead scorpions; (ii) four live, stingless scorpions; or (iii) an equivalent mass of hard-boiled egg, as a control. They did this for three days, and on the fourth day tested the handling abilities of all three pups by providing each with one live, stingless scorpion. They repeated this procedure on six litters in four groups.

The authors found that all pups trained on live scorpions successfully handled the scorpion on the fourth day, whereas those trained on dead scorpions lost the scorpion in two out of six tests, and control pups lost their scorpions in four tests.

Furthermore, in all six trials, the pup trained on live scorpions had the best performance, being either the only pup to handle the scorpion successfully or

[15] Vocal cues can also be relatively easy to present in playback experiments. These experiments are used frequently in the study of vocal learning (e.g., Catchpole and Slater 2008), but have also been used to study the social transmission of other types of traits (e.g., Raihani and Ridely 2008; Davies and Welbergen 2009). See Reader and Biro's (2010) review for more details.

having the fastest handling time (Friedman test[16]: $S = 10.38$, d.f. $= 5$, $p = 0.006$). They also report that all pups not trained with live scorpions were pinched or pseudo-stung (struck by the stingless tail) during the test with a live scorpion, whereas this occurred only once in tests with pups trained on live scorpions (Fisher's test: $p = 0.001$). In combination with their other findings, Thornton and McAuliffe (2006) conclude that there is "strong evidence that the provisioning behavior of meerkat helpers constitutes a form of opportunity teaching, in which teachers provide pupils with opportunities to practice skills, thus facilitating learning."

It is worth noting that to establish social transmission of a trait one also has to establish a causal link between an individual acquiring the trait, and the provision of the relevant social cues to others (see fig. 1.1b). In many cases this will be trivial; for example, a meerkat helper would be unable to provide a pup with a scorpion without first catching scorpions itself. Likewise, to establish social transmission of trait performance, one must establish a causal link between performance of a trait and the provision of relevant social experience to others. Furthermore, to quantify the importance of a given pathway of social transmission in development under natural circumstances, it is not sufficient to estimate the strength of the causal effect of social cues using experimental manipulation. A researcher must also assess how often the relevant social cues are experienced by individuals under such conditions (e.g., Biro et al. 2003).

7.2.3 Translocation studies

The most potent, and arguably most compelling, means of manipulating the developmental experience of natural populations of animals is through the direct transfer of individuals between groups, or the replacement of entire groups. An experimentally satisfying, but rarely utilzed, approach is field translocation experiments; these include the translocation of individuals between populations, or of populations of animals between sites. In principle, such movements could distinguish between alternative explanations for natural traditions (fg. 7.4). For example, if individual animals are introduced into resident populations (manipulation 1 in fig. 7.4A) and over time come to develop the behavior of established residents, this is inconsistent with an explanation in terms of genetic differences between populations, and consistent with some role for social transmission (but also with environmentally evoked behavior). Similarly, if an entire population is replaced (manipulation 2, fig. 7.4B), and the introduced individuals come to exhibit a behavior different from the former inhabitants, this would suggest that the variation does not result from adjusting to divergent ecological conditions. In the latter case, an alternative explanation to social transmission is genetic differences between populations. Both manipulations (or alternative procedures for ruling out other explanations) are necessary to provide convincing experimental evidence for socially transmitted behavior (Laland and Hoppitt 2003).

---

[16] A nonparametric version of a repeated measures ANOVA, for complete block designs.

**Manipulation 1:
ruling out genetic differences**

Environment A   Environment B

(a)

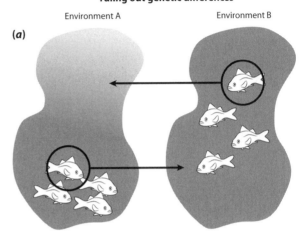

**Manipulation 2:
ruling out ecological differences**

Environment A   Environment B

(b)

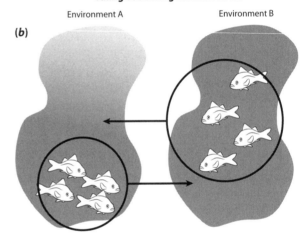

Figure 7.4. Experimental manipulations required to rule out
genetic differences and ecological differences as the cause
of differences in behavior.

The approach has been used successfully to demonstrate culture in fishes. French grunts (*Haemulon flavolineatum*) exhibit traditional daytime schooling sites and twilight migration routes that persist for longer than the lifetime of an individual. Helfman and Schultz (1984) translocated fish between populations and found that, while those French grunts placed into established populations adopted the same schooling sites and migration routes as the residents, control fish introduced into regions from where the residents had been removed did not adopt the behavior of former residents. Helfman and Schultz's study is to this point one of the most elegant demonstrations of animal social transmission and tradition.

Similarly, Caribbean bluehead wrasse (*Thalassoma bifasciatum*) exhibit traditional mating sites, used over generations, which are ecologically similar to many other unused sites (Warner 1988). Removal and replacement of entire populations of these fish established that new sites came into use that were, in turn, used for generations (Warner 1988). This provides strong experimental evidence that their mating sites cannot easily be predicted from knowledge of the local ecology, but rather are maintained as traditions, with youngsters and newcomers learning the routes to sites from experienced residents (Warner 1988, 1990). Interestingly, in a follow up study, Warner (1990) found that when he replaced newly established populations, after just one month the newcomers tended to utilize the same mating sites as their predecessors. This implies that site use is initially based on resource assessment and is not entirely arbitrary; however, it is then preserved by tradition, which generates a "cultural inertia" (Boyd and Richerson 1985) in the face of ecological change.

Unfortunately, this powerful experimentation method is likely to be neither feasible nor ethical for many species, particularly apes. It might be possible to exploit natural movements of animals between populations, or exchange captive animals between holding facilities, in order to isolate instances of social transmission. However, many researchers studying animal cultures or traditions must work with the constraint of having only observational data on which to base their assessments, and so are forced to utilize other approaches such as those described in this book.

7.3 Conclusions

In summary, recent years have witnessed the emergence of a new approach to inferring social learning and transmission by studying the developmental processes that support them. Such studies often set out to document the typical process by which a putative socially transmitted trait is acquired, through detailed longitudinal observations on a number of individuals. A frequently used method aims to identify stages in the development of a character, and to determine the opportunities for social learning. Further insight can be gained by exploring which factors influence the probability that the trait will be acquired, the time at which it is acquired, the rate at which it is performed, or the proficiency with which it is performed. Experimental approaches include the use of seeded and unseeded diffusions, manipulation (e.g., enhancement or depletion) of those social cues thought to be necessary for social transmission and translocation experiments. These approaches are important for at least two reasons. First, they provide powerful means for researchers to establish to what extent the natural behavior of animals is reliant on social transmission. Second, they shed light on how social transmission joins with other influences to affect the development of behavior.

Chapter 8

Social Learning Strategies

8.1 Why Social Learning Is Strategic

Over the last century, it has been widely assumed, albeit often rather tacitly, that learning from others was inherently adaptive. Individuals were thought to benefit by copying[1] others because by doing so they would take a shortcut to acquiring adaptive information, saving on the costs of trial and error. Only recently has it become apparent that this reasoning is overly simplistic. Always copying others, in a mindless or undiscriminating manner, is not a recipe for success.

One way to see this is to think of social learning as a form of information parasitism (Giraldeau et al. 2002). Social learners exploit the information acquired, devised, or discovered through asocial learning but contribute no new information themselves. Asocial learning *produces* information, whereas social learners are information *scroungers*. Game theory models of producer-scrounger interactions reveal that scroungers do better than producers only when fellow scroungers are rare, while at equilibrium the payoffs for producing and scrounging are equal (Barnard and Sibly 1981). This suggests that copying does not inevitably increase payoffs.

The cultural evolution literature reinforces this conclusion. Several theoretical analyses of the evolution of social learning have concluded that social learners have higher fitness than asocial learners only when copying is rare;[2] the reasoning is

---

[1] Our use of the term "copying" refers to any form of social transmission.

[2] The inevitability of this conclusion is contested by some findings from the social learning strategies tournament (see box 8.1) (Rendell et al. 2010), and also by Van der Post and Hogeweg (2009). These findings imply that there are circumstances under which social learning can have higher fitness than asocial learning, even when social learning is widespread. In models, such as the tournament, where individuals possess repertoires of behavior, then social learners are not condemned to perform out-of-date behavior

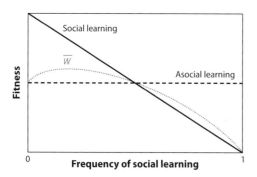

Figure 8.1. The 1988 model by A. R. Rogers assumed that the fitness of asocial learning is constant, while that of social learning is frequency dependent. Social and asocial learners are predicted to reach a mixed equilibrium, at which social learning has not enhanced the mean fitness of the population, a result known as "Rogers' paradox." The dotted line gives the mean fitness of the population ().

that this is because most potential demonstrators are asocial learners who will have sampled accurate information about the environment at some cost (Boyd and Richerson 1985, 1995; Giraldeau et al. 2002; A. R. Rogers 1988; see fig. 8.1). However, assuming that the environment changes over time, as the frequency of social learners increases the value of copying typically declines, if this means that the proportion of individuals (e.g., asocial learners) tracking the environment and producing reliable information is decreasing.[3] At the extreme, all individuals would be copying fellow copiers with no one sampling the environment, which would often make social learning a strategy that has lower fitness than asocial learning. The population is expected to reach a polymorphic equilibrium at which social and asocial learning will be equal in fitness (Boyd and Richerson 1985; A. R. Rogers 1988).

Anthropologist Alan Rogers (1988) first pointed out the "paradox" inherent in the observation that the fitness of social learners at this equilibrium would be no greater than the average individual fitness in a population of asocial learners. His finding is now commonly known as Rogers' paradox (Boyd and Richerson 1985), so called because it contrasts with a commonly held assertion that culture enhances fitness. While Rogers' result is not inherently paradoxical, it appears to conflict with the widely held assumption that social learning underlies the effect of human culture on our ecological success and population growth, which implies an increase in absolute fitness. Other researchers have found that Rogers' general conclusions are robust (Boyd and Richerson 1995).

At first glance, the findings of the social learning strategies tournament (Rendell et al. 2010; box 8.1) might seem to challenge the conclusion that use of social learning should be discriminating. The winning strategy DISCOUNT-MACHINE almost never learned asocially, which might imply that individuals should always copy. However, such a conclusion would be erroneous. The relationship between the amount of copying exhibited by a strategy and the strategy's performance was positive among the top-performing strategies, but

---

if environments change, because they can switch to another behavior in their repertoire. Provided there is an adequate source of behavioral variation (e.g., derived from copy error), social learners can track environmental change effectively, and there is little disadvantage to copying even when others are doing so.

[3] Ibid.

Box 8.1
The social learning strategies tournament

The social learning strategies tournament was a computer-based competition in which entrants submitted a strategy specifying the best way for agents living in a simulated environment to learn. The simulation environment was characterized as a "multiarmed bandit" with, in this case, one hundred possible behavior patterns that an agent could learn and subsequently exploit. Each behavior had a payoff, drawn from an exponential distribution, and the payoff could change over time (the rate of change was a model parameter). This simulated environment contained a population of 100 agents, each controlled by one of the strategies entered into the tournament. In each model iteration, agents selected one of three moves, as specified by the strategy. The first, INNOVATE, resulted in an agent learning the identity and payoff of one new behavior, selected at random. The second, EXPLOIT, represented an agent choosing to perform a behavior it already knew, and receiving the payoff associated with that behavior (which might have changed once the agent learned about it). The third, OBSERVE, represented an agent observing one or more of those agents who chose to play EXPLOIT, and learning the identity and payoff of the behavior the observed agent was performing. Agents could only receive payoffs by playing EXPLOIT, and the fitness of agents was determined by the total payoff received, divided by the number of iterations through which they had lived. Evolution occurred through a death–birth process, with dying agents replaced by the offspring of survivors; the probability of reproduction was proportional to fitness. Offspring inherited the strategy of their parents.

The most important finding was the success of strategies that relied heavily on copying (as in OBSERVE) to learn behavior. Indeed, the winning strategy, called DISCOUNTMACHINE, almost always learned through copying. Social learning proved successful because the exploited behavior patterns available to copy constituted a select subset that had already been chosen for their high payoff. Demonstrators prefiltered behavior, rendering adaptive, high-payoff options available to others.

negative for the poorer strategies. This implies that for social learning to pay, it had to be done well. While the tournament found that social learning was beneficial under a far broader range of conditions than anticipated by previous analytical theory, the copying employed by effective strategies was far from random. Successful strategies timed bouts of copying carefully to coincide with when payoffs dropped, indicative of a change in the environment; hence, behavior that was previously effective no longer proved productive (fig. 8.2). The winner, DISCOUNTMACHINE, engaged in computations to determine, given the agent's age and the rate of environmental change, whether copying was likely to bring new behavior into the agent's repertoire with higher payoffs than its current behavior; DISCOUNTMACHINE then only copied if the expectation was that it would result in new behavior with higher payoffs. Moreover, in most contests, multiple strategies remained in the population, some of which engaged in asocial learning. Part of DISCOUNTMACHINE's success stemmed from its ability to exploit the asocial learning of other strategies parasitically; however, precisely because of the parasitic nature of this relationship, it was unable to exclude other strategies completely. In other words, in tournament simulations there remained a balance of social and asocial learning across the population, and the most effective strategies

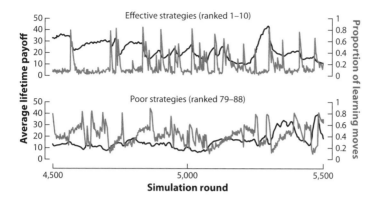

Figure 8.2. The social learning strategies tournament (Rendell et al. 2010) drew attention to the significance of the timing of learning. Successful strategies (top panel) timed bouts of learning (gray line) to coincide with when the environment changed, manifest in a drop in average lifetime payoff (black line), while poorer strategies exhibited no such relationship. See text for further detail. Based on figure 2E in Rendell et al. (2010).

undertook social learning very selectively, reinforcing, rather than challenging, the idea that copying should be strategic.

These theoretical findings suggest that copying others universally and indiscriminately is often not adaptive and would typically not increase the mean fitness of individuals in the population. If social learning is to increase payoffs, and thereby account for the adaptiveness of culture, then individuals must implement it in a more savvy fashion. This leads to the expectation that use of social learning will be *selective,* that individuals will directly sample the environment through their own asocial learning some of the time, and that they will combine social and asocially derived information in a rule-governed manner that is likely to have been shaped by selection (Boyd and Richerson 1995).[4] It is precisely because individuals do not use social learning indiscriminately, and engage in asocial sampling of environments, that social learning is typically adaptive (Galef 1995; Rendell et al. 2010).

Thus, game-theory, population-genetic, and simulation models all lead to the expectation that animals, including humans, ought to be selective with respect to the circumstances under which they rely on social learning and the individuals from whom they learn. A history of natural selection ought to have favored specific adaptive rules in animals, here called "social learning strategies" (Laland 2004; R. L. Kendal, Coolen, et al. 2005; J. R. Kendal, Giraldeau, and Laland 2009), but elsewhere referred to as "transmission biases" (Boyd and Richerson 1985; J. Henrich and McElreath 2003; Richerson and Boyd 2005), that dictate the contexts under which individuals exploit information provided by others. The

---

[4] Alternatively, social learning may prove adaptive because it allows individuals to acquire complex knowledge that would be difficult or impossible to acquire asocially through trial and error (Boyd and Richerson 1995). Such knowledge may be the product of cumulative cultural evolution.

adoption of such strategies would not require that the animals be aware that they are following a strategy, nor that they understand why such strategies may work. Moreover, in accordance with behavioral ecologists' use of the *phenotypic gambit* (Grafen 1984), strategies can be fruitfully studied *as if* the simplest genetic system controlled them. That is, as a reasonable first approximation, research into learning strategies can proceed through functional considerations without any commitment to mechanism, and researchers may assume that it does not matter whether animals adopt such strategies as a consequence of evolved psychological mechanisms, learning, culture, or some combination of processes (although in practice such details may prove important; see Fawcett et al. 2012). Thus, the exhibited strategies may be the product of individual learning and/or cultural evolution with little evolved neural structure (Heyes 2012), and do not imply a commitment to an innate, modular, or adaptationist stance, or imply anything about how such strategies are implemented in the brain.

There are many plausible social learning strategies. Individuals might disproportionately copy when asocial learning would be difficult or costly, when they are uncertain of what to do, when the environment changes, when established behavior proves unproductive, and so forth. Likewise, animals might preferentially copy the dominant individual, the most successful individual, or a close relative. Alternatively, selection might predispose individuals to adopt specific adaptive solutions demonstrated by others, such as a bias toward eating sugar-rich foods, or toward choosing mates according to cues of fertility. Which is the best rule, and how can alternative rules be evaluated?

The term "social learning strategy" was introduced in 2004 (Laland 2004), although the key ideas were implicit in earlier theoretical work (Boyd and Richerson 1985; J. Henrich and McElreath 2003). Rules that govern the use of social information are also referred to as "transmission biases" (Boyd and Richerson 1985; J. Henrich and McElreath 2003; Richerson and Boyd 2005) and "trust" (Harris 2007). In many respects the term "transmission bias" is a more accurate label for the phenomena we describe in this chapter, because we will see that multiple factors may simultaneously influence the likelihood of social transmission. The label "strategy" might imply, misleadingly, that alternative social learning heuristics are mutually exclusive. However, the use of the term "strategy," as a deliberate attempt to equate such social learning heuristics with those strategies subject to analyses by evolutionary game theory (Laland 2004), has proven a real success. In the last decade there has been a rush of experimental studies identifying contexts, or rules, under which a variety of species of animals, including humans, copy others (see R. L. Kendal, Coolen, et al. 2005; R. L. Kendal, J. R. Kendal, et al. 2009; and Rendell, Fogarty, et al. 2011 for recent reviews). These include field-based studies, where social learning strategies have been identified in natural populations of birds and mammals (table 8.1). One reason why the social learning strategies approach has proven productive is that it provides rich possibilities for integrating empirical and theoretical findings. It has the additional advantage in that it helps to draw the field of social learning more closely into a general evolutionary framework. Indeed, some of the theoretical findings described below are based on evolutionary theory (e.g., Boyd and Richerson 1985; Giraldeau and Caraco 2000; Schlag 1998), and there are clear opportunities to develop this theoretical foundation further.

A graphical summary of the social learning strategies discussed in this chapter is given in figure 8.3, while the empirical evidence for some of the better-studied rules is given in table 8.1. We distinguish between *"when" strategies*, which specify the circumstances under which individuals copy others; *"who" strategies*, which identify from whom individuals learn; and *"what" strategies*, which specify what information should preferentially be copied. We also consider the restricted circumstances under which copying randomly might actually be adaptive. In addition, we examine which strategies, if any, might be considered most effective, and how individuals combine strategies. Finally, we look at whether individuals might use metastrategies that specify how to prioritize alternative strategies, or make decisions in the face of conflicting input from competing strategies.

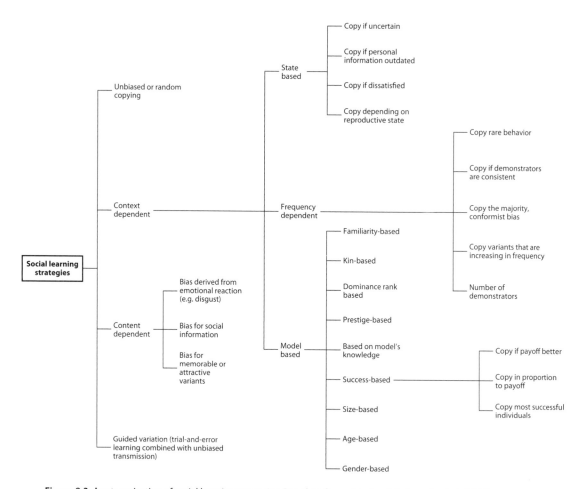

Figure 8.3. A categorization of social learning strategies. Based on figure 1 in Rendell, Fogarty, et al. (2011).

Table 8.1

Strategy	Empirical support
"When" strategies Copy when asocial learning is costly	Public information use in **European starlings** (Templeton & Giraldeau 1996). Acquisition of antipredator behavior in **fishes**, **birds**, and **monkeys** (Chivers & Smith 1995; Kelley et al. 2003; Krause 1993; Mineka & Cook 1988; Suboski & Templeton 1989). Acquisition of costly defense against brood parasites in **birds** (Davies & Welbergen 2009; Campobello & Sealy 2011). **Minnows**, **guppies**, and **sticklebacks** disproportionately copy when asocial learning would be costly (Coolen et al. 2003; R. L. Kendal et al. 2004; Webster & Laland 2008, 2011). Puzzle box solving in **callitrichid monkeys** (Day et al. 2003). Social learning increases with task difficulty in **humans** (Baron et al. 1996; T.J.H. Morgan et al. 2012). Route learning in **fishes**, **dogs**, and **humans** (Laland & Williams 1998; Pongrácz et al. 2003; Reader et al. 2008; Lindeyer & Reader 2010).
Copy when uncertain	Naïve but not experienced **rats** copy the diet preferences of other **rats** (Galef 1996, 2009). Food site preferences copied in naïve but not experienced **guppies** and **sticklebacks** (R. L. Kendal et al. 2004; Van Bergen et al. 2004). Naïve **ants** more likely to follow the chemical trails of others while experienced individuals rely on their own personal experience (Grüter et al. 2011).
"Who" strategies Conformity	Experimental evidence for conformity in **sticklebacks** (Pike & Laland 2010) and **humans** (T.J.H. Morgan et al. 2012). In **humans**, copying increases disproportionately with demonstrator number (Bond 2005) and consistency (T.J.H. Morgan et al. 2012). Conformity observed in the frequency dependent foraging of **guppies** (Day et al. 2001). Social learning increases with demonstrator number in **fishes**, **birds**, **rats**, and **humans** (Lachlan et al. 1998; Laland & Williams 1997; Sugita 1980; Beck & Galef 1989; Chou & Richerson 1992; Lefebvre & Giraldeau 1994; Boyd & Richerson 1985; J. Henrich 2001), but this provides only weak evidence for conformity (see main text).
Copy successful individuals	**Insects** and **birds** copy the nest-site decisions of successful over unsuccessful conspecifics and heterospecifics (Sarin & Dukas 2009; Pasqualone & Davis 2011; Seppanen et al. 2011; Forsman & Seppanen 2011). **Redwing blackbirds** acquire food preferences only from healthy birds (Mason 1988). **Bats** follow successful foragers to food sites (Wilkinson 1992). **Guppies** and **chimpanzees** disproportionately follow successful foragers (Menzel 1974; Lachlan et al. 1998). Dominants inhibit the performance of low-status demonstrators in **monkeys** and **chickens** (Drea & Wallen 1999; Nicol & Pope 1994), while **chimpanzees** and **vervet monkeys** disproportionately copy dominant demonstrators (Horner et al. 2010; Van der Waal et al. 2010). **Children** switch from copying parents to experts as they get older (J. Henrich & N. Henrich 2010; J. Henrich & Broesch 2011). **Adult humans** copy successful individuals (Mesoudi 2008). **Nine-spined sticklebacks** disproportionately copy when the demonstrator receives a higher payoff than they do (J. R. Kendal et al. 2009; Pike et al. 2010). **Indian mynah birds** avoid foraging where conspecific experienced a predator attack (Griffin & Haythorpe 2011).
"What" strategies Content biases	**Humans** exhibit biases for copying more attractive, memorable, or emotionally arousing variants (Heath et al. 2001; Bangerter & Heath 2004; Jones et al. 2007; Little et al. 2008), social information (Mesoudi et al. 2006), cultural stereotypes (Kashima 2000), broad linguistic structure (Mesoudi & Whiten 2004), minimally counterintuitive knowledge (Barrett & Nyhof 2001; Norenzayan et al. 2006), and physical and behavioral traits that are potentially cues of mate quality (Symons 1979; Buss 1989; Jones et al. 2007; Little et al. 2008). Young **songbirds** exposed to conspecific and heterospecific songs disproportionately sing conspecific songs as adults (Marler & Peters 1989; Bolhuis & Gahr 2006). **Rats** differentially adopt, and transmit with higher fidelity, socially transmitted dietary preferences consistent with their prior preferences (Laland & Plotkin 1991, 1993). After social demonstration, **rhesus monkeys** are more likely to acquire a fear of snakes, and **blackbirds** more likely to acquire a fear of predators, than a fear of arbitrary objects (Curio et al. 1978; Mineka & Cook 1988).
Random copying	Forenames, ceramic pottery styles, dog-breed preferences, and other simple, arbitrary choices in **humans** (Shennan & Wilkinson 2001; Herzog et al. 2004; Bentley et al. 2004). Song learning in **chaffinches** (Slater et al. 1980; Slater 1986).

Researchers from the cultural evolution and gene–culture coevolution schools (e.g. Boyd and Richerson 1985; J. Henrich and McElreath 2003; Rendell, Fogarty, et al. 2011) have argued that human minds are able to acquire a wide range of broadly adaptive (or only mildly maladaptive) behavior. As a result, an understanding of why a particular piece of information has been acquired as opposed to its alternatives may be better framed in terms of the questions, "what is the local culture?" or "who is the information source?", rather than "what is the evolved psychological mechanism?" These questions correspond to the when, who, and what strategies, respectively, with the first two as examples of "context biases" (Boyd and Richerson 1985; J. Henrich and McElreath 2003). The idea of a context bias is simply that the social context determines what is learned. These researchers maintain that some quite general evolved rules (such as "conform," or "copy the most successful individual") could have been favored by selection, and do not greatly prespecify informational content.

In contrast, researchers from the evolutionary psychology tradition maintain that content biases shape what we eat, how we structure our world to be ergonomically rewarding, and many aspects of our social behavior (Cosmides and Tooby 1987; Tooby and Cosmides 1989, 1992; Buss 2008). The idea of a content bias is that animals are fashioned to acquire specific adaptive information, largely irrespective of the type of social context, or who the demonstrator is. Indeed, some researchers from this perspective suggest that, when social transmission occurs, information is not so much transmitted as reconstructed by the observer, with the demonstrator better regarded as a trigger to individual learning (Sperber 1996). The appearance of replication occurs, it is argued, only because individuals possess the same evolved structure, termed "cognitive attractors," in their mind (Sperber 1996). Thus, even among those researchers who agree that the acquisition of socially transmitted knowledge must be reliant on evolved mechanisms that detect and filter incoming information, there is disagreement as to whether the knowledge is correctly viewed as a major causal influence on behavior or just another environmental cue. It is only through careful empirical research that the relative importance of when, who, and what strategies will be established.

8.2 "When" Strategies

8.2.1 Copy when established behavior is unproductive

The simplest "when" strategy is perhaps to copy when established behavior is unproductive. Here, "established behavior" could refer to unlearned behavior or to learned solutions to related problems. We have already described one piece of theoretical support for this strategy; namely, the social learning strategies tournament (Rendell et al. 2010), where successful strategies were found to target their learning to coincide with periods when average population payoffs dropped, indicating a change in the environment that had rendered a previously productive

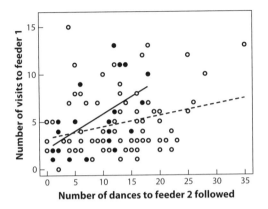

Figure 8.4. Grüter and Ratnieks (2011) found a relationship in honeybees between the number of waggle dances followed and the number of times F1 foragers visited the familiar but now empty F1 feeder. Each circle represents one F1 forager. Black circles show the foragers that were successfully recruited to the F2 feeder, which was advertised by waggle dances; white circles show foragers that were not recruited. Based on figure 2 in Grüter and Ratnieks (2011). The lines represent the significant positive relationships between the number of F2 dances followed and the number of times an F1 forager revisited the F1 feeder; these are for F1 foragers recruited (*continuous line*) and nonrecruited (*dashed line*) to the F2 feeder.

behavior less profitable (fig. 8.2). The issue of when to break off exploiting current knowledge in order to invest in further knowledge gain (the exploitation/exploration tradeoff) had not been incorporated into previous theory in this field. It nonetheless fits well with some empirical findings, suggesting that animals disproportionately copy when current knowledge is proving unsuccessful. For example, Grüter et al. (2010) and Grüter and Ratnieks (2011) observed that as long as the food source of foraging honeybees remains reasonably profitable, the worker honeybees will rely on their memories and return to the same place again and again (fig. 8.4), equivalent to agents in the tournament exploiting behavior already in their repertoire. However, if the profitability deteriorates sufficiently, foraging bees start seeking social information; they then start paying more attention to waggle dances when foraging at familiar locations becomes unsuccessful as a result of environmental changes.

Further support comes from studies investigating when animals switch between producing and scrounging. Lefebvre and Palameta (1988) conducted an investigation of the spread of a food-finding behavior in populations of pigeons, in which the birds were required to learn to peck open a carton containing seed. The pigeons would scrounge (take food from others) when possible; it was only when there were so few birds producing food that scrounging became unproductive and some scroungers then switched the food-finding behavior (i.e., they scrounged information when they could no longer scrounge food). The suggestion that this reflects a strategy of learning only when there is no easier option is supported by the observations that scroungers and producers switch strategy to maintain a frequency-dependent balance, and that the proportion of scroungers diminishes as the producers share increases (Giraldeau and Beauchamp 1999).[5] The establishment by earlier experimental analyses that asocial learning of the task was unlikely suggests that, in this instance, social learning underpinned adoption of the producer strategy (Lefebvre and Palameta 1988).

Implicit in this account are the assumptions that social learning can facilitate the adoption of a producer strategy but that no learning is required to scrounge food. The hypothesis that scrounging food is a first or preferred choice and learning to produce, whether socially or asocially, is a last resort used only when scrounging food is unprofitable or costly is supported by evidence that the opportunity to scrounge blocks the learning of a food-producing tactic. Giraldeau and Lefebvre (1987) reached such a conclusion after finding that pigeons that obtained food as a result of a demonstrator's removal of a stopper to a container of seed were then poor at learning the stopper-removal behavior; this was in comparison with birds that received no food during the demonstration. Similar observations have been reported in zebra finches (Beauchamp and Kacelnik 1991) and tufted capuchins (Fragaszy and Visalberghi 1990). Giraldeau and Lefebvre (1986) observed that scrounger pigeons that switched to producing when producers had been removed from the population, then switched back to scrounging when the producers were returned, and there is evidence that dominant individuals (who presumably have the choice) are more likely to be scroungers than subordinate individuals (Barta and Giraldeau 1998; Giraldeau and Caraco 2000). These findings are consistent both with the conclusion that scrounging is a preferred strategy, and with the suggestion that individuals adopt whatever strategy has the highest payoff at the time. Several recent studies in pigeons, ravens, meerkats, and callitrichid monkeys report that scrounging food can promote acquisition of novel food-finding techniques (Giraldeau and Lefebvre 1987; Fritz and Kotrschal 1999; Caldwell and Whiten 2003; Thornton and Malapert 2009). Such findings are not inconsistent with a *copy-when-established-behavior-is-unproductive* strategy, provided individuals had no opportunity to scrounge on their return to the task.

8.2.2 Copy when asocial learning is costly

Theoretical explorations of the adaptive advantages of social learning have led to a consensus that greater reliance on social learning could be favored as the costs of asocial learning increase (Boyd and Richerson 1985, 1988; Feldman et al. 1996). Potential costs include the energy expended in searching for and processing valuable resources; the risk of unreliable (asocially acquired) information; as well as the viability deficits associated with hazardous asocial learning, such as the learning of antipredator behavior through direct experience. These considerations imply that a *copy-when-asocial-learning-is-costly* strategy might be

---

[5] Here the use of the *copy-when-established-behavior-is-unproductive* strategy refers to the initial *acquisition* of the producing behavior. Whether or not individuals continue to produce or switch back to scrounging depends largely on the respective payoffs, although there is evidence that prior experience also affects reliance on producing and scrounging (Katsnelson et al. 2008; Morand-Ferron and Giraldeau 2010).

adaptive, a suggestion consistent with Boyd and Richerson's (1985) *costly information hypothesis.*[6] This hypothesis focuses on the evolutionary tradeoff between acquiring accurate but costly information versus less accurate but relatively low-cost information. Individuals should copy when the costs of learning through trial and error are high, taking advantage of the relatively cheap information provided by others. A similar argument was proposed by Bandura (1977, 12), who stated, "the more costly and hazardous the possible mistakes, the heavier is the reliance on observational learning from competent examples."

The assumption that some tasks are so difficult that they are unlikely to be learned asocially can frequently be observed in the social learning literature. For example, Byrne (1999) and Byrne and Russon (1998) argued that the elaborate, hierarchically organized food-processing techniques exhibited by gorillas when consuming plants with physical and chemical defenses are likely to be acquired through program-level imitation (see section 4.1.6.2), since they are too complex for an individual to acquire asocially. Byrne and Russon (1998) write, "Learning by individual experience is not completely disproven by these data, but it becomes a contrived alternative" (676).

Copy when asocial learning is costly is perhaps the best supported of all social learning strategies, with a host of experimental findings implying that diverse species of animals copy in this context (table 8.1). This is almost certainly because it is easy to envisage circumstances under which information might be costly to acquire asocially (e.g., learning the identity of predators or predator evasion tactics). It is, perhaps, no coincidence that there is considerable evidence for the social learning of antipredator behavior in animals, including fishes, birds, and monkeys (table 8.1). The same reasoning applies to the acquisition of costly defense against brood parasites in host bird species. Campobello and Sealy (2011) reasoned that reed warblers may benefit by acquiring antiparasite responses from conspecifics instead of through trial and error, because of their high nesting density and consistent cuckoo-specific responses that also recruit conspecifics. Using treatments that included presentations of taxidermic mounts, clutch manipulations, and playbacks, the authors tested the effect of conspecific defense on the response intensity of nesting reed warblers. Exposure to social cues resulted in an increase of cuckoo-specific nest defense responses, whereas experience with natural or experimental parasitism did not produce any change in defense intensity (see also Davies and Welbergen 2009).

While several studies have produced data consistent with this hypothesis, there have been few attempts to test it directly. An exception is a pair of experiments by Webster and Laland (2008), who gave European minnows (*Phoxinus phoxinus*) a choice between a socially demonstrated (i.e., conspecifics present) and a nondemonstrated (i.e., conspecifics absent) prey patch under conditions of low, indirect, and high simulated predation risk; the authors could then assess

---

[6] Although the costly information hypothesis lays emphasis on the costs of *acquiring* personal information, the same reasoning holds with respect to the costs of *using* personal information (Kendal, Coolen, et al. 2005).

individuals' reliance on social cues (fig. 8.5a). Whether they had no relevant prior experience or conflicting prior experience, subjects were observed to spend more time in the demonstrated patch than in the nondemonstrated patch when simulated risk was high compared to when it was low (i.e., that is, under conditions where individual learning is costly (fig. 8.5b).

Support for this strategy also comes from studies of public-information use in birds and fishes. Coolen et al. (2003) examined the propensity of wild-caught three-spined (*Gasterosteus aculeatus*) and nine-spined (*Pungitius pungitius*) sticklebacks to use public information about the profitability of food patches. Individual fish were restricted to a central compartment of an aquarium from where they could see two equivalent-sized shoals of conspecifics, each feeding at one of two feeders. Following observation, the demonstrators and all food were removed from the tank; the observers were released and their choice of feeder monitored. Solely on the basis of the demonstrators' success (i.e., how much food the demonstrators ate), observers were required to choose the richer of the two feeders. At test, nine-spined but not three-spined sticklebacks preferentially chose the goal zone that had formerly held the rich feeder, suggesting that they were able to exploit public information, while further experiments ruled out alternative explanations (Coolen et al. 2003; Coolen et al. 2005; Webster and Hart 2006). The collection of personal information in open water is costlier for nine-spines than for three-spines because the former have inferior structural antipredator defenses and are consumed preferentially by piscivorous fish (Hoogland et al. 1957). Because of these costs, nine-spines may forego the opportunity to collect reliable personal information and favor vicarious assessment of foraging opportunities through observational learning. Thus, public-information use in sticklebacks may reflect the differential costs of personal information acquisition.

Similarly, Templeton and Giraldeau (1996) found that European starlings used public information concerning the foraging successes and failures of conspecifics only when accurate information about patch quality was difficult or costly to acquire via personal sampling alone. Likewise, R. L. Kendal et al. (2004) report that guppies ignore social information when reliance on personal information would not incur costs; however, they switch to utilizing social information when these costs are substantive, even where such information conflicts with their personal experience.

There are relatively few empirical studies that directly explore whether and how the probability of social learning is influenced by task difficulty. An exception are the experiments conducted by R. L. Kendal and colleagues (Day et al. 2003; R. L. Kendal, J. R. Kendal, et al. 2009), who presented a series of novel puzzle box tasks to captive populations of callitrichid monkeys. Judging by the mean time required for the monkeys to solve them, the tasks varied significantly in difficulty. Each puzzle box could be opened in one of two ways (e.g., by opening one of two doors) to access food, with the alternatives differing in location and color but being otherwise equivalent. Although the monkeys learned all the tasks, application of the option-bias method (box 6.1) revealed evidence that the means of opening a more difficult puzzle box, and not the easiest one, was learned socially

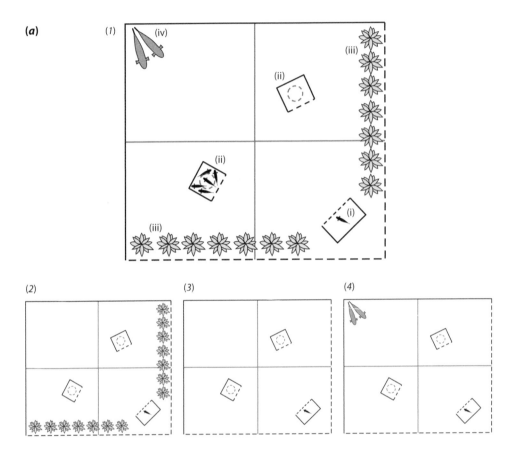

(R. L. Kendal, J. R. Kendal, et al. 2009). Presumably, easy tasks could be solved through asocial learning at little personal cost (in terms of time and energy) to the solver, whereas the solutions of more complex puzzles were associated with a sufficiently large cost to render social learning adaptive. Similar observations have been reported in human subjects, who were found to imitate more as task difficulty increased (Baron et al. 1996; T.J.H. Morgan et al. 2012).

In circumstances in which asocial learning is associated with significant costs, doing what others do may be adaptive for an individual even if the population's behavior is suboptimal. In game theoretical terms, arbitrary and even maladaptive traditions may emerge as Nash equilibria if each individual is reinforced for doing what others are doing, or penalized for breaking the convention (Boyd and Richerson 1985, 1992; Giraldeau et al. 2002). For example, Laland and Williams (1998) found that guppies will swim an energetically costly long route to feed when a short route is available, provided that conspecifics take the long but not

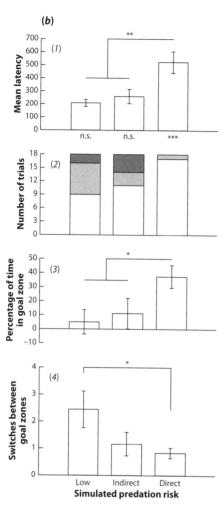

Figure 8.5. Webster and Laland (2008) found that minnows copy only when using private information would be costly. (*a*) The experimental setup: (*1*) the test tank, configured for (*2*) low-, (*3*) indirect-, and (*4*) direct-predation risk treatments. Here (*i*) is a plexiglass box start shelter, (*ii*) are stimulus chambers, simulating food patches, (*iii*) is cover, in the form of artificial vegetation, and (*iv*) shows the location of two model predators. (*b*) Experimental results: (*1*) Mean latency (s.e.) of focal fish to emerge from the start box shelter. (*2*) The number of trials in which fish first entered the demonstrated (*white*) and nondemonstrated (*gray*) goal zones. Fish that failed to enter either are shown in dark gray. (*3*) The mean difference between time spent in the demonstrated and nondemonstrated goal zones (s.e.). (*4*) The mean number of switches (s.e.) made between goal zones by the focal fish. The symbols *, **, *** indicate statistical significance at 5%. 1% and 0.1% respectively; "n.s." indicates no statistically significant difference. Based on figures 1 and 2 in Webster and Laland (2008).

the short route. Solitary fish learn to take the short route very quickly, but fish that are members of a group that chose the long route take a much longer time to adjust to the shorter route. Pongrácz et al. (2003) and Reader et al. (2008) report virtually identical findings for route learning in dogs and humans. If one regards the ecological environment (e.g., food sources in the environment) as the sole source of reward, it is difficult to explain why individuals exposed to the tradition should take longer than solitary individuals to adopt the short route. But if one recognizes the rewards of the social environment (e.g., the benefits of aggregation for effective predator defense in fishes) and the costs of asocial learning (e.g., an elevated risk of predation), then it becomes possible to envisage animal traditions that are inherently self-perpetuating.

The only clear failure to find support for this strategy that we are aware of is Galef and Whiskin's (2006) assessment of Norway rats' (*Rattus norvegicus*) reliance on social information while foraging in risky situations. Following a

30-minute interaction with a demonstrator rat that imparted cues on its breath as to whether it had been eating either cinnamon- or cocoa-flavored food, naïve rats were provided with both food flavors in a single location that afforded them little cover/refuge and entailed traversing an open space (from the safety of the nest box) to reach it. The potential cost of consuming these unfamiliar foods was increased for the rats by placing two cats in a cage in the housing room for either (i) 4 hours, (ii) 24 hours, or (iii) allowing the cats to roam freely. Rats did not eat more of the diet eaten by their demonstrator as potential predation risk increased. However, the cost the rats were facing was largely in traveling, over open ground, to the food site (where both foods were presented in close proximity), rather than in choosing between the two foods. As the two foods were essentially in the same location, individual rats may have concluded that both were equally safe to eat. A follow-up study (Galef and Yarkovsky 2009) also failed to find an increase in reliance on socially acquired information when foraging in risky situations, but since the foods were again presented in close proximity, it suffers from the same deficiencies (see R. L. Kendal, J. R. Kendal, et al. 2009 for discussion).

8.2.3 Copy when uncertain

Boyd and Richerson (1988) developed a model exploring the advantages of reliance on social and asocial learning in a temporally variable environment, in which animals have to make decisions as to which of two environments they are in and choose the appropriate behavior (fig. 8.6). Behavior 1 is appropriate in environment 1, and behavior 2 in environment 2, while performing the wrong behavior results in a fitness cost. The animals base their decision on the magnitude of a continuous parameter (x) representing the outcome of direct observation. If x has high values above a threshold value d, the animals "know" they are in environment 1 and perform behavior 1. If x has low values (below $-d$) they "know" they are in environment 2 and perform behavior 2. If x has intermediate values ($-d < x < d$) animals are uncertain as to which environment they are in, and it is assumed that they will copy the behavior of others.

There is considerable empirical support for Boyd and Richerson's assumption that individuals copy when uncertain (R. L. Kendal, J. R. Kendal, et al. 2009). For example, B. G. Galef and E. E. Whiskin describe an experiment (reported in Galef 2009) in which they fed one group of observer rats both cinnamon-flavored food and cocoa-flavored food for 24 hours, while another group of observer rats were given unflavored food. The researchers then allowed each member of the groups to interact with a demonstrator rat fed either the cinnamon- or cocoa-flavored diet before offering each observer a choice between the diets. Galef and Whiskin found that demonstrators had significantly greater influence on the food choices of rats choosing between unfamiliar than familiar foods. Likewise, Forkman (1991) and Visalberghi and Fragaszy (1995) have reported, respectively, that Mongolian gerbils (*Meriones unguiculatus*) and capuchin monkeys (*Cebus apella*) increased their consumption of food in the presence of feeding conspecifics when the food was unfamiliar, but not when it was familiar. Coolen et al. (2003) found

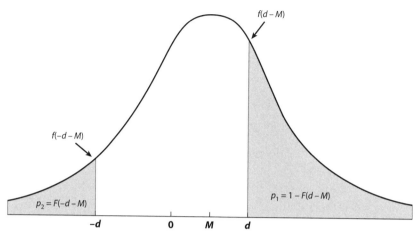

Figure 8.6. Boyd and Richerson (1988) specify a model in which organisms copy when they are uncertain. The adaptive problem each individual faces is to identify which of two habitats they are in, with a different behavior required in each. Boyd and Richerson assume that the results of each individual's experience can be quantified in terms of a single, normally distributed random variable x. If the environment is in state 1, the mean value of x is M; if it is in state 2, the mean value is $-M$. If the individual's estimate of x based on its personal experience is in the tails of the distribution, then it can be sure what habitat it is in (state 1 if x is large, state 2 is x is small), and consequently does not need to copy others. Conversely, where x takes on intermediary values it is unclear what the state of the habitat is, and the individual relies on copying. The range of circumstances under which individuals copy ($-d < x < d$) is specified by the parameter d. Based on figure 2.1 in Boyd and Richerson (1988).

that nine-spined sticklebacks that did not have personal information copied the patch choices of others, whereas Van Bergen et al. (2004), testing the same species in an identical setup, found that fish would ignore social information when they had relevant personal information. Similarly, naïve ants are more likely to follow the chemical trails of others, compared to experienced individuals who rely on their own personal experience (Grüter et al. 2011).

Giraldeau et al. (2002), and Bikhchandani et al. (1992, 1998) proposed that individuals may use social information not only because their personal information is unreliable but because the accumulated knowledge of conspecifics potentially represents a source of information with even greater reliability. In spite of this, they predict that in specific instances, reliance on the decisions of others can lead to arbitrary or even maladaptive traditions in animals (Giraldeau et al. 2002). Theoretical work regarding reliability and the value of information in communication and mating systems (Sirot 2001; Koops 2004) suggests that, even if the costs of misinformation are high, animals should still use information, provided that it is usually reliable. This requires animals to be able to assess the relative reliability of personal versus social information correctly, and to use social information when they are uncertain that personal information is reliable.

In a study of nine-spined sticklebacks, Van Bergen et al. (2004) manipulated the reliability of personal experience concerning the profitability of two foraging patches. Fish were allocated to one of three conditions, where they received

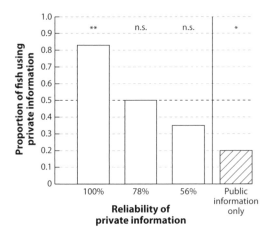

Figure 8.7. Van Bergen et al. (2004) manipulated the reliability of personal experience concerning the profitability of two foraging patches for nine-spined sticklebacks, and then subjected them to a conflicting public demonstration (where the richer patch in their personal experience was the poorer patch in the public demonstration), followed by a test. They found that fish increasingly relied on the social information provided by their demonstrators as their certainty in their personal experience diminished. Based on figure 2a in van Bergen et al. (2004). The symbols * and ** indicate a statistically significant difference from 0 at 5% and 1% respectively; "n.s." indicates no statistically significant difference.

100%, 78%, or 56% reliable personal experience as to which of two feeders was "rich" and which "poor," corresponding to increasingly noisy information, then received the conflicting social information of watching other fish feed at a higher rate at what had been the personal-poor feeder. Finally, they were tested individually for their feeder preference. Following this demonstration, only fish in the 100% reliable condition continued to prefer the feeder that was "rich" according to their personal information, as fish in the other conditions exhibited no preference (fig. 8.7). The study suggests that fish increasingly relied on the social information provided by their demonstrators as their certainty in their personal experience diminished.

8.2.4 Copy when prior information is outdated

Some theoretical models predict that individuals should *acquire* personal information, and ignore social information, when the latter is likely to be outdated as in, for example, changing environments[7] (Boyd and Richerson 1985, 1988; Feldman et al. 1996; J. Henrich and Boyd 1998; Moscarini et al. 1998; Doligez et al. 2003). Van Bergen et al. (2004) manipulated the degree to which personal information, regarding the relative profitability of two foraging patches, was outdated, and explored how this prior experience affected individuals' subsequent acquisition of public information. Nine-spined sticklebacks were allocated to one

---

[7] The models of Van der Post and Hogeweg (2009) and Rendell et al. (2010) contest this conclusion.

of four conditions, where they received personal information as to which of two feeders was "rich" and which was "poor," either 1, 3, 5, or 7 days prior to receiving conflicting public information, and a test. Fish with only a 1-day delay between receiving personal and public information ignored the social information and first visited the feeder that was "rich" according to their personal information. Fish with delays of 3 and 5 days showed no feeder preference, and those experiencing a 7-day delay preferred the feeder that was "rich" according to the public information. Van Bergen et al. argue that as personal information becomes increasingly outdated, nine-spined sticklebacks become increasingly reliant upon socially acquired information. They maintain that personal information was not forgotten after 7 days, but rather that fish in the 7-day condition ignored their personal information in favor of the public information.

8.2.5 Copy when dissatisfied

Mathematical analyses by Schlag (1998, 1999) established that an imitation rule termed "proportional reservation," but which might more intuitively be called "copy when dissatisfied," was a highly effective strategy. Here, the size of the pay-off of their current behavior determines an individual's satisfaction as a linear function. The individual retains its current behavior with a probability equal to this satisfaction and otherwise copies the action of a randomly chosen demonstrator. This strategy has the advantage that it is potentially simple to implement, because it does not require individuals to assess the payoff received by a demonstrator or to make any judgment as to the relative profitabilities of alternative behavior patterns.

Galef et al. (2008) manipulated rats' dissatisfaction with their diet (experiment 1) or environment (experiment 2). In the first experiment, rats were maintained for one week on either an unpalatable low caloric diet, requiring increased handling time to maintain their health (dissatisfied condition), or a palatable and relatively high caloric diet (satisfied condition). Following this, the subjects were exposed for 30 minutes to a demonstrator rat carrying cues on the breath as to whether it had eaten a marjoram- or anise-flavored diet. In the following 24 hours the subjects were exposed to both diets and "dissatisfied" rats ate more of the food that their demonstrator had eaten than did "satisfied" rats. In the second experiment, as essentially crepuscular burrowing animals, rats were either housed on a hard substrate with no cover, constant light, and an overly warm ambient temperature (1: dissatisfied condition); or with bedding, cover, a 12:12 light:dark cycle, and appropriate ambient temperature (2: satisfied condition). Again, "dissatisfied" subjects ate more of the diet indicated by their demonstrator than did "satisfied" rats. Galef et al.'s experiments are consistent with the strategy of "copy when dissatisfied," although the latter experiment exemplifies "dissatisfaction" more broadly than the former. In experiment 2, rats appear to be generally dissatisfied with their lot, while in experiment 1, they are specifically dissatisfied with their feeding regime—that is, they were dissatisfied in the same domain as the social information. There is a case for characterizing these as two separate strategies.

8.2.6 Is copying a first or last resort?

Enquist et al. (2007) explored the relative merits of two conditional strategies, which they call the *critical-social-learner* and *conditional-social-learner* strategies. For the *critical-social-learner* strategy, individuals are assumed to copy first, and to rely on trial and error learning only if copying fails to produce satisfactory outcomes. The *conditional-social-learner* refers to strategies that involve individual learning first, with a switch to social learning only if individual learning fails to provide satisfactory rewards. Enquist and colleagues conclude that the critical-social-learner strategy is always superior to pure social learning, and typically results in a higher fitness than pure individual learning. Critical social learning is an *evolutionarily stable strategy* (ESS) (Maynard-Smith 1982), unless cultural transmission has low fidelity, the environment is highly variable, or social learning is much more costly than individual learning. Similarly, because dissatisfied individuals abandon behavior with low rewards to copy other individuals that have retained behavior with high rewards, like *copy-when-dissatisfied*, conditional social learning generates and maintains adaptive behavior in a population. A population of *conditional-social-learners* cannot be invaded by *critical-social-learners* if the cost of individual learning is sufficiently small, while *conditional-social-learners* cannot invade a population of *critical-social-learners* if individual learning is sufficiently costly, because conditional social learners pay too great a cost up front. Enquist and colleagues find that the *critical-social-learner* is a superior strategy to *conditional-social-learner* over a broader range of conditions.

Enquist et al.'s (2007) findings are consistent with the hypothesis that social learning should be tried first when a behavior is either too costly or too complex to be invented by a single individual, a condition that applies to much of human culture. However, copying first may not be best in highly variable environments, where asocial learning is a more effective first recourse, or in cases where asocial learning is very low risk. Enquist et al.'s analysis suggests the interesting possibility that social learning strategies may be better characterized as specifying when social learning is not to be used than as specifying when it is.

However, Rendell et al. (2009) report that in a spatially varying environment, the *conditional social learning* strategy is favored over the *critical social learning* strategy under most conditions. In circumstances in which social learning is effective, then *critical social learning* will be at an advantage, because it will tend to pay the cost of asocial learning less than *conditional social learning*. Conversely, if social learning is ineffective, then the reverse is true. Rendell et al. conclude that since increased environmental spatial variation decreases the effectiveness of social learning, it will favor *conditional social learning*.

Data consistent with use of a conditional social learner strategy is observed in rats. Norway rats find it easier to detect a sodium rather than a protein deficiency in food, an observation that Galef and Whiskin (2008b) exploited to test a prediction, derived from Boyd and Richerson's (1988) model, that aspects of diet that are difficult to learn should be differentially copied. Galef and Whiskin induced dietary deficiencies in rats by feeding them either a sodium-deficient or protein-deficient diet for seven consecutive days. They then offered observers a choice between cinnamon- and cocoa-flavored foods rich in both protein and sodium; these observers had previously interacted with a demonstrator rat that had eaten either cinnamon-flavored or cocoa-flavored food. They found that, whether severely or moderately deprived, protein-deprived observer rats relied more heavily on socially acquired information than did sodium-deprived observer rats, confirming Boyd and Richerson's (1988) prediction.

8.3 "Who" Strategies

A primary focus of the formal theory underlying social learning strategies is on "who strategies," which specify learning from particular individuals or collections of individuals (Boyd and Richerson 1985; J. Henrich and Boyd 1998; Schlag 1998, 1999; Wakano and Aoki 2007; J. R. Kendal, Giraldeau, and Laland 2009). These strategies are also receiving attention from researchers in a wide variety of academic disciplines with interests in the experimental analysis of social learning, cultural transmission, and cultural evolution (Kameda and Nakanishi 2002, 2003; McElreath et al. 2005; Efferson et al. 2008; Mesoudi and O'Brien 2008; Horner et al. 2010; Dindo et al. 2009). One relatively well-studied class of rules are frequency-dependent strategies, such as "conformity" and "anticonformity" (Cavalli-Sforza and Feldman 1981; Boyd and Richerson 1985; J. Henrich and Boyd 1998; Wakano and Aoki 2007; J. R. Kendal, Giraldeau, and Laland 2009), which involve individuals selectively adopting traits based on how common they are. "Who" strategies also encompass payoff-based rules, where copying depends on the return to the observed individual (Schalg 1998, 1999; J. R. Kendal, Giraldeau, and Laland 2009), and model-based biases, such as a prestige bias (J. Henrich and Gil-White 2001).

There is, of course, a long-standing interest among social psychologists in how and why individuals adopt the decisions of others (Bond 2005; Bond and Smith 1996; Latané 1981; Latané and Wolf 1981; Nowak et al. 1990; Coultas 2004; Tanford and Penrod 1984; Mullen 1985). Social impact theory clearly relates to social learning strategies, and has been extended to consider its effect upon population-level belief patterns (Nowak et al. 1990; Latané 1996). However, such work typically focuses on the adoption of arbitrary or incorrect group decisions (Asch 1955), and by doing so limits consideration of the evolutionary, population-level consequences of such behavior, and of whether the use of social information leads to adaptive behavior.

Other investigations of this topic derive from the animal behavior literature. Coussi-Korbel and Fragaszy (1995) introduced the concept of *directed social learning*; according to the concept, the identity and characteristics of demonstrator and observer critically affect the probability of social learning. Coussi-Korbel and Fragaszy suggested that the social rank, sex, age, patterns of association, and other characteristics of demonstrator and observer frequently influence the likelihood of social learning. As a result, information may be transmitted through subsections of animal societies at different rates. The authors' argument was based primarily on observations of nonhuman primates, but their ideas apply more generally, and are well supported in other taxa.

8.3.1 Frequency-dependent biases

Following Boyd and Richerson (1985, 206), we define conformist transmission as positive frequency-dependent copying. That is, it entails the disproportionately likely adoption of the variant exhibited by most individuals (which may not be the most commonly performed behavior if individuals exhibit the behavior multiple times; Haun et al. 2012). Here the frequency data relates to the behavior of individuals within a sample, with each individual contributing a single datum. Theoretical analyses have revealed that in most circumstances in which natural selection favors reliance on social learning, this type of conformity is also favored (Boyd and Richerson 1985; J. Henrich and Boyd 1998). Conformist transmission is of theoretical interest both because it is an effective means of acquiring adaptive information in a spatially variable environment, and because it can maintain differences between populations (Boyd and Richerson 1985; Richerson and Boyd 2005).[8] It was established by J. Henrich and Boyd (1998) that an even broader range of conditions favors conformist transmission than social learning, including temporally and spatially variable environments. These findings suggest that much animal social learning should involve individual adoption of the behavior of the majority—that is, a *copy-the-majority* strategy.[9] Boyd and Richerson (1985), Richerson and Boyd (2005), and their collaborators (e.g., J. Henrich and Boyd 1998, 2001; J. Henrich 2001; J. Henrich and McElreath 2003, 2007; McElreath et al. 2005; Efferson et al. 2007; N. Henrich and J. Henrich 2007) have consistently placed emphasis on conformist transmission as an important mechanism of cultural evolution, stressing the evidence for conformity found in the social psychology literature (Asch 1955; Coultas 2004). However, this emphasis has recently received criticism from other cultural evolution theoreticians, whose analysis reveals that conformity hinders cumulative cultural evolution, and as a consequence would be selected against in many circumstances (Eriksson et al. 2007). These differences may relate in part to different assumptions about patterns of environmental variation, and the cumulative or noncumulative nature of cultural change. Kandler and Laland (2009) found that a strong conformist bias can hinder the spread of a beneficial variant, supporting Eriksson et al.'s (2007) argument that natural selection will not favor conformity if it prevents cumulative cultural evolution. Conversely, Kandler and Laland (2009) found that weak conformity

---

[8] To coin a useful distinction from social psychology (Deutsch and Gerard 1955), here we are interested in *informational* rather than *normative* conformity, the latter being concerned with gaining positive appraisal from others rather than useful knowledge.

[9] The 1998 model by J. Henrich and Boyd is an extension of an earlier model by Boyd and Richerson (1988); the former implements a *copy-when-uncertain strategy* (see fig. 8.6 in J. Henrich and Boyd [1998]). Strictly speaking, J. Henrich and Boyd's model provides support for a conditional form of conformist bias, along the lines of *copy-the-majority-when-uncertain*.

typically increases the frequency of the most beneficial variant at equilibrium, which can be viewed as consistent with J. Henrich and Boyd's (1998) conclusion that selection will favor conformity over a broad range of circumstances. Hence, one reading of the literature is that the strength of conformity is thus likely to be a key issue. To the extent that human cultural change can be regarded as manifestly cumulative (Enquist et al. 2007; Ghirlanda and Enquist 2007), conformist transmission might be expected to be comparatively weak. However, conformity may not act alone. Another reading is that conformity, to be adaptive, must be combined with another strategy, such as asocial learning, prestige bias, or payoff-based copying (Boyd and Richerson 1995; J. Henrich and Boyd 1998). As we discuss below, this suggestion garners both empirical and theoretical support.

Empirical evidence of conformity has proved surprisingly elusive (McElreath et al. 2005; Efferson et al. 2008; Eriksson et al. 2007; Eriksson and Coultas, 2009), and some work has suggested that changes in frequency may be more salient than absolute frequencies (Toelch et al. 2010). McElreath et al. (2005) had groups of people repeatedly play a computer-based task (planting one of two crops with different yields), either where asocial learning alone was possible, or with the opportunity to view the previous choice of one randomly selected group member (allowing social learning), or with the opportunity to view the previous choices of all group members (allowing conformity). The use of social information increased when individual learning was relatively inaccurate, supporting the strategy of "copy when uncertain." Similarly, although models indicated that conformity was the best strategy under all conditions, it was used only when the environment fluctuated, hence when there was a cost to individual learning. Thus, to the extent that McElreath et al. find support for conformity, it is a strategy used when individuals are uncertain.

Similarly, in another computer-based study, Efferson et al. (2008) found considerable heterogeneity, as yet unexplained, in the extent to which humans follow a strategy of "copy the majority," even when doing so would be in their interests. A fraction of the population appeared to conform but the rest, which Efferson et al. characterized as "mavericks," relied on asocial learning. Eriksson et al. (2007) found that the likelihood of conforming is highly context-dependent. People reported that, when presented with vignettes pertaining to either a scenario of novel food choice (different soups) or punishment of social defectors (defection consisted of undertaking large print jobs on a communal machine), they would conform in regard to the former but not the latter.

Strong experimental evidence for conformity can be found in a series of experiments investigating factors affecting when and how humans use social information across several computer-based tasks (T.J.H. Morgan et al. 2012; fig. 8.8). These authors (T.J.H. Morgan et al.) found that both the number of demonstrators and consensus among demonstrators increased subjects' use of social information in an adaptive fashion. In three of four experiments, the rate of copying increased disproportionately with the number of demonstrators. As majority decisions of larger groups are known to be more likely to be correct (i.e., less error prone) than those of smaller groups (King and Cowlishaw 2007), this tendency is

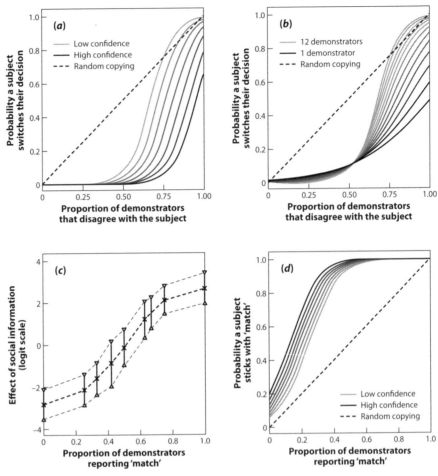

Figure 8.8. The 2012 experimental studies by T.J.H. Morgan et al. of social learning in humans provide evidence for conditional conformity (a, b). Here subjects had to judge whether two shapes (differently rotated in spaces) were the same. They had an initial, asocial choice, and then chose again after receiving social information about the choices of others. The proportion of demonstrators that disagree with the subject's initial, asocial, choice strongly affects the probability that subjects will change their decision. (a) Decreases in subject's confidence based on asocial information (number of demonstrators = 12), and (b) increases in demonstrator number (subject confidence = low) increases the likelihood of a switch following conflicting social information. Subject behavior, when uncertain and given many demonstrators, is conformist in that subjects are disproportionately likely to switch their decision when faced with a strong opposing majority. The black dashed lines portray the expected result of random copying. (c, d). The proportion of demonstrators reporting that the displayed shapes match strongly affects the probability that subjects' final decision will be that the shapes match. (c) In line with a conformist response to social information, when unconstrained (i.e., modeled as categorical), intermediate levels of consensus have a disproportionately large effect on decision making. The y-axis is the change to the linear predictor of the model, on which a change of magnitude four could alter the probability that a subject decides displayed shapes match by as much as 76%. Accordingly, without other influences, social information is likely to have a dramatic effect on subject behavior. (d) Subject behavior, however, is also strongly affected by prior information and their confidence in it (the lines shown are for cases where subjects already believe that the shapes match). Thus, although subject behavior may not be conformist, their response to the social information alone was conformist. The black dashed line portrays the expected result of random copying. Based on figure 2 in T.J.H. Morgan et al. (2012).

likely to prove adaptive. Demonstrator consensus also strongly affected the rate of copying. In an additional theoretical analysis, T.J.H. Morgan et al. showed, using probability theory, that the greater the consensus, the more likely that the majority opinion is correct, provided individuals make independent decisions and perform above chance levels.

The data suggested a conditional implementation of a conformist social learning strategy, dependent on the number of demonstrators as well as the subject's confidence in their own judgment. Rather than simply copying the majority under all circumstances, individuals copied the majority when they were unsure (as predicted by Henrich and Boyd [1998], and as found by McElreath et al. [2008]), and when there were a large number of demonstrators behaving consistently (fig. 8.8).

Conformity results in the disproportionate adoption of popular traits at the expense of rare traits, producing an S-shaped relationship between trait frequency and probability of adoption. This S-shaped curve was found to occur only when there were many demonstrators, and subjects expressed low confidence (T.J.H. Morgan et al. (2012). Furthermore, while formal models set the threshold for disproportionate copying at 0.5 (Boyd and Richerson 1985), T.J.H. Morgan et al. observed a higher threshold. While this data is consistent with a conformist use of social information, as subjects' behavior was the result of both social and asocial influences, the resultant behavior was not always conformist and depended on the magnitude of the asocial influence. Like Efferson et al. (2008), T.J.H. Morgan et al. found that individuals varied in their sensitivity to social influences. Nonetheless, the restricted conditions under which conformist behavior was realized suggest a tendency to rely on asocial over social information as a first resort. Increased subject confidence strengthens this tendency, making subjects more willing to go against the consensus. Such behavioral flexibility may explain the contradictory findings of previous work.

Social learning in which the probability of adopting a pattern of behavior increases with the proportion of demonstrators occurs in other animals, including guppies (Lachlan et al, 1998; Laland and Williams 1997; Sugita 1980), rats (Beck and Galef 1989; Chou and Richerson 1992), and pigeons (Lefebvre and Giraldeau 1994). Strictly speaking, this work provides only possible evidence of conformity, since the experiments would need to demonstrate a *disproportionate* tendency to adopt the behavior of the majority rather than a linear association between the probability of learning and the proportion of demonstrators exhibiting the behavior. Day et al. (2001) found indirect evidence of conformity in a series of experiments in fish in which the effect of shoal size on foraging efficiency was investigated. In a first experiment, they presented a hidden food source to shoals of guppies in open water. Large shoals were found to locate food faster than small shoals, probably because fish in large shoals have more shoal mates from which to acquire information, and large numbers of individuals at a food site attract conspecifics more rapidly than small aggregations. However, in a second experiment, the fish had to swim through a hole in an opaque maze partition to get to a food source. In this situation, the opposite result was found; smaller shoals located food faster than larger shoals. The seemingly conflicting findings of these experiments make sense in light of the observation that guppies have a preference to join large over small shoals, which implies that individuals ought to be more willing to leave smaller rather than larger shoals. Swimming through an opaque partition to locate food involved breaking visual contact with the shoal and, hence, effectively leaving it. Under such circumstances, conformity, which is the result of the natural shoaling tendency of these fish, leads to greater reluctance to acquire a novel behavior in large than in small shoals. This interpretation was supported by the findings of a third experiment, which replicated the second except that it made use of a transparent partition. In this experiment, individuals in large shoals once again located the food faster than those in small shoals. Here, visual contact between fish was maintained because the partition was transparent, so fish passing through it were not leaving the shoal. Hence, social transmission of foraging information was not hindered by conformity, and large shoals were at an advantage.

Perhaps the clearest evidence for conformity outside of humans is provided by Pike and Laland's (2010) investigation of conformist transmission in the nine-spined stickleback. These authors manipulated the number of demonstrator fish at two feeders (one with a large amount of food [rich] and the other with a small amount [poor]) during a demonstration phase; they evaluated how demonstrator number affected the likelihood that focal fish subsequently copied the demonstrators' apparent patch choices after the demonstrators were removed. Pike and Laland observed a significantly increased level of copying with increasing numbers of demonstrators at the richer of the two feeders, with copying increasing disproportionately, rather than linearly, with demonstrator number. Control conditions with nonfeeding demonstrators showed that this was not simply the result of a preference for shoaling with larger groups, implying that nine-spined sticklebacks can also adopt a strategy of *copy the majority*.

Whiten et al. (2005) also claim evidence of conformity in an experimental study investigating the spread of alternative tool-using techniques by foraging chimpanzees. Unobserved by group mates, the experimenters first trained a high-ranking female from each of two groups of captive chimpanzees to adopt one of two different techniques for obtaining food from the same apparatus. They then reintroduced each female to her respective group. All but 2 of 32 chimpanzees mastered the new technique under the influence of their local expert, whereas none did so in a third population into which no trained expert was introduced. Most chimpanzees adopted the method seeded in their group, and these traditions were maintained over time. The authors claimed that a conformity bias was shown. However, this reasoning is based on a different definition of conformity to that given above,[10] and is not unambiguously a manifestation of positive frequency-dependent social learning.

---

[10] Whiten et al.'s (2005) claim of conformity is based on the observation that a small number of individuals changed their behavior from the minority to the majority variant.

In theory, the reverse pattern of negative frequency dependence might occur where novelty confers a selective advantage, a finding referred to as "anti-conformity." In the same way that males with rare phenotypes can have a mating advantage in populations of Drosophila (Spiess 1968), rare behavior patterns may be disproportionately adopted. One possible example is interspecific vocal mimicry in birds, such as starlings, parrots, and mynahs. At the extreme, male European marsh warblers copy the sounds of an average of 77 other species (Dowsett-Lemaire 1979). Catchpole and Slater (1995) suggest that the most likely explanation for male birds habitually copying the vocalizations of heterospecifics is sexual selection, due to females favoring males with large repertoires. They point out that the most striking cases of vocal mimicry occur in species with very elaborate songs.

8.3.2 Success biases

Intuitively, learning from others would be most effective if animals disproportionately adopted the behavior of the more successful members of their population. There is no point in copying someone whose behavior is manifestly unproductive. There are a number of related strategies that could be adopted, the utility of which has been explored using theoretical models. One such strategy is to copy successful individuals by directly monitoring their payoffs (Schlag 1998). An alternative approach is to identify successful individuals indirectly using cues such as health, wealth, or reproductive success. Boyd and Richerson (1985) labeled this latter strategy an "indirect bias," and analyses suggest that it leads to adaptive behavior (Boyd and Richerson 1985).

An indirect bias has the advantage of being relatively easy to put into practice, but the disadvantage is that it is not always clear which of a successful individual's many traits is the major source of its success. Pop and film stars do not make their millions as a result of their political views, yet they frequently exert an influence on the political beliefs and values of their fans. Mathematical analyses suggest that this strategy may be favored by natural selection even though it may sometimes allow neutral and maladaptive traits to hitchhike along with those traits that engender success (Boyd and Richerson 1985).

Animals provide evidence for copying on the basis of both direct and indirect assessments of success, although it is often unclear which is operating. Clear evidence for a success bias is found in studies of foraging behavior in chimpanzees, including Menzel's (1973a, 1974) reports that naïve chimpanzees were more likely to follow an informed individual with knowledge of the whereabouts of hidden food than an uninformed conspecific. A similar phenomenon has been described in guppies (Lachlan et al. 1998). Likewise, bats that are unsuccessful in locating food alone follow previously successful bats to feeding sites, using cues indicative of recent feeding, such as defecation (Wilkinson 1992). Insects and birds copy the nest-site decisions of successful rather than unsuccessful conspecifics and heterospecifics (Sarin and Dukas 2009; Pasqualone and Davis 2011; Seppanen et al. 2011; Forsman and Seppanen 2011). For example, fruit flies (*Drosophila melanogaster*) that experienced novel food together with other females who had laid eggs on that food, subsequently exhibited a stronger preference for laying eggs on that food over another novel food; this behavior was compared with focal females that experienced the food alone (Sarin and Dukas 2009). Similar behavior is observed in the walnut-infesting fruit fly, *Rhagoletis suavis* (Pasqualone and Davis 2011). Likewise, pied flycatchers (*Ficedula hypoleuca*), who naturally have a partially overlapping ecology with resident titmice species, adopted the breeding site choices of successfully breeding coal tits and blue tits, but avoided sites where tits had few chicks in their nests (Seppanen et al. 2011; Forsman and Seppanen 2011). In addition, Indian mynah birds avoid foraging where conspecifics experienced a predator attack (Griffin and Haythorpe 2011).

The probability that redwing blackbirds will acquire a food preference through social learning is affected by whether the demonstrator bird becomes sick or remains well (Mason 1988), a manifestation of indirect bias. Similarly, nine-spined sticklebacks differentially copy the food site choices of larger compared to smaller fishes; this may occur because size is an indicator of prior foraging success (Duffy et al. 2009). Finally, R. B. Payne (1985) suggested that village indigobirds may copy the individual with the greatest mating success so that his songs are both shared and persistent. These examples suggest that animals will sometimes rely on indirect cues of success in cases where immediate success is difficult to assess directly.

Henrich and Gil-White (2001) suggest that the evolution of a *copy-successful-individuals* strategy could explain the formation of prestige hierarchies, since highly skilled individuals will be at a premium. Individuals adopting this strategy may be selected to pay deference to successful individuals in exchange for preferential access and assistance in learning. Henrich and Gil-White suggest that such deference benefits may take many forms, including coalition support, gifts, and caring for offspring. Eventually, such deference behavior itself becomes a reliable cue as to which individuals possess adaptive knowledge. The authors J. Henrich and N. Henrich (2010) found that, in traditional Fijian populations, food taboos that lead pregnant and lactating females to avoid consumption of toxic fish are initially transmitted through families, but as individuals get older they preferentially seek out local prestigious individuals to refine their knowledge (see also J. Henrich and Broesch 2011). However, in contrast to J. Henrich and Gil-White's (2001) expectations, in a test of predictions from the prestige-bias model in the Bolivian Tsimane, Reyes-Garcia et al. (2008) found weak evidence that prestige was associated with ethno-medicinal plant knowledge, and no evidence that prestige was associated with age.

Most human social learning experiments have a very simple structure, requiring individuals to select one of two options. A more realistic task would require individuals to choose between options that vary along a continuous scale and in many different dimensions. Mesoudi and O'Brien (2008) carried out such an investigation based on the cultural transmission of prehistoric arrowheads. In their computer game, participants had to design and test "virtual arrowheads" in "virtual hunting environments," and were given the option to use information about the success of other peoples' arrowheads. Arrowheads could vary with respect to their overall shape, width, thickness, length, and color, and the virtual environments were set up so that there were several different combinations of traits that conferred high hunting returns (in evolutionary terms, there were multiple adaptive peaks). Mesoudi and O'Brien found that despite experiencing repeated hunts in which they gradually refined their own designs, individuals would frequently abandon their arrowhead and copy the designs produced by others. Moreover, participants tended to copy the arrowhead design of the single most successful hunter, even with respect to nonfunctional features, such as color. The results are very consistent with Boyd and Richerson's (1985) model of indirect bias and can potentially explain the actual patterns of prehistoric arrowhead variation observed in the archaeological record (Bettinger and Eerkens 1999).

In the context of animal studies, the closest variable to prestige is dominance or social rank, although Henrich and Gil-White (2001) emphasize that these are not the same (prestige relates to influence not power, and esteem for prestigious individuals is freely given rather than coerced). Nonetheless, we can ask whether there exists a corresponding dominance or power bias, such that animals are more likely to copy high-ranking rather than low-ranking conspecifics. Surprisingly few attempts have been made to address this question. Horner et al. (2010) report that, when given opportunities to watch alternative solutions to a foraging problem performed by two different models of their own species, chimpanzees preferentially copied the method shown by the older, higher-ranking individual with a prior track record of success. However, evidence is beginning to emerge from studies of birds and monkeys that low-status demonstrators may be ineffective relative to high-status demonstrators because of the inhibiting effects of the presence of conspecifics on the performance of behavior by low-status demonstrators (Drea and Wallen 1999; Nicol and Pope 1994). This might result in an apparent tendency to copy dominants as an artifact of lower rankers not performing behavior that could be copied.

A standard assumption of theoretical models of human decision making is that individuals are able to evaluate the payoffs associated with behavioral alternatives directly (Gintis 2000; Schlag 1998), but it is not clear to what extent animals are able to make such judgments. The collection of information about relative patch quality by monitoring the success of others has been called *public information use* (Valone 1989). Public information use has been reported in the selection of food patches and breeding sites in birds (Doligez et al. 2002; Templeton and Giraldeau 1996) and in assessment of the relative value of mates and competitors in fish (McGregor et al. 2001). For example, starlings can use the foraging success of other birds to assess patch quality, and they exploit this information in their judgments to stay or switch patches (Templeton and Giraldeau 1996). The aforementioned work on public information use in sticklebacks (Coolen et al. 2003), where nine-spines differentially chose the richer of two food patches demonstrated by other fishes, would also seem to be consistent with this strategy.

Another heuristic is *copy if better*, whereby individuals switch strategy if the returns of the behavior adopted by the demonstrator exceed those of their own behavior (Schlag 1998). Schlag's game theoretical analyses reveal that when information concerning the success of others is unreliable and noisy, a *copy-if-better* strategy outperforms a *copy-the-most-successful-behavior* strategy. However, Schlag reports that in risky environments *always* copying all individuals that seem to be reaping greater returns can lead the entire population to choose the alternative with the lowest expected payoff. A much better rule, which Schlag calls "proportional imitation," is one by which observers copy an individual that performed better than they did, with a probability that is proportional to how much better that individual performed. Schlag found that this version of the *copy-if-better* strategy always leads the population to the expected payoff-maximizing action.

Schlag's (1998) analysis has uncovered two further strategies that appear to be equally effective as "proportional imitation," but may be easier for animals to use. One such strategy, which he terms "proportional observation," requires individuals to copy the behavior of a demonstrator with a probability proportional to the demonstrator's payoff. Thus, once again, animals have to make a judgment as to the profitability of another individual's behavior, but this rule seems less complicated than the proportional imitation rule, since a comparison between self and other is not required. There is evidence that nine-spined sticklebacks disproportionately copy when the demonstrator receives a higher payoff than they do (J. R. Kendal, Giraldeau, and Laland 2009; Pike et al. 2010). It was established that nine-spined sticklebacks are able to switch foraging patch preferences to exploit a more profitable food patch if the returns to "demonstrator" fish are greater than their own, but are less likely to copy when low-profitability patches are demonstrated (J. R. Kendal, Giraldeau, and Laland 2009). The results are consistent with these fish employing a proportional (rather than absolute) implementation of a *copy if better* social learning strategy (Laland 2004); however, the results are ambiguous as to which rule is being deployed. Pike et al. (2010) extended this work by conducting an experiment that differentiates among these alternative rules, finding that sticklebacks exploited a socially demonstrated food patch under restricted and predicted circumstances, copying the patch choice of demonstrator individuals with a probability proportional to the demonstrators' payoff (i.e., by Schlag's proportional observation rule). Similar findings were reported for human subjects by T.J.H. Morgan et al. (2012).

Another possibility is a rule that Schlag termed "proportional reservation," but which might be called the *copy-if-dissatisfied* strategy. Here, the size of the payoff to an individual's current behavior determines its satisfaction, in a linear function. The individual retains its current behavior with a probability equal to this satisfaction. Hence, if the level of satisfaction is low, there is a large probability that it will copy the action of a randomly chosen demonstrator, whereas if its satisfaction is high the probability is small. This strategy has the advantage that it is potentially simple to put into practice, because it does not require individuals to assess the payoff to a demonstrator or to make any judgments as to the relative profitability of alternative behavior patterns. Schlag's analysis suggests that the proportional-imitation, proportional-observation, and proportional-reservation tactics will all have equal fitness at equilibrium. However, the proportional-reservation rule would appear to be easier to implement than the other two strategies. Thus, we might anticipate that a *copy-if-dissatisfied* strategy is likely to be widespread in nature (e.g., Grüter and Ratnieks 2011; fig. 8.4).

8.3.3. Kin and age biases

An extremely common observation in the social learning literature is that individuals often copy kin. A meta-analysis by Reader (2000) found disproportionate reports of learning from mothers in nonhuman primates. Any kin bias may reflect nothing more than the fact that individuals spend considerably more time in the presence of kin than they spend in the presence of those who are not kin. However, there are two reasons to suspect that selection may have favored a disproportionate degree of learning from kin. First, social learning, particularly in a variable environment, is of use only to the extent that demonstrator and observer experience the same environment and reap the same rewards (Boyd and Richerson 1985, 1988). This may be more likely among kin because, by virtue of their shared genes (and other sources of similarity), kin may be more likely than nonkin to reproduce similar behavior variants and to experience the same affective sensations in reinforcement. Second, in situations in which information transmission is costly individuals may have more to gain by providing reliable information to kin than to nonkin, and less to gain from deceiving them. Conceivably, Hamilton's (1964) rule may apply to social learning, with the probability of social learning being directly proportional to the coefficient of relatedness of observer and demonstrator. It may be no coincidence that one of the most sophisticated cases of animal communication that results in social learning—the famous dance of the honeybee—occurs among female worker bees that share up to three quarters of their genes. Similarly, animal teaching occurs disproportionately among cooperative breeders (Thornton and Raihani 2008; Hoppitt et al. 2008), perhaps because they often exhibit high relatedness (Cornwallis et al. 2010). Similarly, if "friends" are regarded as individuals with whom one trades altruistic acts (Trivers 1971), by similar lines of reasoning we might expect more social learning among friends than among nonfriends in a *copy-friends* strategy. Although fish may not have "friends," they do express preferences for shoaling with familiar individuals (Griffiths 2003), and guppies have been reported to acquire foraging information more effectively from familiar than from unfamiliar demonstrators (Swaney et al. 2001).

Another possible strategy, *copy-older-individuals*, has been assumed in theoretical analyses (Kirkpatrick and Dugatkin 1994) on the basis of reports of age-biased mate-choice copying in female guppies. This theoretical work implies that age-biased copying might generally be adaptive (Kirkpatrick and Dugatkin 1994). It is well established that younger fish copy the mate-choice decisions of older individuals (Dugatkin and Godin 1993). Duffy et al. (2009) found that this extended to nine-spined sticklebacks in a foraging context. Younger birds also have been found more likely to copy than older birds (Biondi et al. 2010), and to benefit from foraging with more experienced birds (Kitowski 2009). Biro et al. (2003, 2006) report that young chimpanzees were far more likely than older individuals to attend to the novel nut cracking or leaf use in water holes of group members (see section 7.1.1).

8.4 "What" Strategies

"What" strategies specify that individuals copy behavior or knowledge according to its content;this is also known as a "content bias." For example, there is evidence that humans exhibit biases for copying more attractive, memorable, or emotionally evocative variants (Heath et al. 2001; Bangerter and Heath 2004; Jones et al. 2007; Little et al. 2008). Heath et al. (2001) proposed that infectious ideas, like rumors and urban legends, spread because they undergo a kind of *emotional selection*—they are selected and retained in the social environment in part based on their ability to tap emotions that are common across individuals. These authors found that people say they would be more likely to pass along stories that evoke more disgust, compared to less disgusting alternatives. A classic example is the rumor, which spread rampantly, that a customer once found a rat in their Kentucky fried chicken. It is easy to imagine that a disgust bias might have an evolutionary origin, since things that we find disgusting, such a rotting and infested food, might well be harboring disease.

Evolutionary theories concerning the origins of human intelligence suggest that cultural transmission might be biased toward social rather than nonsocial information (Humphrey 1976; Byrne and Whiten 1988; Dunbar 1998). Mesoudi et al. (2006) tested this by passing social and nonsocial information along multiple chains of participants. A first experiment found that gossip, defined as information about intense third-party social relationships, was transmitted with significantly greater accuracy and quantity than equivalent nonsocial information concerning individual behavior or the physical environment. A second experiment replicated this finding while controlling for narrative coherence, and additionally found that information concerning everyday (not involving gossip) social interactions was transmitted just as well as the interactions that did consist of gossip. Mesoudi et al. concluded that human cultural transmission is biased toward information concerning social interactions over equivalent nonsocial information.

In chapter 3, we described a study by Barrett and Nyhof (2001), which found that descriptions of living creatures or physical objects that were "minimally counterintuitive" were passed along transmission chains with higher fidelity than entities with either no counterintuitive elements or with extremely counterintuitive (i.e., bizarre) material. Norenzayan et al. (2006) tested this idea with Brothers Grimm fairytales, and confirmed that the most successful stories were those with two to three counterintuitive elements, rather than those with either no such elements or with a great number. Anthropologists Pascal Boyer (1994) and Scott Atran (2002) have suggested that supernatural beliefs (e.g., in ghosts, spirits, gods) are prevalent around the world precisely because they are minimally counterintuitive. This means they are rendered more memorable by violating some common beliefs, but are consistent with others. Religions are composed of a compelling blend of commonsensical and counterintuitive elements that give them a mnemonic advantage, which in turn facilitates retention and social transmission relative to stories with either too many or no counterintuitive beliefs (Atran 2002).

Perhaps the most commonly given example of a content bias is the widely recognized human preference for sugar-, fat-, and salt-rich foods. Plausibly valued as rare commodities by our ancestors, modern food-production methods have solved the problem of scarcity for many of us, but we continue to crave these now abundant foods to such an extent that we consume quantities that are excessive and sometimes lead to disease. Seemingly, we are predisposed to acquire from others a preference for hamburgers, fries, and chocolate cake. However, any human dietary bias is unlikely to be restricted to salt-, sugar-, and fat-rich foods, which have garnered attention primarily because of their negative consequences for our health, and not because they are the only way in which our diet has been constrained by selection. Moreover, this dietary bias may be extremely common in nature. For example, rats exhibit prior preferences for certain flavored foods over others, and the favored variants are transmitted with greater fidelity (Laland and Plotkin 1993).

Other possible content biases include the acquisition of common human fears, such as the fear of spiders, snakes, heights, or the dark (Nesse 1990; Buss 2012), and biases toward transmitting cultural stereotypes (Kashima 2000), broad linguistic structure (Mesoudi and Whiten 2004). Physical and behavioral traits that are potentially cues of mate quality comprise an additional content bias (Symons 1979; Buss 1989; Jones et al. 2007; Little et al. 2008).

In nonhuman animals, adaptive specializations in social learning provide further evidence for content biases. Young male songbirds appear predisposed to acquire conspecific over heterospecific song (Marler and Peters 1989). Similarly, rhesus monkeys are prepared to acquire a fear of snakes from others, but not flowers (Mineka and Cook 1988), and Curio et al. (1978) found that observer blackbirds acquired a significantly higher level of fear when the stimulus paired with conspecific alarm cries was a stuffed predator, rather than a plastic bottle.

8.5 Random Copying

A theme of this chapter has been that social learning is unlikely to generally be random, and animals are expected to copy in a highly selective and discriminating manner. In spite of this general pattern, cultural evolution theory has established that there are a relatively narrow range of circumstances in which pure, unbiased social learning outcompetes both individual learning and conditional strategies, while also increasing fitness (Rendell et al. 2010). The conditions for this exist when individual learning is challenging (e.g., very costly in time) but there are a range of viable alternatives available to copy, any of which might produce a reasonably effective, if not globally optimal, solution. Interestingly, these conditions seem to fit well with some examples of human cultural evolution that are best described by the kind of drift dynamics expected under unbiased (or random) copying (Bentley et al. 2004).

A growing body of evidence suggests that random copying may be a feature of human cultural evolution. This should not surprise us; after all, the random copying of genetic material, known as random genetic drift, is acknowledged to be a major process underlying biological evolution. Models of genetic drift are sometimes described as 'the neutral model' because the variants are considered to be neutral with respect to the success of the individual. Neutral models work for some cases of animal social learning, such as birdsong learning; in species such as chaffinches, one performed song variant is as good as another, and random copying describes the observed patterns of change (Slater 1986; Slater et al. 1980). In recent years, cultural evolutionists have suggested that a range of phenomena, from the popularity of dog breeds to the decorative patterns on pottery, can be understood in these terms (Bentley et al. 2004).

Cavalli-Sforza and Feldman (1981) carried out the first theoretical analysis of the properties of random cultural drift, specifying how random copying will affect the distribution of cultural variants. For example, if the forenames given by parents to their babies were chosen through a random copying process, we would expect to see a large number of very uncommon names and a very small number of highly popular names. This distribution of data is known as a "power-law distribution," and it is exactly what is observed. Hahn and Bentley (2003) showed that the frequency distributions of baby names used in the United States in each decade of the twentieth century, for both males and females, obey a power law that is maintained for more than one hundred years; even though the population is growing, names are being introduced and lost every decade, and large changes in the popularity of specific names are common. Similar distributions with this power-law property characterize a range of other cultural phenomena, including the decorative motifs on pottery excavated from early German farming settlements (5300–4850 BC), the frequency of citation of patented inventions in the United States over a 40-year period, and the citation of scientists in academic journals (Bentley et al. 2004; Shennan and Wilkinson 2001; Simkin and Roychowdhury 2003).

One problem with the detection of "random copying" by testing for consistency with a power-law distribution is that it is not clear that this is necessarily a very sensitive test, and it may not be able to discriminate random copying from cases where there is a weak transmission bias (Mesoudi and Lycett 2009). Indeed, a recent analysis suggests that some features of the dog breed and baby name distributions are inconsistent with random copying (Acerbi et al. 2012). Of relevance here is an issue that we discuss in the final section of this chapter—the distinction between strategies as psychological rules or population-wide practices. It is entirely possible that people choose their babies' names according to a wide range of entirely idiosyncratic strategies, biases, and other criteria that, since they are inconsistent, do not sum up across individuals, and at the population level appear to be random copying. There may be circumstances where it is important to distinguish between individuals genuinely exhibiting random (unbiased) copying and individuals behaving inconsistently but according to the dictates of a wide range of different strategies. If strategies are best characterized, as we suggest, at the psychological level, then ultimately researchers will require tools that allow them to discriminate between alternative rules at this level. A further concern with the power-law approach is that such a test involves no null hypothesis, and therefore does not really find positive evidence for random copying, so much as the absence of evidence for some strong forms of cultural selection. Fortunately, there are methods for isolating specific social learning strategies now available, to which we turn in the following section.

8.6 Statistical Methods for Detecting Social Learning Strategies

Most of the work described here, to the extent that it has attempted to distinguish between alternative strategies, has done so by experimentally exposing subjects to various treatments in which they observe, for example, different classes or numbers of individuals choosing each available option. Treatments are chosen such that the relative response to each is predicted to differ across strategies. Observers are then tested for their own response and the strategy being used is inferred accordingly.

A more general approach is to fit models, representing different strategies, to the data and determine which strategy, or combination of strategies, explains the data best (perhaps by using AIC; see box 5.3). This has the advantage that it can be used to infer strategies when the experimenter does not have control over the types or numbers of demonstrations that the subjects observe, for example, in observational field studies, or in more naturalistic laboratory studies in which subjects can freely interact.

McElreath et al. (2008) have provided a framework that enables this to be done for a number of strategies, including frequency-dependent copying, a payoff-biased copying strategy, and hierarchical combinations of the two (see box 8.2 for mathematical details). McElreath et al.'s approach can be extended to consider any strategy that a researcher wishes to consider, so long as the probability that each option will be chosen on a given "trial" is a function of information the researcher has available (including cases in which the relevant variables are experimentally controlled). For example, a dominance bias could be fitted to the data if it includes how many times each subject has observed the options chosen by others of different ranks, or an estimate of this based on a sample. The approach also offers an alternative means by which to compare different developmental hypotheses of the relative rate of trait performance, such as Perry's (2009) study of alternative foraging techniques in capuchin monkeys (see section 7.1.6).

Box 8.2

McElreath et al.'s approach for estimating and comparing strategies

McElreath et al.'s (2008) approach is intended to be a general approach for inferring social learning strategies in cases where an experimenter does not have complete control over the social information received by subjects, as in freely interacting groups of individuals. They applied their method to experimental data in which human participants were in control of a virtual farm and had to make decisions as to which of two "crops" to plant each season. Each participant was informed of the "yield" resulting from each choice, and also had access to social information about the choices of others in their group, as well as their resulting yields.

For their data McElreath et al. considered five strategies: (i) individual learning, (ii) frequency-dependent social learning, (iii) compare means (a payoff-biased strategy), (iv) a hierarchical combination of ii and iii, and (iv) a hierarchical combination of i and iii. These strategies were formulated as probabilistic models allowing McElreath et al. to fit them to the data using maximum likelihood, and assess the relative fit of each using AIC (see box 5.3):

i. Individual (asocial) learning

The attraction score of a particular option i, in season $t+1$, is given as:

$$A_{i,t+1} = (1-\phi)A_{i,t} + \phi\pi_{i,t},$$

where $\pi_{i,t}$ is the payoff received for option i in season t (0 if it was not sampled), and $0 \le \phi \le 1$ is a parameter determining how much weight is given to more recent information. Individuals then choose option i with probability:

$$p(i\,|\,A_t,\Theta)_{t+1} = \frac{\exp(\lambda A_{i,t})}{\exp(\lambda A_{1,t}) + \exp(\lambda A_{2,t})},$$

where Θ is a vector of all parameters (which are selected to maximize likelihood), and $0 \le \lambda \le \infty$ is a parameter determining the extent to which choices are determined by the attraction scores rather than at random. When $\lambda = 0$, choices are completely random as $\lambda \to \infty$, they become more deterministic.

ii. Frequency dependent social learning

The frequency dependent social learning strategy allows for the fact that subjects may combine social information with their individual information:

$$p(i\,|\,A_t,\Theta)_{t+1} = (1-\gamma)\frac{\exp(\lambda A_{i,t})}{\exp(\lambda A_{1,t}) + \exp(\lambda A_{2,t})} + \gamma\frac{n_{i,t}^f}{n_{1,t}^f + n_{2,t}^f},$$

where $0 \le \gamma \le 1$ is the weight given to social information relative to individual information, $n_{i,t}$ is the number of individuals observed to have chosen option i in period t; and f determines how nonlinear frequency dependence is. When $f = 1$, copying is unbiased. When $f > 1$, more common options have exaggerated chances of being copied, resulting in positive frequency dependence. When $f < 1$, frequency dependence is negative, and more commonly observed options are less likely to be copied.

iii. Compare means

This strategy has a similar form to ii:

$$p(i\,|\,A_t,\Theta)_{t+1} = (1-\gamma)\frac{\exp(\lambda A_{i,t})}{\exp(\lambda A_{1,t}) + \exp(\lambda A_{2,t})} + \gamma\frac{\overline{\pi}_{i,t}^{f}}{\overline{\pi}_{1,t}^{f} + \overline{\pi}_{2,t}^{f}},$$

where $\overline{\pi}_{i,t}$ is the observed mean payoff for option i in season t. McElreath et al. (2008) fixed $f = 100$, ensuring the threshold behavior predicted by their theory, though in principle it could be fitted as a parameter.

iv. Hierarchical compare means/frequency-dependent social learning (HCMFD)

Here we have

$$p(i\,|\,A_t,\Theta)_{t+1} = (1-\gamma)\frac{\exp(\lambda A_{i,t})}{\exp(\lambda A_{1,t}) + \exp(\lambda A_{2,t})} + \gamma\left((1-Y)\frac{\overline{\pi}_{i,t}^{100}}{\overline{\pi}_{1,t}^{100} + \overline{\pi}_{2,t}^{100}} + Y\frac{n_{i,t}^{f}}{n_{1,t}^{f} + n_{2,t}^{f}}\right),$$

where Y determines the relative weight given to compare means and frequency-dependent information. However, the idea was that frequency-dependent social learning is used when copy means fails to provide a clear signal, so we have:

$$Y(\delta,\overline{\pi}_{1,t},\overline{\pi}_{2,t}) = \frac{2}{1 + \exp(\delta\,(\overline{\pi}_{1,t} - \overline{\pi}_{2,t})^2)} \equiv Y$$

such that less weight is given to compare means as the difference between observed means decreases, where δ is a parameter that determines how quickly reliance on payoffs decreases, as the difference in observed means increases.

The hierarchical compare means/individual learning (HCMINDIV) strategy was defined in a similar manner (see McElreath et al. 2008 for details).

The log-likelihood is determined in the usual manner:

$$\log(L) = \sum \log[p(D_j\,|\,A_{t-1},\Theta)_t],$$

where D_j gives the choice (data) for the jth decision of the dataset, and the summation is across all seasons and individuals. For each strategy the parameters are chosen to maximize the likelihood using optimization algorithms, such as the optim function in R (R Core Development Team 2011). The log-likelihood for the fitted strategy can then be used to calculate AIC or AIC_c (see box 5.3), and the Akaike weights calculated to quantify the relative support for each strategy. McElreath et al. found overwhelming support for the HCMFD strategy (Akaike weight ≈ 1) for their data, which fit with their theoretical predictions. For this model, f was estimated at approximately 2, indicating that the frequency-dependent component of the strategy was subject to positive-frequency dependence.

8.7 Meta-strategies, Best Strategies, and Hierarchical Control

There is much work to be done before researchers can be said to have a good understanding of social learning strategies. Nonetheless, the concept is proving to be a productive conceptual tool with which to explore how animals, including humans, learn. The fact that specific strategies can be delineated and subjected to both theoretical and empirical investigation, with the findings of each type of study informing the other, is a major plus associated with the approach. However, such research is revealing that the actual strategies used in nature may be more complex, and context dependent, than originally conceived.

For example, in the course of less than a decade's investigation into public information use in nine-spined sticklebacks in a single laboratory, we witness experimental evidence consistent with these fish engaging in no fewer than six separate social learning strategies. True, not *all* strategies were supported; for example, there are no data consistent with sticklebacks pursuing a random copying strategy, nor is there evidence found for *copy if dissatisfied* (Pike et al. 2010). Nonetheless, the diversity of supported strategies negates any overly simple notion of social learning strategy and it undermines a research agenda dedicated to working out *the* strategy used by an animal. Rather, it implies that animals may switch between strategies, or combine them in a complex decision tree, utilizing the available internal and external cues in a flexible and adaptive manner. Indeed, at this stage we cannot exclude the possibility that the apparent use of strategies is an artifact of a single, more complex decision-making process that employs multiple cues.

The study by T.J.H. Morgan et al. (2012) presents a similar picture for human social learning. Their experiments provide conditional or strong support for the use of nine social learning strategies predicted by the theoretical cultural evolution literature, including conformity, payoff-based copying, *copy-when-asocial-learning-is-costly* and *copy-when-uncertain*. Importantly, these various influences appear to operate simultaneously, and interact to produce behavior leading to effective decision making and higher payoffs. Wood et al. (2012) reach a similar conclusion. This renders the job of determining the context specificity of social learning even more challenging. It is not sufficient to work out which rule is being implemented; one must also work out the rules that specify which rule should be implemented.

We envisage three logically distinct possibilities for how strategy use may be functionally organized. First, we can conceive of evolved *meta*-strategies that dictate social learning strategy use in a context-specific manner, identifying the circumstances under which individuals switch between one strategy and another. This first possibility implies that under normal circumstances, each animal would base any given decision on a single strategy. While the aforementioned findings in humans and fishes imply that multiple strategies are used simultaneously, these analyses are conducted across a sample, rather than on an individual-by-individual basis; they leave open the possibility that individual strategy use may be comparatively simple. Nonetheless, the observed conditional use of strategies militates against this interpretation; indeed, T.J.H. Morgan et al. (2012) are able to rule out the possibility that the appearance of multiple strategy use is solely the result of a mixed population. For these reasons, we regard this explanation as the least likely of the three possibilities, at least in humans.

A second, more plausible, account is that strategy use is dictated by hierarchically organized decision trees with multiple input variables. The aforementioned stickleback work implies that this is how alternative strategies might be integrated. Sticklebacks rely on up-to-date and unambiguous personal information when it's available, but they use social information when they lack relevant personal information or where it is outdated or ambiguous (Coolen et al. 2003; Van Bergen et al. 2004). But what social information should they use? When sticklebacks have information about the payoff to demonstrators, they use it preferentially (Coolen et al. 2005), exhibiting public-information use, which subsequent study has shown operates through a proportional observation rule (Pike et al. 2010). Conversely, when information concerning demonstrator payoff is missing, the fish switch to reliance on conspecific numbers as an indirect cue indicating patch quality (Coolen et al. 2005), probably utilizing a conformist learning strategy (Pike and Laland 2010).

It is plausible that strategy use may be hierarchically organized (Laland 2004). There are good evolutionary reasons for anticipating this sort of organization of behavior (Dawkins 1976), and hierarchical control has been reported for a great deal of human and animal behavior (Byrne and Russon 1998). The findings of T.J.H. Morgan et al. (2012) are consistent with the idea of hierarchical control—for example, the decision as to whether or not to use social information appears to logically precede the decision whether to implement a conformity rule. However, even here, it is clear that the decision trees employed by individuals are likely to be more complex than originally envisioned (Laland 2004).

A third type of organization would involve individuals simultaneously utilizing all of the available information to make their decisions. From this last perspective, it may be better to think of "strategies" as factors or weighted biases that collectively influence reliance on social learning; this perspective is consistent with statistical models like multiple regression. It is also possible that some so-called strategies may be patterns that emerge as outcomes of this decision-making process. In principle, it would be possible to sort between these three models of how social learning is functionally organized in the brain using the statistical tools described in the previous section.

Theoretical models exploring the merits of reliance on alternative social learning strategies rarely find that a single strategy will dominate under all conditions. Rather, different conditions favor different strategies. Future research would benefit from paying greater attention to the contexts under which different rules prove effective. Unfortunately, analytical models are rarely able to compare more than a very small number of strategies simultaneously, and are obviously restricted to those that occur to the investigator. The social learning strategies tournament provided one vehicle for comparing the relative merits of a large number of strategies, and had the additional advantage that diverse strategies could be proposed by any of the hundreds of individuals that entered the competition. The winning strategy implemented a rule that can be paraphrased as "copy when doing so will bring new behavior into your repertoire with higher expected fitness than current behavior." Such a rule possesses some of the properties one might expect in a meta-strategy. Rendell et al. (2010) draw attention to the parallels between the tournament findings and a rule that emerged from a population genetic analysis by J. R. Kendal, Giraldeau, and Laland (2009). A rule formulated by J. R. Kendal, which is analogous to Tilman's (1982) R rule from ecology,[11] predicted that among competing social learning rules, the fittest rule will be the one that can persist with the lowest frequency of asocial learning. This led these authors to expect, over evolutionary time that is punctuated by the repeated influx of different social learning rules, a continuous reduction in the frequency of asocial learning would continue. If correct, this may help to explain our species' extraordinary reliance on culture.

Finally, we note that much of the literature on social learning strategies fails to distinguish between strategies as sets of psychological rules and strategies as population-level outcomes of natural selection or cultural transmission. Rather, the literature tended to assume that implementation of particular psychological rules would generate characteristic population-level signatures, the detection of which potentially allows strategy identification. Two sets of findings miligate against this position, and suggest that the assumption is likely to prove untenable. First, the realization that multiple strategies may be used by animals in identical contexts, and even at the same time, means that population-level patterns may not be easily decipherable as the signatures of underlying psychological rules. Second, we have seen discordance between psychological rules and population-level outcomes. For example, T.J.H. Morgan et al. (2012) find evidence for conformist social learning, but it rarely results in conformist behavioral outcomes, because other biases are operating. Conversely, we consider above the possibility that random copying at the population level may not be indicative of the employment of a random copying strategy at the individual level. For these reasons, we advocate that the term "social learning strategy" be restricted to the individual-rule, or psychological-rule, level,[12] which means that evidence for the use of a particular strategy requires confirmation that individuals implement particular copying rules.

---

[11] Tilman's (1982) R rule states that among competitors for a resource, the dominant competitor will be the species that can persist at the lowest resource level.

[12] The situation may be more complex still, because in principle there need not be a simple correspondence between the psychological processes underlying strategy usage and the behavioural outcomes of a single individual.

Chapter 9

Modeling Social Learning and Culture

9.1. Introduction

In chapter 2 we described how, since the 1970s, researchers from a variety of academic disciplines have been developing mathematical models to explore how aspects of culture change over time. In subsequent chapters we sketched statistical models of social learning of, for example, diffusion data. In this chapter we look more closely at the various ways in which social learning and culture have been formally modeled, focusing primarily on the methods rather than the scientific findings, but nonetheless using examples to illustrate the value of mathematical modeling. We begin by considering the merits of a theoretical approach in general, touching on ways in which challenging and rich concepts such as "culture" or "tradition" can be operationalized through modeling. We then go on to describe a number of modeling approaches to this topic.

9.1.1. Why model?

Researchers draw on mathematical modeling because it is a valuable aid to understanding complex phenomena. To be useful, modeling inevitably requires simplifying assumptions. For example, population biologists often make accurate predictions about field data despite assumptions such as random mixing, discrete generations, and no inbreeding (Haldane 1964). Similarly, physicists have made enormous progress by assuming that bodies can be treated as particles with no friction or air resistance. Much theoretical work that employs mathematical modeling has been conducted in the belief that reducing the analysis of the processes under consideration to a few relevant parameters, and formalizing their relationships to capture their essence, can greatly aid our understanding. A model allows researchers to investigate how the interactions between key parameters will play out, which is not always easy to do without the help of mathematics. Models inform us about the likely outcomes of complex interactions between processes, often revealing unanticipated equilibria or surprising dynamics. Models can also help make sense of challenging data sets by isolating key underlying processes. However, modeling can be very useful even in the absence of relevant data by providing researchers with access to processes that are inaccessible through empirical analysis. Researchers cannot, for example, conduct breeding experiments using human beings in order to test hypotheses about our evolution. They can, however, develop mathematical models of such processes, analyze them, and use the results to test the feasibility of their hypotheses; they can then reject unreasonable hypotheses.

We understand that not everybody is mathematically minded, and that some researchers are even skeptical of the value of a modeling approach, or worry that some theory is not sufficiently well-grounded in actual data. It is certainly true that if a model ignores vital processes or important parameters, then it may generate misleading answers to the questions being addressed. At the same time, it would be naïve to criticize a model for not incorporating every parameter that may play a role, however small. Substituting a complex model for a complex world does not aid our understanding. Conversely, when a simple model accurately predicts aspects of a complex world, we have gained some insight into what the important parameters may be. A mathematical model may be simple compared with the larger world, but it is rarely simple compared with the intuitive "models" we can construct in our heads. It is precisely because our minds cannot juggle many variables or processes together and work out exactly how they will influence each other that constructing a model to do this for us can aid our understanding. Unlike verbal theory, the formalization of framing processes in mathematical terms forces the researcher to be explicit about the assumptions and precise about the predictions of a theory, thus clarifying thinking and facilitating discussion of the relative merits of alternative hypotheses.

A model is only as good as its assumptions. Alternative assumptions generate alternative models, which differ in their predictions, and then the researcher can embark on model selection procedures to ascertain the best model for a given data set. To gain powerful predictability from a model, it is important that the assumptions be valid, and this usually means they must be grounded by relevant empirical data. Theory is often of greatest value where it is tightly integrated with empirical research. Modeling potentially yields both valuable insights and hypotheses that can feed back to generate predictions, which in turn can be subject to further experimentation. Progress comes from repeated bouts of empirical and theoretical research, each informing the other recursively. However, even in the absence of data, alternative models clarify the likely ramifications of different processes. Some models are built more as conceptual tools to establish the likely consequences of the operation of one or more processes, rather than to predict patterns in data sets.

The model-building exercise typically starts with a set of assumptions about the key processes to be explored, and the nature of their relations. These assumptions are then translated into the mathematical expressions that constitute the model. The operation of the model is then investigated, typically using a combination of analytical mathematical techniques and simulation, to determine relevant outcomes, such as the equilibrium states or patterns of change over time.

The following examples of the modeling of cultural transmission illustrate, using a variety of methods, how theoretical approaches can make an enormous difference to understanding complex cultural phenomena. For more on mathematical modeling in related fields, see Otto and Day (2007), McElreath and Boyd (2007), or Kokko (2007).

9.1.2. Operationalizing the culture concept

A leading nineteenth-century anthropologist famously characterized culture as "that complex whole which includes knowledge, belief, art, morals, custom, and any other capabilities and habits acquired by man as a member of society" (Tylor 1871, 1). While anthropology and the other social sciences have moved on considerably in the intervening period, something like Tylor's definition still captures what many of us think of as "culture." For theoretically minded researchers this is a problem, since amorphous definitions like Tylor's are not particularly conducive to scientific analysis. Human culture has proven a difficult concept to pin down. Not only is there little definitional consensus regarding "culture" within the social sciences, there appears to be little appetite for definition (Kroeber and Kluckhohn 1952; Keesing 1974; Durham 1991).

In this vacuum, cultural evolutionists, eager to explore how aspects of culture change over time using quantitative tools, have taken pragmatic steps toward operationalizing culture. For researchers like Luca Cavalli-Sforza, Marc Feldman, Robert Boyd, and Peter Richerson—pioneers of the mathematical modeling of human culture—"culture is information capable of affecting individuals' behaviour that they acquire from members of their species through teaching, imitation, and other forms of social transmission" (Richerson and Boyd 2005, 5). Such "information" includes knowledge, beliefs, values, and attitudes. The advantage of this stance is that a population's culture can be characterized as comprised of a series of "traits" or "characters," each consisting of discrete packages (e.g., the presence or absence of dairy farming), or simple distributions (i.e., willingness to cooperate with non-relatives), of socially learned information. This greatly facilitates exploration of the processes by which different variants of this information, or measures of its central tendency and variation, change in frequency in a population.

Cultural evolution is typically modeled as a broadly Darwinian process comprising the selective retention of favorable socially transmitted variants, as well as a variety of nonselective processes, such as drift, migration, and invention (Boyd and Richerson 1985; Cavalli-Sforza and Feldman 1981). However, cultural evolution models depart from those of biological evolution in significant ways to take account of important differences between these processes (e.g., transmission between nonrelatives). Rather than attempting to describe the entire culture of a society simultaneously, culture is broken down into specific traits, allowing their frequencies to be tracked mathematically. This allows, for example, different social learning rules, or patterns of cultural change, to be investigated in a straightforward manner.

This broad characterization does not capture some features of culture that many social anthropologists consider important, such as shared intentions and values, ethnic boundaries and moral codes, or social institutions. However, for cultural evolutionists, what such definitions lack in nuance they make up for in practicality, because they allow simple models to be formulated and tested. These models can then be, and indeed have been, applied to explore the evolution of socially constructed features, such as ethnic markers and shared norms in human populations (Boyd and Richerson 1985). A broad characterization of culture also allows cultural evolution to be formally investigated in other animals, as in models of birdsong learning (Lachlan and Slater 1999; Beltman et al. 2003, 2004).

9.1.3. Parallels between biological and cultural evolution

Cultural evolutionists have long noted similarities between biological and cultural evolution (for overviews, see Mesoudi et al. 2004, 2006; Mesoudi 2011). For example, many aspects of culture appear to be comprised of discrete heritable units, which are in many respects akin to genes; examples are words, concepts, or beliefs. What is critically required for the Darwinian process is the existence of variants that compete with each other. Human culture also frequently exhibits this characteristic. For example, there are 6800 languages spoken around the world (Grimes 2002) and over 10,000 different types of religious belief, 150 of which have more than a million followers (Barrett et al. 2001). Darwin (1871) himself argued that natural selection occurs among words in a language, with favored ones being retained and less popular ones lost. The "struggle" in cultural evolution involves not so much the effect of the character on the ability of individual animals to survive and reproduce (although this can affect the frequency of cultural traits in some cases), but rather the relative competitiveness of functionally equivalent solutions to specific problems; the solutions would be adopted by individuals, or passed on to others, through differential social learning. For example, archaeologists have tracked increasing frequencies of one artifact (e.g., types of arrowheads) and the corresponding decreasing frequencies of competing ones (O'Brien and Lyman 2000), while similar patterns are reported for bird song dialects (Catchpole and Slater 2008).

New ideas are introduced into cultural evolution through behavioral and cognitive *innovation* in a manner functionally equivalent to the way that new variants are introduced into biological evolution through mutation (Cavalli-Sforza and Feldman 1981; Plotkin 1994; Simonton 1999). Random copying is formally equivalent to random genetic drift, and can be modeled in a similar manner using neutral models (Cavalli-Sforza and Feldman 1981; Bentley et al. 2004). The splitting of conceptual lineages resembles speciation, while extinction is like lineage loss (e.g., the loss of a language). Once again, these processes can be modeled using tools similar to those used by biologists, including phylogenetic methods (Mace and Holden 2005; Pagel et al. 2007; Gray and Atkinson 2003).

More generally, the argument that culture exhibits a number of key Darwinian properties is well supported (Mesoudi et al. 2004, 2006; Mesoudi 2011). The significance of such parallels is that they justify the application of models and methods devised for the study of biological evolution to the study of cultural change. Cultural evolutionists argue that borrowing Darwinian concepts and methods, suitably adjusted to the structural peculiarities of human culture, is a sensible route to a reasonable theory of human culture and thus to an improved understanding of human behavior (Boyd and Richerson 1985). However, the two processes are obviously not identical, and biological methods and models are rarely applied to cultural phenomena without consideration of potential differences. While some anthropologists have been critical of this borrowing, such criticism is usually based on distortion and misunderstanding—for example, an overly simple notion of genetics is contrasted with a complex portrayal of culture (Bloch 2000; Ingold 2007). Critics (e.g., Bloch 2000; Gould 1991; Pinker 1997) commonly reject outright any evolutionary analysis of culture by appealing to putative differences that are frequently illusory or unfounded (Mesoudi et al. 2006). See Richerson et al. (2005) and Mesoudi et al. (2004, 2006) for robust defenses of the use of evolutionary models to study human culture.

In the sections below, we describe various approaches by which cultural evolution has been modeled. In each case we endeavor to give the reader some insight into the methods used, referring them to more detailed treatments where available.

9.2. Theoretical Approaches to Social Learning and Cultural Evolution

9.2.1 Population-genetic style models of cultural evolution

As described in chapter 8, social learning researchers and cultural evolutionists have described a number of learning rules, commonly known as "transmission biases" (Boyd and Richerson 1985) or "social learning strategies" (Laland 2004), that specify when and how an individual should copy and from whom they should learn (fig. 8.3). Many mathematical studies have focused on establishing the theoretical viability of a given strategy or a small number of strategies, and have explored the conditions under which each is expected to prosper (Boyd and Richerson 1985; Henrich and Boyd 1998; Enquist et al. 2007; Wakano and Aoki 2007; Nakahashi 2007; J. R. Kendal, Giraldeau, and Laland 2009). Typically, each type of social learning strategy requires its own model with which researchers can explore the full ramifications of the application of the decision rules contained in the strategy. Some of the key ideas are discussed below.

Given a choice between two alternative behavior patterns, individuals may be more likely to adopt one variant than another (Cavalli-Sforza and Feldman 1981).

Boyd and Richerson (1985) refer to this as *biased cultural transmission*, as opposed to the adoption of variants in proportion to their frequency, which is known as "random copying" or "unbiased cultural transmission" (see below). Various types of bias may exist (see table 9.1). Through *direct bias*, individuals choose one of two or more alternative traits to adopt according to some intrinsic quality of the trait (also known as a "content bias"). A direct bias might result from a genetic predisposition to favor certain types of information, similar to Lumsden and Wilson's (1981) "epigenetic rules," or to how evolutionary psychologists think about learning. Stanford University anthropologist Bill Durham (1991) has argued that the individual choices underpinning these cultural processes are guided, but not determined, by genetic predispositions, prior knowledge, and other experiential factors; this characterization is fairly representative of the views of the cultural evolution community.

The cultural traditions that an individual picks up will often depend upon who else in the population has adopted that tradition. In the case of a *frequency-dependent bias*, the commonness or rarity of a trait affects the probability of information transmission (fig. 8.3). We saw in chapter 8 that individuals can be disproportionately predisposed to adopt the behavior of the majority, and that this positive frequency-dependent bias may generate *conformity*. People may also use cues about one trait, such as wealth, to choose which individuals to observe to acquire information about another trait, such as fashions in clothing. This form of learning is called *indirect bias* by Boyd and Richerson (1985), and it can be regarded as one of several types of *model-based biases*, which also include copying according to the prestige, relatedness, or success of the demonstrator (here "model" is used as a synonym for our term "demonstrator") (see fig. 8.3).

The cultural traditions of a population may change over time if individuals alter the cultural information that they receive before passing it on. Boyd and Richerson (1985) discuss *guided variation*, which refers to a process by which individuals acquire information about any behavior from others and then modify the behavior based on their personal experience (fig. 8.3). Here cultural variation is guided by individual experience, which may allow behavioral traditions to evolve gradually toward the optimal behavior for that environment, as human behavioral ecologists envisage.

Social transmission can occur *vertically* (i.e., from parents to offspring), *obliquely* (from the older to the younger generation, as in learning from teachers or religious elders), or *horizontally* (within generation transmission, such as learning from friends or siblings) (Cavalli-Sforza and Feldman 1981). Of course, genetic inheritance is primarily vertical and hence, as social transmission frequently occurs through some combination of these modes of information transmission, cultural evolution may commonly exhibit quite different properties from biological evolution. Methods are available to model each of these modes of transmission, alone or in combination, in addition to other forms of cultural transmission (see Boyd and Richerson 1985; Cavalli-Sforza and Feldman 1981).

These models are usually formulated as systems of recursions that track the frequencies of cultural (and sometimes also genetic) variants in a population, often with fitness defined by the match between a behavioral phenotype and the environment. Systems range from those containing only two possible discrete behavioral variants, through multistate models, to traits that vary continuously along one or more dimensions, with evolutionarily stable strategy (ESS) and population genetic analyses applied to these models. Typically, dynamics are described at the population level, often making assumptions like well-mixed populations, random mating, discrete generation, and so forth, although methods also exist for relaxing these assumptions (Cavalli-Sforza and Feldman 1981; Boyd and Richerson 1985). In box 9.1, we illustrate the methods using models of unbiased, directly biased, and frequency-dependent biased cultural transmission.

9.2.2. Population-genetic style models of gene-culture coevolution

The notion that genes and culture might coevolve was first proposed over 30 years ago by pioneers of the field of "gene-culture coevolution," a branch of theoretical population genetics (Feldman and Cavalli-Sforza 1976; Boyd and Richerson 1985; Feldman and Laland 1996; Richerson and Boyd 2005). These researchers view genes and culture as two interacting forms of inheritance, with offspring acquiring both a genetic and a cultural legacy from their ancestors. Genetic propensities, expressed throughout development, influence what cultural organisms learn. Culturally transmitted information, expressed in behavior and artifacts, spreads through populations, modifying selection pressures that feed back onto populations. While social scientists rarely consider that an individual's genotype may affect which cultural traits they adopt, this should not be regarded as contentious. There are many examples, particularly related to agriculture, food production, and dietary habits, where such interactions have been documented (Laland et al. 2010; Richerson et al. 2010).

Gene-culture coevolutionary analyses typically build on conventional population genetic theory. In addition to tracking how allele or genotype frequencies change in response to evolutionary processes, such as selection and drift, the analyses incorporate cultural transmission, with cultural transmission modeled as described in the preceding section. A theoretical investigation of the evolution of conformist transmission, by Henrich and Boyd (1998), provides a useful illustrative example of the gene-culture modeling approach (see box 9.2). Gene-culture analyses explore how learned behavior coevolves with alleles that affect the expression and/or acquisition of the behavior, or whose fitness is affected by the cultural environment. The approach has been used to explore the adaptive advantages of reliance on learning and culture (Boyd and Richerson 1985; A. R. Rogers 1988; Feldman and Laland 1996; Enquist et al. 2007); to investigate the inheritance of behavioral and personality traits (Cavalli-Sforza and Feldman 1973; Laland et al. 1995a; Feldman and Otto 1997); and to examine specific topics in human evolution, such as language or cooperation (Aoki and Feldman 1987, 1989; Boyd et al. 2003, 2010).

Gene-culture coevolutionists view culture as a dynamic process that can shape the material world (Boyd and Richerson 1985; Feldman and Laland 1996;

Box 9.1

Modeling cultural evolution

Consider a cultural trait with two alternative variants, denoted c and d, acquired through social learning. Boyd and Richerson (1985) propose different models that track the spread of c in the population as a result of different learning strategies. Here the proportion of the population with c is denoted by p, (where $0 < p < 1$). Each individual in the population is exposed to three randomly selected cultural role models; thus, from probability theory (binomial distribution), the probability of having i role models with trait c, given p, is $M(i|p) = \binom{3}{i} p^i (1-p)^{3-i}$. To model cultural transmission with frequency-dependent bias, the strength of which is D, expressions for the probability that an individual acquires c when i role models have c are given in table 9.1.

Table 9.1. Probability that an individual acquires trait c given its frequency in the set of cultural role models

Number of role models with c	Probability that a focal individual acquires c
0	0
1	$1/3 - D/3$
2	$2/3 + D/3$
3	1

Unbiased transmission—that is, random copying—is the special case when $D = 0$. The entries in table 9.1 can be combined with expressions, derived from probability theory, for the probabilities that individuals will have 0 to 3 c-carrying role models; when summed, these give a recursion for the expected frequency of c in the population in the next generation: $p' = p + Dp(1-p)(2p-1)$. A direct learning bias can be modeled by assuming that some feature of trait c renders it inherently more likely to be copied. The variable B is the strength of this direct bias, and the recursion expression is $p' = p + Bp(1-p)$. These equations can be used to compare the fate of trait c over time under different transmission biases or strategies, and show that the different individual level learning strategies produce different outcomes at the population level (fig. 9.1).

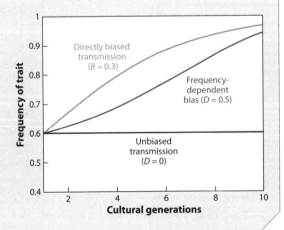

Figure 9.1. Individual-level transmission biases produce different outcomes at the population level. The figure shows the time course of trait c when different biases are operating. Based on figure 1 in Rendell, Fogarty, et al. (2011).

Richerson and Boyd 2005; Laland 2008). Their models have established that cultural processes can dramatically affect the rate of change of allele frequencies in response to selection, sometimes speeding it up and sometimes slowing it down. Recent estimates of the coefficient of selection associated with selected human genes exposed to culturally modified selection reveal unusually strong responses. The lactose-tolerance allele has spread from low to high frequencies in less than 9000 years since the inception of farming (Bersaglieri et al. 2004). Such observations have led to the argument that culturally derived selection pressures may frequently be stronger than noncultural ones, for two reasons. First, cultural processes occur by virtue of acquired knowledge carried in human brains that is often reliably transmitted between individuals. Although the constancy of selection pressures varies from trait to trait, there is evidence that culturally modified selective environments are capable of producing unusually strong natural selection that is highly consistent in directionality over time (Bersaglieri et al. 2004). Second, many genes are favored as a result of coevolutionary events triggered by phenotypic changes in other species, or in response to other gene-frequency changes in their genomes. When changes in one genetic trait drive changes in a second, the rate of response in the latter depends in part on the rate of change in the former, which, as a rule, is not fast. In comparison, new cultural practices typically spread more quickly than a genetic mutation, simply because cultural learning operates at faster rates than biological evolution (Feldman and Laland 1996). If a cultural practice modifies selection of human genes, the larger the number of individuals exhibiting the cultural trait, the greater is the intensity of selection on the gene. A rapid spread of the cultural practice leads very quickly to maximal intensity of selection on the advantageous genetic variant(s). The effect of these factors has been repeatedly demonstrated by gene-culture coevolutionary models, which consistently report more rapid responses to selection than conventional population genetic models (Boyd and Richerson 1985; Feldman and Laland 1996; Feldman and Cavalli-Sforza 1989; Laland 1994; Laland, Kumm, and Feldman 1995). This may help to explain the argument that culture has "ramped up" human evolution (Hawks et al. 2007).

However, equally important is the observation that cultural selection pressures may frequently arise and cease to exist faster than the time required for the fixation of the associated beneficial allele(s). In this case, culture may drive alleles only to intermediate frequencies (Coop et al. 2009). Another possibility is that cultural processes can buffer selection (Feldman and Laland 1996). For example, imagine a population exposed to changes in temperatures that under normal circumstances would engender bouts of selection for genes favored in hot or cold climates. However, if humans can put on or take off clothes, build fires, find caves and develop means of cooling, they effectively counteract the changing selection pressures. The temperature changes actually experienced by the population are dampened relative to the external environment, and as a consequence selection is weak. One prediction from this cultural mitigation of selection is that we now expect more (of what would otherwise be) deleterious alleles in the human gene pool than we would in the absence of cultural activities (Laland et al. 2010).

The evolution of conformist transmission: An example of gene-culture coevolution (adapted from J. Henrich and Boyd 1998)

The researchers J. Henrich and Boyd (1998) study the evolution of conformist transmission with a model that assumes environments vary in both time and space. The focal population is subdivided into a number of large subpopulations, each of which experiences one of two environmental states, labeled 1 and 2. Each individual also can acquire one of two behaviors, similarly labeled 1 and 2. Behavior 1 is favored (meaning it gives the individual a better chance of surviving) in environment 1 and behavior 2 is favored in environment 2. Thus, selection will favor learning mechanisms that make individuals more likely to acquire the favored behavior in the current environment.

The model of J. Henrich and Boyd's is divided into four stages: cultural transmission (which includes conformist transmission), individual learning, migration, and natural selection. First, individuals acquire their initial behavior by copying members of the previous generation. Genetic variation among individuals affects both the extent to which individuals rely on social learning and the degree of conformism in that social learning. The probability that an individual without any propensity for conformism (unbiased transmission) acquires a behavior is the same as the frequency of that behavior in the previous generation, while those with a conformist psychology have a higher probability of acquiring the most common behavior. During cultural transmission individuals experience both unbiased transmission and conformist transmission from three individuals drawn randomly from the subpopulation.

The structure of social learning is described by two parameters, L and Δ. L measures the relative reliance on social learning versus individual learning, while the strength of the conformist effect is measured by the parameter Δ, which ranges from 0 to 1. The authors assume a haploid genetic system with 400 alleles (at a single locus) that vary in both the degree of reliance on social learning and the strength of conformist transmission that they produce, with 20 alleles covering the range in each of these two dimensions. Initially, populations consist mostly of individuals possessing the genotype for exclusively asocial learning ($L = 0$) and no conformist effect ($\Delta = 0$), with a very small portion uniformly distributed over all the remaining genotypes.

A three (cultural) parent model is used by J. Henrich and Boyd, drawing on Boyd and Richerson (1985). Hence, following cultural transmission, the frequency of behavioral trait 1 and genotype jk in subpopulation i, u'_{1ijk}, is given by:

$$u'_{1ijk} = \underbrace{u_{1ijk}}_{\substack{\text{unbiased} \\ \text{transmission}}} + \underbrace{u_{1ijk}q_i(1-q_i)(2q_i-1)\Delta}_{\substack{\text{conformist} \\ \text{effect}}} ,$$

where q_i is the frequency of behavior 1 in subpopulation i, Δ is the strength of the conformist effect for allele type k, and u_{1ijk} is the frequency of the same phenogenotype in the previous time step (i.e., prior to cultural transmission). An equation similar to the above equation can be written for u_{2ijk}.

Next, all individuals try to learn the best behavior for the current environment (asocial learning). Each individual acquires information from the environment that allows them to infer which trait is currently adaptive. Because environmental information is imperfect, individual learning does not always produce the correct behavior for the current environment. The variables p_{1j} and p_{2j} represent the probability of acquiring behavior 1 or 2 through individual learning, respectively, and

$$u''_{1ijk} = u'_{1ijk} - \underbrace{u'_{1ijk}p_{2j}}_{\substack{\text{those that learned} \\ \text{behavior 2 individually}}} + \underbrace{u'_{2ijk}p_{1j}}_{\substack{\text{those that learned} \\ \text{behavior 1 individually}}}$$

After the cultural transmission and individual learning phases, each subpopulation experiences migration and selection. During the migration phase, individuals move between subpopulations, with the parameter m representing the proportion of each subpopulation that has moved to a new one. Immigrants with a heavy reliance on social learning (high L) and a strong conformist effect ($\Delta \approx 1$) can more rapidly learn the common behavior in their new environment through imitation. This is represented by the following:

$$u'''_{1ijk} = u''_{1ijk} - \underbrace{mu''_{1ijk}}_{\substack{\text{migration} \\ \text{out}}} + \underbrace{m\bar{u}''_{1jk}}_{\substack{\text{migration} \\ \text{in}}}$$

where,

$$\bar{u}''_{1jk} = \frac{\sum_i^n u''_{1ijk}}{n}$$

and $n =$ the number of subpopulations.

Finally, selection influences the process because individuals with the nonadaptive behaviors are less likely to survive than those with the adaptive behavior. During each generation, a subpopulation may experience a sudden shift in the environment from the current state to the alternative state, with fixed probability.

$$u_{1ijk} = \frac{u'''_{1ijk}(W + D)}{\bar{W}}$$

$$u_{2ijk} = \frac{u'''_{2ijk} W}{\bar{W}}$$

where

$$\bar{W} = (W + D)\sum_i \sum_j \sum_k u'''_{1ijk} + W \sum_i \sum_j \sum_k u'''_{2ijk}$$

and $W = 1$, the baseline fitness, and $D = 0.01$, the strength of selection.

The analysis of their model led J. Henrich and Boyd to the view that conformist transmission is favored under a very broad range of conditions, broader in fact than the range of conditions that favor a substantial reliance on (any form of) social learning. They demonstrate that conformist transmission can evolve even in temporally fluctuating environments. It is favored when the accuracy of environmental information is poor (leading to ineffective asocial learning) and when conformist social learning is fairly accurate. They show that natural selection favors increasing the reliance on social learning, and that there is no case in which a strong reliance on social learning evolves but conformist transmission does not. If J. Henrich and Boyd are right, the fact that humans are capable of sophisticated and effective forms of social learning implies that conformism also is likely to be an important component of human social learning.

Although Lumsden and Wilson's (1981) book *Genes, Mind, and Culture* was the first extended treatise on gene-culture coevolution, it had little lasting influence. Conversely, Cavalli-Sforza and Feldman's (1981) *Cultural Transmission and Evolution* and Boyd and Richerson's (1985) *Culture and the Evolutionary Process* laid the theoretical foundations for the field. Today, the study of gene-culture co-evolution brings together researchers from a variety of disciplines, including evolutionary biology, genetics, anthropology, and economics (Richerson and Boyd 2005; Laland and Brown 2011). The field has developed in a variety of ways. One class of models investigates the inheritance of behavioral and personality traits (Cavalli-Sforza and Feldman 1973; Otto et al. 1995). Other models explore the adaptive advantages of learning and culture (Boyd and Richerson 1985; Enquist et al. 2007; Feldman et al. 1996; A. R. Rogers 1988). The methods have also been applied to address specific cases in which there is an interaction between cultural knowledge and genetic variation that influences the prevalence of cultural knowledge. These include the evolution of language and other lateralized characters in the body and brain (e.g., handedness) (Aoki and Feldman 1987, 1989; Lachlan and Feldman 2003; Laland, Kumm, Van Horn, and Feldman 1995; Laland 2008); changes in the genetic sex ratio in the face of sex-biased parental investment (Kumm et al. 1994); the spread of agriculture (Aoki et al. 1996); the coevolution of hereditary deafness and sign language (Aoki and Feldman 1991); the emergence of incest taboos (Aoki and Feldman 1997); the effect of cultural niche construction on human evolution (Laland et al. 2001; Laland et al. 2010); and the evolution of human mating systems (Mesoudi and Laland 2007). In addition, an extensive gene-culture coevolutionary literature on the evolution of cooperation has developed (Boyd and Richerson 1985; Fehr and Fischbacher 2003; Gintis 2004).

9.2.3 Neutral models and random copying

Most cultural evolution theory assumes that individuals copy in a highly selective and discriminating manner. However, as outlined in chapter 8, a growing body of evidence suggests that random copying may also be a feature of human cultural evolution, under restricted circumstances. We described how, at the population level, random copying is expected to generate a "power-law distribution," but cautioned against inferring from this distribution that individuals are necessarily employing the psychological rule of copying at random. For example, Hahn and Bentley (2003) claimed, through conformity with the power-law distribution, that in spite of all of the thought and care that individual parents put into choosing their child's name, as a group, parents behave in a manner that is equivalent to the case in which they copy names at random. The reason for this is nothing more than the fact that common names are more likely to be observed and considered by parents than obscure names, and the likelihood that a name is chosen is roughly proportional to its frequency at the time. Models of random copying are sometimes described as the "neutral model" because the variants are considered to be neutral with respect to their probability of being reproduced.

The neutral model has also been used as a null hypothesis, against which to detect departures from random copying. For example, researchers have suggested that cultural drift can explain changes in the popularity of dog breeds in the United States over a 50-year period; these changes also conform to a power law distribution (Herzog et al. 2004). This match between theory and data suggests that the frequency of each particular breed fits with a pattern of random copying by owners. This finding conflicts with the widely held, but unsupported, view that a breed taking the "best in show" award at prominent dog shows witnesses a short-term increase in their popularity. However, the study did note one exception in the data that did not fit the random copying model. In the eight years after the release of the 1985 version of the Disney movie *101 Dalmatians*, new Dalmatian registrations showed a sixfold increase, a change that far exceeds typical fluctuations in breed popularity under the neutral model. It would seem that the exposure of would-be dog owners to Dalmatians through the movie far exceeded those that occurred through seeing actual Dalmatians, dramatically boosting the breed's sales, albeit temporarily. In practice, it may often be difficult to discriminate cases with random copying from those with weak transmission biases (Mesoudi and Lycett 2009). In both the baby name and dog breed examples, a recent analysis suggests that the observed patterns of change actually contradict the expectations of the neutral model in important respects (Acerbi et al. 2012).

As its name implies, the neutral model was devised to apply to neutral traits—that is, characters that have little impact on the survival or reproduction of their carrier. This sits naturally with applications of the theory to cases like the decorative motifs on pottery or the popularity of dog breeds, since it is easy to imagine such decisions are essentially arbitrary. It also works for some cases of animal social learning, such as birdsong learning, where one song variant is as good as another for species like chaffinches, and random copying describes the observed patterns of change (Catchpole and Slater 2008). However, it is more challenging to imagine that traits with a potential impact on Darwinian fitness may be copied at random. Yet that is exactly what Tanaka et al. (2009) proposed with respect to the copying of many treatments for diseases, such as complementary medicines.

Tanaka et al. noted that complementary medicines, traditional remedies, and home cures are used extensively worldwide, in spite of serious doubts about the efficacy and safety of many treatments. To make sense of this, they constructed stochastic models, which revealed that the most efficacious treatments are not necessarily those most likely to spread (see box 9.3). Ironically, superstitious remedies or even maladaptive practices sometimes spread well because their very ineffectiveness results in longer, more salient demonstrations, precisely because the demonstrators are sick for longer, leading to a larger number of converts, which sometimes compensates for greater rates of abandonment (see fig. 9.2). These authors argue that random copying fits the data well, because this explains why ineffective treatments are common; conversely, copying that is biased toward acquiring effective treatments does not explain the data, because the latter rule would not lead to ineffective treatments spreading. Making decisions about the efficacy of treatments adopted by others is challenging, because people frequently

Box 9.3

Complementary medicine and the spread of ineffective treatments for disease

Tanaka et al. (2010) develop mathematical models of the spread of self-medicative treatments for medical conditions in order to explore both the factors that lead to treatments becoming widespread and how a treatment's efficacy affects its rate of spread. They show the treatments that spread are not necessarily those most efficacious at curing the ailment, and explain how "superstitious treatments" with little efficacy and even maladaptive practices can spread under broad conditions.

Tanaka et al. derive expressions for the cultural fitness (mean number of converts to the treatment resulting directly from observation of a given demonstrator) Φ, and probability of spread, P(spread), of new treatments. Treatments are acquired through social learning, but their spread depends on a variety of factors, including the efficacy of the practice, and rates of conversion, death, recovery, and abandonment of treatment (see fig. 9.2). They propose a continuous-time branching process to model the cultural transmission of self-medicative practices, and track individuals who practice the treatment and therefore demonstrate the cultural trait.

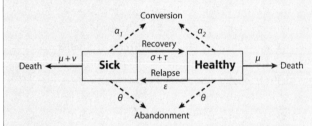

Figure 9.2. This figure illustrates the processes through which demonstrators of a treatment can change health state in Tanaka et al.'s (2009) model. Based on figure 1 in Tanaka et al. (2009). The symbols α_1 and α_2 give the rates at which sick and healthy individuals, respectively, are converted to using the treatment, while θ is the rate of abandonment of the treatment.

Each individual can be either ill or healthy. Ill demonstrators convert observers to using their treatment at rate α_1, and healthy demonstrators at rate α_2. If only ill individuals demonstrate the treatment, then $\alpha_2 = 0$. Judgments about the efficacy of the treatment are made by those who have adopted it, with individuals abandoning the treatment at a rate ρ, which was set as a decreasing function of the total rate of recovery ($\tau + \sigma$), where σ is the natural rate of recovery and τ is the rate of recovery due to the treatment. Tanaka et al. defined a superstition to be a trait for which $\tau \approx 0$, that is, the trait is largely ineffective. If $\tau < 0$, the treatment is maladaptive, hindering recovery. The time U spent by an individual being sick before recovery, death, or abandonment is exponentially distributed with rate parameter $\lambda = \mu + v + \theta + \tau + \sigma$, where μ is the baseline death rate and v is the additional rate due to disease.

The time W spent in health before dying or abandoning the treatment is exponentially distributed with the parameter $\zeta = \mu + \theta + \varepsilon$. The number of converts per demonstrator then follows a Poisson distribution conditional on the time spent in each state, with mean $\alpha_1 U + \alpha_2 W$. The behavior of the models is well illustrated by a general expression for the mean cultural fitness of the treatment (Φ), defined as the mean number of converts to the treatment resulting directly from observation of a given demonstrator, or

$$\Phi = \frac{\alpha_1}{\lambda} + \frac{\alpha_1 (\tau + \sigma)}{\lambda \zeta},$$

which is simply α_1/λ when only ill individuals demonstrate the treatment. In a generalized model, recovered individuals relapse to illness at rate ε, leading potentially to multiple episodes of illness, and

$$\Phi = \frac{1}{\psi}\frac{\alpha_1}{\lambda} + \left(\frac{1}{\psi} - 1 + \frac{\tau + \sigma}{\lambda}\right)\frac{\alpha_2}{\zeta}.$$

Across a broad range of conditions, the most effective treatments are not necessarily those most likely to spread, and superstitious treatments with no effectiveness, or even involving maladaptive practices, frequently have the highest cultural fitness. The spread of treatments depends critically on the rates of recovery from illness and abandonment of the treatment, with high recovery/low-abandonment favoring superstitious/maladaptive treatments, and low-recovery/high-abandonment favoring effective treatments (fig. 9.3). Factors that precipitate low abandonment, such as social norms favoring traditional remedies, or treatments that are costly to learn, potentially facilitate the spread of superstitions/maladaptive traits, particularly in chronic cases. The analysis can explain the ineffectiveness of many prominent complementary and traditional medicines.

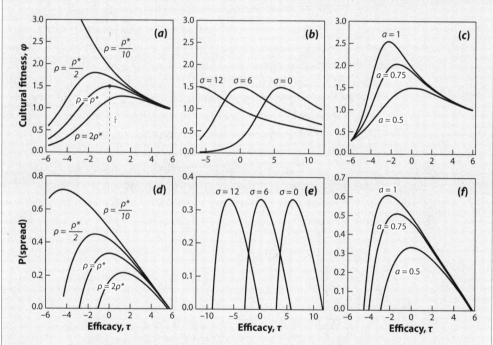

Figure 9.3. Tanaka et al.'s (2009) mathematical models revealed that the most efficacious treatments for disease are not necessarily those most likely to spread. Indeed, purely superstitious remedies, or even maladaptive practices, spread more readily than efficacious treatments under specified circumstances. Low-efficacy practices sometimes spread because their very ineffectiveness results in longer, more salient demonstration and a larger number of converts, which more than compensates for greater rates of abandonment. The figure shows effects for a single episode of illness with treatment demonstration only during illness. The cultural fitness (a–c) and probability of spread (d–f) of self-medicative treatments, plotted as a function of treatment efficacy τ, when there is a single episode of illness and demonstration occurs only during illness. Left (a and d), effect of varying maximum rate of abandonment ρ. Middle (b and e), effect of varying rate of recovery α. Right (c and f), effect of varying rate of decay in treatment abandonment. Based on figure 3 in Tanaka et al. (2009).

recover irrespective of treatment, are poor about making judgments about what led to recovery, and different people offer conflicting advice. According to Tanaka et al. (2009), in spite of our best efforts to find the cure to what ails us, we end up copying at random.

9.2.4 Social foraging theory

Another theoretical tradition that has investigated social learning is social foraging theory (Giraldeau and Caraco 2000), a subfield of behavioral ecology. Social foraging models characterize individuals as making decisions on the basis of economic information concerning the fitness consequences of behavioral alternatives, recognizing that individuals acquire this information through learning. While individuals can acquire useful knowledge through interacting directly with the environment, sometimes foragers utilize information generated through the foraging of other animals; this information might concern the richness of a food patch, or a method of food processing. Where this occurs, foraging models need to take account of social learning.

Giraldeau et al. (1994), motivated by observations of social foraging in pigeons, developed a mathematical model to explore how the costs and benefits of this learning are affected by scrounging. They concentrated on within-generation social learning of a trait that enhances resource production, assuming both frequency-dependent asocial learning (which decreases due to scrounging, because an animal that scrounges reduces its opportunity for learning through its own experiences) and frequency-dependent social learning (which increases with the number of demonstrators). The acquired trait results in an increased ability to find resource clumps relative to a baseline rate. Giraldeau et al. found that social learning increased the expected number of individuals foraging at the elevated rate relative to asocial learning, and with no social learning there was a significant fitness cost to group foraging. They hypothesized that the adaptive function of social learning may be to allow individuals to circumvent some of the inhibitory effects that scrounging has on asocial learning of a foraging skill, and thereby to learn to produce.

Arbilly et al. (2011) also introduced social learning into a producer-scrounger game, exploring how it affected decision making under risk. As food items with high nutritional value are rarer and are likely to be depleted first, searching for them may involve repeated failures even though on average they are optimal choices. For asocial learners, this can generate risk aversion. Arbilly et al. found that in an environment where the most productive resources occur with the lowest probability, socially acquired information is strongly favored over individual experience. The advantage of social learning in a risky environment stems from the fact that attending to the successes of others allows patch status to be updated. This leads to the best food source being revisited even when individuals have themselves experienced repeated failures, thereby circumventing risk aversion.

For an introduction to social foraging theory, see Giraldeau and Caraco (2000).

9.2.5 Spatially explicit models

Early models of the evolution of social learning commonly assumed a perfectly mixed population, meaning that each individual was equally likely to interact with any other. While this was a sensible place to start, and led to important insights, the spatial structure of most real populations should give pause to the assumption (Cavalli-Sforza and Feldman 1981). There are several reasons why a spatially explicit approach is often appropriate for social learning analyses. Studies of social behavior in other contexts have shown that spatial factors can profoundly affect evolutionary outcomes (Nowak and May 1992; Kerr et al. 2006; Silver and Di Paolo 2006). Some human cultural phenomena, such as agricultural practices, are physically grounded in space (Durham 1991). Aspects of spatial structure have also been usefully incorporated into simulation models of cultural evolution and gene-culture coevolution—for example, in explorations of the appearance of modern human behavior (Powell et al. 2009), and the evolution of lactase persistence (Gerbault et al. 2012). Moreover, social learning is widespread in animals, many of which are sedentary and/or territorial. In such cases, an analysis that recognizes individuals are often more likely to learn from their near neighbors is appropriate. More generally, by comparing well-mixed and spatially structured populations, analyses can characterize the extremities of a range of unstructured to structured populations. Learning in a structured population is a special case of bias in social learning, where nearby individuals are preferred as models to distant ones, and in this respect, spatially explicit analyses are more generally instructive with respect to the effects of bias (Kameda and Nakanishi 2002).

Rendell et al. (2010) show that spatial structure affects the evolution of social learning. As described in chapter 8, A. R. Rogers (1988) pointed out the "paradox" inherent in the observation that the fitness of social learners at equilibrium would be no greater than the average individual fitness in a population of asocial learners; this result seemingly conflicts with the observation that social learning underlies the effect of human culture on our ecological success and population growth. Rendell et al. (2010) introduced spatial structure into analysis of this problem, applying a cellular automata model. This meant that an individual's interactions were restricted to their immediate neighbors. They found that spatial structure further exacerbated Rogers' paradox, because pure social learners could actually continue to increase in numbers even while driving fitness below that of the original, purely asocial learner population. The key to this surprising result is the observation that the social learning genotype was buffered from the invasion of (globally) more fit asocial learners in spatial clusters of social learners (fig. 9.4). Here a "cluster" is defined as a contiguous group of individuals who have at least one immediate neighbor who shares their particular strategy. Social learners behaved parasitically, gleaning useful information from asocial learners, but suffering little or no cost for doing so. This free information, however, was recycled again and again, becoming outdated as the environment changed. As social learners clustered together, the individuals at the centre of the cluster, surrounded by other information parasites, quickly declined in fitness in a changing

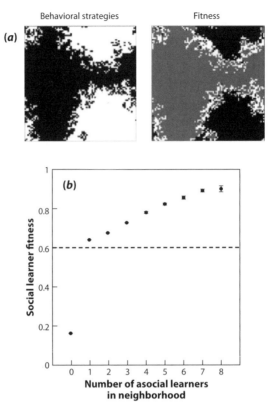

Behavioral strategies   Fitness

(a)

(b)

Social learner fitness

Number of asocial learners
in neighborhood

Figure 9.4. Rendell et al. (2010) explored the evolution of reliance on social and asocial learning using a spatially explicit stochastic model. They found that the spatial structure introduced a new paradox, which is that social learning can spread even when it decreases the average fitness of individuals below that of asocial learners. (*A*) Snapshot of a running simulation with local learning and dispersal and a spatially uniform environment, taken within a few iterations of an environmental state change. The *left panel* shows spatial distribution of learning strategies (black = asocial learner, white = social learner). The *right panel* shows the fitness of each individual; the lighter the cell, the higher the fitness value. Asocial learners have a fixed fitness (1 − *ca*), which appears as gray. Social learners in the border regions have the highest fitness (appearing white in the right panel) even though most of the social learners have much lower fitness (appearing black). (*B*) Mean fitness of social learners plotted against the number of asocial learners in their neighborhood over the same simulation (error bars show 95% CI). Dashed line shows fitness of asocial learners. Based on figure 2 in Rendell et al. (2010).

environment. However, as soon as one individual learner appeared inside the cluster, by mutation, the fitness of its social learning neighbors jumped above the population average for asocial learners, since they were immediately able to learn accurate information from it. This meant that asocial learners would struggle to reinvade once social learning was established, even though the global fitness of social learners was lower than that of asocial learners. Added to this were edge effects around the clusters themselves. The social learners at the edge of the clusters interacted with social learners inside the cluster and asocial learners outside, allowing them to ferry new, accurate information from asocial learners into the social learning cluster. The leading edge of the social learning clusters therefore had extremely high fitness, which allowed the spread of social learning despite its low fitness at the cluster centers (fig. 9.4). Rendell et al.'s analysis suggests that spatial structure increases the possibility that social learning can propagate maladaptive information.

9.2.6. Reaction-diffusion models

Several authors have used reaction-diffusion models to investigate cultural evolution. These include studies of the spread of farming (Aoki et al. 1996), language competition (e.g., Kandler and Steele [2008] and Pinasco and Romanelli [2006]), cultural hitchhiking (Ackland et al. 2007), and prestige bias (Ihara 2008). The models comprise reaction, diffusion, and competition components, which collectively are well suited to capture many aspects of the spread of an innovation in a finite population. Such models describe the spread dynamics of n competing (mutually exclusive) variants of a cultural trait within a population. Such variants might represent alternative beneficial subsistence techniques, technologies, religious beliefs, or languages. Using a continuous differential equation based approach, researchers can determine the temporal and spatial changes in the frequencies of the variants.

Cultural variant frequency change is determined by two main components—diffusion and growth. The diffusion component models the spread of a cultural variant from a specific location in space as a random walk, with density-dependent mixing, equivalent to the random spread of an innovation through direct contact between individuals. The propensities of variants to spread out in space vary and are described by the "diffusion coefficients." The reaction term describes the increase in frequency of each variant among naïve individuals (individuals that have not yet adopted the trait), according to a specific "growth" parameter, where a conventional logistic growth is often assumed; the reaction term also captures the effects of competition. Variants are assumed to differ in their growth propensities, depending on the benefits each conveys to its adopters.

Kandler and Laland (2009) applied reaction-diffusion models to investigate how both independent invention and the modification and refinement of established innovations affect cultural diversity. They found the introduction of new variants typically increased cultural diversity substantially in the short term; however, it could actually decrease diversity, depending on the frequency of existing variants and on how diversity was measured, if the new variant had high fitness and thereby drove out less fit variants. Significant rates of innovation were typically required for innovation to reliably increase diversity. They also found that if each variant was an improvement on the previous variants, innovation through modification of existing variants could generate oscillating trends in diversity. Provided there was sufficient average time between innovations for the most beneficial variants to reach high frequency, a cycling pattern of diversity emerged. Novel invention generally supported higher levels of cultural diversity than modification or refinement of existing variants. Moreover, while equilibrium levels of cultural diversity typically increased with innovation rate, this increase could be surprisingly modest, particularly when innovation occurred through refinements of earlier variants, or when conformist social learning was operating. For example, they found that a doubling of the innovation rate was typically associated with an increase in long-term cultural diversity of 6% to 32%.

9.2.7. Agent-based models

Another approach that has proved useful in the study of social learning is individual- or agent-based modeling. Researchers specify a population of agents that are programmed to behave according to prespecified individual-level rules. These rules specify whether and how the individuals will aggregate, for example, or whether, when, and how they will forage or engage in agonistic encounters. In spatial models, individuals are typically able to move around a two-dimensional grid, which can be set up to provide resources, such as food, distributed in a naturalistic manner.

Van der Post and Hogeweg (2008) used this approach to investigate the emergence of dietary traditions in animals. They showed that grouping behavior (approaching and following others) by itself is a sufficient social influence on individual learning for supporting the inheritance of diet traditions. A similar conclusion was reached by Franz and Matthews (2010), also applying an individual-level model; they determined that social enhancement processes can support both adaptive and maladaptive traditions. These studies imply that the traditions observed in many animals could result from very simple learning processes, such as local and stimulus enhancement.

Van der Post and Hogeweg also found that grouping was sufficient to generate "cumulative" group-level learning through which groups increased diet quality over the generations by gradually homing in on high-profitability foods over time. They found that whether "traditions" or "progressive change" dominates depends on foraging selectivity (that is, to what extent less preferred resources are consumed). A later paper by the same authors (Van der Post and Hogeweg 2009) concludes, like Rendell et al. (2010), that social learning is likely to be adaptive in a highly changing environment. This finding contrasts with analytical population-genetic models, which consistently predict that high rates of environmental change render social learning maladaptive because it leads to inheritance of outdated information (e.g., Boyd and Richerson 1985, 1988; Stephens 1991; Feldman et al. 1996). The main reason for this difference is that in the simulation models, outdated preferences are unlikely to be inherited. Other factors reinforce these effects, including that asocial learning is not assumed to automatically track environmental change. In the case of Van der Post and Hogeweg's (2009) study, cultural inheritance can only take place during collective foraging in shared contexts, and this cannot happen once resources have disappeared. Arguably, the greater realism of the agent-based models might suggest that their conclusions are the more likely to prove robust, at least for the specific circumstances being modeled.

9.2.8. Phylogenetic models

Another recently devised approach to modeling cultural processes and testing historical hypotheses has come from the application of phylogenetic methods to interpret aspects of human cultural variation. These methods, again borrowed from evolutionary biology, reconstruct the history of cultural traits by mapping diverse populations onto a tree of relatedness, generally based on slowly changing linguistic data; alternatively, the methods construct a cultural phylogeny based on the trait of interest. They provide a means for researchers to explore the macroevolutionary aspects of cultural change.

There are reasons for expecting phylogenetic methods, devised for biological evolution, to also work for cultural phenomena at least some of the time. Events responsible for genetic differentiation between populations are equally very likely to generate cultural diversification. For example, the physical separation of a single population into two populations is expected to reduce contact between them, which will contribute to both linguistic and genetic divergence. It is well known that geographical distance between populations is a good predictor of decreasing exchange of individuals and also of genetic diversity, and the same is true for cultural diversity (Cavalli-Sforza et al. 1992; Cavalli-Sforza and Wang 1986). However, phylogenetic methods typically assume vertical transmission, and culture may be transmitted horizontally to a greater extent than genes (Borgerhoff Mulder et al. 2006), at least in animals. For example, in Europe the replacement of Celtic languages by Latin in the territories of the Roman Empire, and of Latin by Anglo-Saxon in England, are well documented. Such "borrowing" across historical lineages potentially creates problems for this approach, which assumes that the data are inherited in a treelike manner.

Cultural evolutionists have asked whether cultural data exhibit treelike structures (irrespective of whether these map onto genetic trees). If this is the case, then phylogenetic methods used in evolutionary biology can legitimately be used to reconstruct the historical relations among cultural traits and test hypotheses concerning past demographic and dispersal events. Conversely, if cultures change primarily through mixing, borrowing, and horizontal transfer between lineages, then treelike structures will not be apparent, and other methods of analysis will be more appropriate.

Gray and Jordan (2000) conducted a pioneering study of language evolution, illustrating how phylogenetic methods applied to culture can be used to test hypotheses about human prehistory. They focused on the Austronesian languages spoken in Southeast Asia and the Pacific Islands. Previous researchers had documented sets of words, or lexical items, for each of the 77 languages spoken in this area of the world. Gray and Jordan applied established phylogenetic statistical techniques, regularly used within biology, to ask whether the languages fit a treelike structure, and to identify the best tree. The tests established that the language dataset had a significant phylogenetic signal to it; that is, it had the treelike property characteristic of biological evolution rather than the interwoven matrix that excessive mixing of language elements would generate. Gray and Jordan then used this tree to test two competing hypotheses for the colonization of the Pacific by Austronesian-speaking farmers from China and Taiwan. In one hypothesis, the farmers expanded rapidly, with little mixing with earlier hunter-gatherer settlers (the "express-train hypothesis"); in the other, the expansion was slower and with more intermixing (the "entangled bank model"). Gray and Jordan used archaeological data to predict the sequence of colonization from Taiwan to New Zealand and Hawaii, showing that it had occurred in an ordered geographical pattern. They then compared this pattern to the language tree. A good correspondence between the two would require few colonization steps to map the geographical evidence onto the language tree, whereas a poor fit would require many more steps. In other words, a rapid expansion of the population across the region will be expected to have produced a distribution of languages that mirrored the geographical pattern, and hence the archaeological data, whereas a slower expansion that included mixing of languages with those of native populations would have produced a more complex picture, with a less clear fit between the archaeological data and the language tree. The

Figure 9.5. Projectile points from the southeastern United States and placed on a phylogenetic tree illustrating their divergence from a single common ancestor. Redrawn from Michael J. O'Brien, John Darwent, R. Lee Lyman, "Cladistics Is Useful for Reconstruction Archaeological Phylogenies: Palaeoindian Points from the Southeastern United States," *Journal of Archaeological Science* 28: 1115–1136, copyright 2001, with permission from Elsevier.

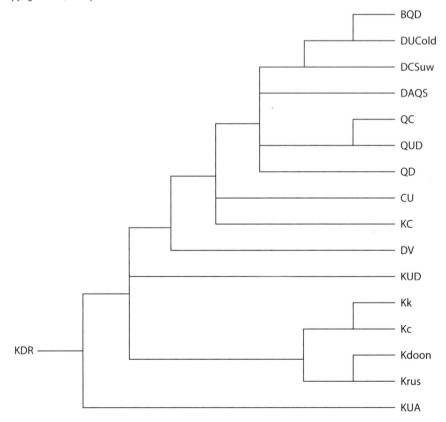

obtained fit was close, indicating that the express-train model fit the language tree exceptionally well and significantly better than would be expected by chance. The authors found no support for the entangled-bank model. Similar methods have since addressed a number of hypotheses about human prehistory, including the expansion of Indo-European languages with the spread of agriculture from Anatolia approximately 8000 to 9500 years BP (Gray and Atkinson 2003), and the evolution of languages in punctuational bursts (Atkinson et al. 2008).

In addition to proving a useful framework within which to integrate data from ethnography, archaeology, linguistics, and genetics (Gray and Jordan 2000), phylogenetic methods can also shed light on the mechanisms of cultural evolution. For example, Pagel et al. (2007) showed how the frequency with which words are used within modern languages predicts their rate of replacement. Frequently used words evolve at a slower rate, whereas infrequently used words evolve more rapidly. Speaking words a great deal appears to act like a purifying mechanism that prevents changes from occurring.

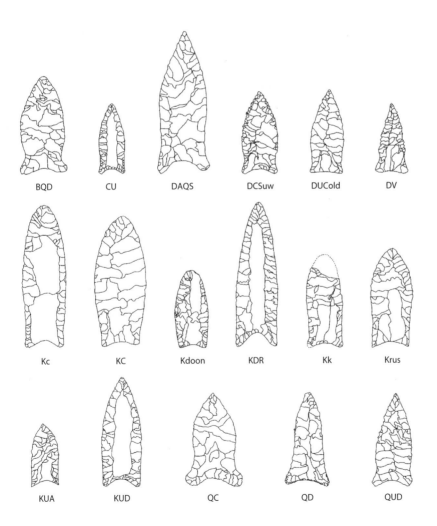

The historical relations among variants of cultural traits can be investigated by constructing a cultural tree of the trait itself. For example, O'Brien et al. (2001) and O'Brien and Lyman (2003) carried out a phylogenetic analysis of 621 Paleo-Indian projectile points from the southeastern United States (see fig. 9.5), illustrating their patterns of divergence from the ancestral projectile point. Tehrani and Collard (2002) used similar methods to reconstruct the history of Turkmen textile patterns. Interestingly, they found that these fabric designs are better characterized in a diverging, treelike manner rather than the converging pattern expected if there had been extensive borrowing across lineages. They draw on Durham's (1991) argument that the apparent prevalence of warfare, language barriers, within-group marriage, and ethnocentrism in human societies may act in a similar way to reproductive isolating mechanisms in biology, with both generating a branching pattern for cultural traits. In contrast, Jordan and Shennan (2003) found branching processes to be much less dominant in Californian basketry traditions.

Phylogenetic tools can also be used to evaluate the past dependency of one cultural trait on another, to explore whether they are best described by independent or dependent evolution. For example, Holden and Mace (2003) established that the acquisition of cattle by Bantu farmers led to a change from matrilineal to patrilineal inheritance of wealth. The same methods can be used to explore gene-culture coevolution, with the prediction that allele frequencies will covary with the cultural trait. Holden and Mace (1997) conducted such an analysis on the coevolution of dairy farming and the lactose tolerance allele in Africa, finding that dairy farming evolved first, creating the conditions under which the tolerance alleles spread. See Mace and Holden (2005) for a more detailed treatment.

9.3. Conclusions

It can be seen that a wide variety of mathematical tools can be deployed to investigate social learning, tradition, and culture. We emphasize that individual researchers can, and do, utilize several of these approaches, and need not adhere to a single methodology. Moreover, the above account only scratches the surface of this rich and multifaceted literature, and it would take a book-length treatment to review this body of theory comprehensively (e.g., McElreath and Boyd 2007). Our objective here has been more modest; we have sought only to illustrate the principal methods available to social learning researchers, rather than review them in detail, in the hope that it will encourage others to ultilize them. Even in this respect, our summary must be regarded as superficial, because there are a number of other mathematical tools, including information cascade models (Kübler and Weizsäcker 2004; Goeree et al. 2007), adaptive dynamics (Beltman et al. 2003), network models (e.g., Fowler and Christakis 2008), and economic models of how social norms affect preferences (e.g., Benhabib et al. 2011a, 2011b). All of these have been used to study various aspects of social learning. Nonetheless, our treatment is sufficient to make two important points. First, it is clear that mathematical modeling is an extremely useful tool to the social learning researcher, and that numerous fascinating and important insights have emerged from theoretical analyses. Second, mathematical modeling can feed back to direct and inspire further empirical research by, for example, providing testable hypotheses. There can be no doubt that a big part of the growth in social learning research can be attributed to the vibrant interplay of theoretical and empirical research in this field.

Chapter 10

Conclusions

The study of social learning is currently receiving unprecedented scientific attention, spread across the sciences, social sciences, and humanities. We set out to summarize and extend new developments in this burgeoning field, render the new tools more accessible and heighten awareness of this exciting area of research. Our primary goal has been to provide a practical guide to the methods for research into social learning, and we review empirical approaches for use in both the laboratory and the field, as well as the primary theoretical tools. In many areas, there is much room for further development of this methodology. In this final, brief chapter, we close by summarizing our conclusions from each of the preceding chapters, and suggest routes forward in the development and application of research methods.

In chapter 1, we laid the groundwork for the rest of the book by defining some key terms: social learning, social transmission, imitation, and innovation. Definitions have been contentious for all of these terms, and this is especially pertinent for "social learning," which has been used with varying degrees of generality. We advocate using this term in a very wide-ranging way: "learning that is facilitated by observation of, or interaction with, another individual or its products," which is a slight modification of a commonly used definition (Heyes 1994). This definition has the function of delineating the field, with different terms, such as "social transmission," being used for more specific phenomena.

In chapter 2, we provided a brief history of social learning research. It is important that researchers new to social learning consider the history of the field, particularly given its heterogeneous origins, because doing so will help them make sense of the current state of affairs. For example, it is only in the light of history that the eclectic terminology of social learning researchers can be understood.

We believe that inclusion of this brief history at the start of the book set the stage for us to take a more contemporary and forward-looking approach in the remaining chapters.

Chapter 3 covered methods for studying social learning in the laboratory, including the more traditional demonstrator-observer experiments, as well as recent transmission chain and diffusion studies, advances in the study of innovation, and experimental investigations of the biological bases of social learning. Arguably, this last area offers the most scope for development of new methodology, perhaps drawing on the methods already successfully used to elucidate the neural basis of learning in birdsong.

In chapter 4, we presented a number of mechanisms that have been proposed by the social learning community, and we discussed the various means by which they can result in social transmission. We also described how they can be distinguished experimentally. We conclude that in many cases it is very difficult to unambiguously assign a mechanism (or set of mechanisms) to a case of social transmission. This led us to propose a way of characterizing mechanisms of social transmission based on the information usually available to a researcher. One approach that shows much promise is to use statistical model fitting to identify the mechanisms of social learning in the field, where experimental control is limited (e.g., see box 4.2). Such studies have the advantage that they can inform researchers of how important different social learning mechanisms are in groups of animals freely interacting in their natural environment, thereby indicating what relationship each mechanism has to the emergence of traditions and culture (Hoppitt et al. 2012). Finally, we suggested that "teaching" might be viewed as another dimension to social learning, rather than as an alternative to social learning mechanisms in the observer. We hope that this perspective will allow further progress to be made in determining the conditions under which teaching evolves (e.g., Fogarty et al. 2011).

Recent debates about possible animal "cultures" and "traditions" have led to a call for methods that can detect social transmission in natural circumstances (R. L. Kendal et al. 2010), and these are considered in chapters 5 to 7. In chapter 5, we focused on statistical methods for diffusion data, where the researcher has data on the pattern of spread of a behavioral trait. We argue that the traditional method for analyzing such data, the "diffusion curve analysis," is flawed, and we focus on an alternative "network-based diffusion analysis" (NBDA) (Franz and Nunn 2009), which infers social transmission when the pattern of spread follows a social network. Other analyses have used the spatial spread of a trait to infer social transmission; again, this is a key area for development of new methods.

In chapter 6, we turn to repertoire-based data, where the researcher has a snapshot of the behavioral traits in different individuals or groups' repertoires. Recent debate in this area has concerned the usefulness of the "method of exclusion," where a role for social transmission is inferred when a pattern consistent with social transmission (such as within-group homogeneity) cannot be explained by alternative genetic and ecological hypotheses. Some progress can be made in assessing the method of exclusion—for example, by distinguishing between "basic" and "advanced" versions. Previous work assessing the genetic and ecological hypotheses can then be seen as either refuting an argument by the basic method of exclusion, or supporting an argument by the advanced method of exclusion, depending on the null hypothesis adopted.

While much progress has been made on the "advanced" method of exclusion, we believe that such work would benefit from being integrated into a more general model-fitting approach, where the focus is on quantifying the contributions of social transmission, genetics, and ecology in shaping behavioral repertoires. We discuss a matrix regression method suggested by Whitehead (2009), but future work might incorporate more specific models of genetic inheritance and ecological influence, such as those considered by Krützen et al. (2005) for sponging in bottlenose dolphins (see also box 6.5). Finally, we suggested that repertoire-based data might be addressed by an even broader causal modeling framework, and show how the previously used approaches fit into this framework.

In chapter 7, we discuss developmental approaches, broadly defined as any approach that aims to elucidate the role of social influences in the development of a behavioral trait. We discuss methods for utilizing observational data on the developmental process, focusing on case studies, but suggest some extensions and refinements of the methods employed. We then discussed experimental manipulations that can be made in a field context (Reader and Biro 2010), including diffusion studies, manipulations of social experience, and translocation experiments. In the future, field studies might be combined with statistical techniques such as those described in chapter 5, box 4.2, and box 8.2 to extract further information about the pattern of spread of a trait, and the mechanisms and strategies being used.

Chapter 8 described how game-theory, population-genetic, and simulation models have all led to the expectation that animals, including humans, ought to be selective with respect to the circumstances under which they rely on social learning and the individuals from whom they learn. Natural selection ought to have favored specific adaptive copying rules, called "social learning strategies," and we presented some of the theoretical and empirical support for the most prominent of these. The fact that specific strategies can be clearly delineated and subject to both theoretical and empirical investigation is a major plus associated with the strategies approach, and partly explains the growth of interest in social learning. However, such research is revealing that the actual strategies deployed in nature may be more complex, and context dependent, than originally conceived, with evidence for multiple strategies being used simultaneously. Future work will shed light on how the various social and asocial sources of information are combined in effective decision making.

Chapter 9 reviewed some of the mathematical tools that can be applied to investigate social learning, tradition, and culture. Historically, the field has been dominated by cultural evolution, and gene-culture coevolutionary and neutral models in the population genetics tradition, largely through the pioneering work of Cavalli-Sforza and Feldman (1981) and Boyd and Richerson (1985). Here too, however, recent years have witnessed a diversification of methodology, with new bodies of theory emerging from social foraging theory and phylogenetics, and from the application of other tools to social learning, such as agent-based modeling. The benefits of a strong theoretical foundation to social learning research go far beyond the novel insights that are gained from such analyses. A key emphasis throughout this book has been that mathematical and statistical modeling is at its most powerful when tightly integrated with empirical research. In this way, formal theory can feed back to direct and inspire further empirical research, and to provide testable hypotheses and clarify arguments; at the same time, empirical findings can be used to inform the model-building exercise, and ensure that the insights gained are well grounded and plausible. We appreciate that this theoretical emphasis does not always make our book easy reading, nor does it necessarily make the job of the social learning researcher any easier. However, we hope to have convinced the reader that it is worth persevering with these theoretical tools, because of the powerful insights they can bring.

This is an exciting time for social learning researchers, with the recent rush of new people, methods, and theory into the field. Hopefully, this book will have proven useful as a resource for getting to grips with the broad and often confusing methodology, terminology, and theory, not only for those readers just starting out, but also for those long involved who wish to retrain and upgrade their own toolboxes, as well as for others with more peripheral interests. We hope that we have made some useful contributions to the future direction of research in the field.

References

Acerbi, A., S. Ghirlanda, and E, Magnus. 2012. The logic of fashion cycles. *PLoS ONE* 7: e32541, doi:10.1371/journal.pone.0032541.

Ackland, G. J., M. Signitzer, K. Stratford, and M. H. Cohen. 2007. Cultural hitchhiking on the wave of advance of beneficial technologies. *Proc. Natl. Acad. Sci. USA* 104: 8714–8719.

Acosta-Calderon, A., and H. Hu. 2004. Robot imitation: A matter of body representation. *Proceedings of the International Symposium on Robotics and Automation, Querétaro, Mexico*, 25–27 August, pp. 137–144..

Adams, J. 1984. Learning of movement sequences. *Psychol. Bull.* 96: 3–28.

Akins, C. K., and T. R. Zentall. 1996. Imitative learning in male Japanese quail using the two-action method. *J. Comp. Psychol.* 110: 316–320.

———. 1998. Imitation in Japanese quail: The role of demonstrator reinforcement. *Psychon. Bull. Rev.* 5: 694–697.

Akins, C. K., N. Levens, and H. Bakondy. 2002. The role of static features of males in the mate choice behaviour of female Japanese quail (*Cortunix japonica*). *J. Exp. Psychol. Anim. Behav. Process* 58: 97–103.

Allport, G. W., and L. Postman. 1947. *The Psychology of Rumor*. New York: Henry Holt.

Amaral, D. G. 2003. The amygdala, social behavior, and danger detection. *Ann. NY Acad. Sci.* 1000: 337–347.

Ammerman, A. J., and L. L. Cavalli-Sforza. 1984. *The Neolithic Transition and the Genetics of Populations in Europe*. Princeton, NJ: Princeton University Press.

Anisfeld, M. 1991. Neonatal imitation: Review. *Dev. Rev.* 11: 60–97.

———. 1996. Only tongue protrusion modeling is matched by neonates. *Dev. Rev.*16: 149–161.

Aoki, K. 1987. Gene-culture waves of advance. *J. Math. Biol.* 25: 453–464.

Aoki, K., and M. W. Feldman. 1987. Toward a theory for the evolution of cultural communication: Coevolution of signal transmission and reception. *Proc. Natl. Acad. Sci. USA* 84: 7164–8716.

———. 1989. Pleiotropy and pre-adaptation in the evolution of human language capacity. *Theor. Popul. Biol.* 35: 181–194.

———. 1991. Recessive hereditary deafness, assortative mating, and persistence of a sign language. *Theor. Popul. Biol.* 39: 358–372.

———. 1997. A gene-culture coevolutionary model for brother-sister mating. *Proc. Natl. Acad. Sci. USA* 94: 13046–13050.

Aoki, K., M. Shida, and N. Shigesada. 1996. Travelling wave solutions for the spread of farmers into a region occupied by hunter-gatherers. *Theor. Popul. Biol.* 50: 1–17.

Apesteguia, J., S. Huck, and J. Oechssler. 2003. Imitation: Theory and experimental evidence. *CESifo Working Paper Series* No. 1049.

———. 2005. Imitation: Theory and experimental evidence. Unpublished manuscript, accessed April 2005. University of Bonn.

Arbilly, M., U. Motro, M. W. Feldman, and A. Lotem. 2011. Evolution of social learning when high expected payoffs are associated with high risk of failure. *J. R. Soc. Interface* 7: 1604–1615.

Asch, S. E. 1955. Opinions and social pressure. *Sci. Am.* 193: 31–35.

Atkinson, Q. D., A. Meade, C. Venditti, S. J. Greenhill, and M. Pagel. 2008. Language evolves in punctuational bursts. *Science* 319: 588.

Atran, S. 2002. *In Gods We Trust: The Evolutionary Landscape of Religion*. New York: Oxford University Press.

Atton, N., W. Hoppitt, M. Webster, B. Galef, and K. N. Laland. 2012. Information flow through threespine stickleback networks without social transmission. *Proc. R Soc. Lond. B.* DOI: 10.1098/rspb.2012.1462.

Aunger, R., ed. 2000. *Darwinizing Culture: The Status of Memetics as a Science*. Oxford, U.K.: Oxford University Press.

Bagehot, W. 1873. *Lombard Street: A Description of the Money Market*. London: Henry S. King.

Baldwin, J. M. 1896. A new factor in evolution. *Am. Natl.* 30: 441–451.

Balleine, B., and A. Dickinson. 1998. The role of incentive learning in instrumental outcome revaluation by sensory-specific satiety. *Anim. Learn. Behav.* 2: 46–59.

Bandura, A. 1977. *Social Learning Theory*. Englewood Cliffs, NJ: Prentice Hall.

———. 1986. *Social Foundations of Thought and Action: A Social Cognitive Theory*. Englewood Cliffs, NJ: Prentice Hall.

Bangerter, A. 2000. Transformation between scientific and social representations of conception: The method of serial reproduction. *Br. J. Soc. Psychol.* 39: 521–535.

Bangerter, A., and C. Heath. 2004. The Mozart effect: Tracking the evolution of a scientific legend. *Br. J. Soc. Psychol.* 43: 605–623.

Barnard, C. J., and R. M. Sibley. 1981. Producers and scrounders: A general model and its application to captive flocks of house sparrows. *Anim. Behav.* 29: 543–550.

Baron, R. S., J. A. Vandello, and B. Brunsman. 1996. The forgotten variable in conformity research: Impact of task importance on social influence. *J. Pers. Soc. Psychol.* 71: 915–927.

Barrett, J. L., and M. A. Nyhof. 2001. Spreading non-natural concepts: The role of intuitive conceptual structures in memory and transmission of cultural materials. *J. Cognit. Cul.* 1: 69–100.

Barrett, L., R. Dunbar, and J. Lycett. 2001. *Human Evolutionary Psychology*. London: Macmillan.

Barta, Z., and L.-A. Giraldeau. 1998. The effect of dominance hierarchy on the use of alternative foraging tactics: A phenotype-limited producing-scrounging game. *Behav. Ecol. Sociobiol.* 42: 217–223.

Bartlett, F. C. 1932. *Remembering.* Oxford, U.K.: Macmillan.

Barton, R. A. 2006. Primate brain evolution: Integrating comparative, neurophysiological, and ethological data. *Evol. Anthropol.* 15: 224–236.

Basalla, G. 1988. *The Evolution of Technology.* Cambridge: Cambridge University Press.

Bateson, P., and P. Gluckman. 2011. *Plasticity, Robustness, Development and Evolution.* Cambridge, U.K.: Cambridge University Press.

Bateson, P., and P. Martin. 2000. *Design for a Life: How Behavior and Personality Develop.* New York: Simon and Schuster.

Beauchamp, G., and A. Kacelnik. 1991. Effects of the knowledge of partners on learning rates in zebra finches (*Taeniopygia guttata*). *Anim. Behav.* 41: 247–254.

Beaumont, M. 2010. Approximate Bayesian computation in evolution and ecology. *Annu. Rev. Ecol. Evol. System.* 41: 379–406.

Beaumont, M., W. Zhang, and D. Balding. 2002. Approximate Bayesian computation in population genetics. *Genetics* 162: 2025–2035.

Beck, B. B. 1974. Baboons, chimpanzees and tools. *J. Hum. Evol.* 3: 509–516.

Beck, M., and B. Galef. 1989. Social influences on the selection of a protein-sufficient diet by Norway rats (*Rattus norvegicus*). *J. Comp. Psychol.* 103: 132–139.

Benhabib, J., A. Bisin, and M. Mackson, eds. 2011a. *Handbook of Social Economics, Volume 1A.* Amsterdam: North-Holland.

———, eds. 2011b. *Handbook of Social Economics, Volume 1B.* Amsterdam: North-Holland.

Behrens, T., L. Hunt, M. Woolrich, and M. Rushworth. 2008. Associative learning of social value. *Nature* 456: 245–249.

Beltman, J. B., P, Haccou, and C, ten Cate. 2003. The impact of learning foster species' song on the evolution of specialist avian brood parasitism. *Behav. Ecol.* 14: 917–923.

———. 2004. Learning and colonization of new niches: A first step toward speciation. *Evolution* 58: 35–46.

Bentley, R. A., M. W. Hahn, and S. J. Shennan. 2004. Random drift and culture change. *Proc. R. Soc. B* 271: 1443–1450.

Berger, S. M. 1962. Conditioning through vicarious instigation. *Psychol. Rev.* 69: 450–466.

Bersaglieri, T., P. C. Sabeti, N. Patterson, et al. 2004. Genetic signatures of strong recent positive selection at the lactase gene. *Am. J. Hum. Genet.* 74: 1111–1120.

Bettinger, R. L., and J. Eerkens. 1999. Point typologies, cultural transmission, and the spread of bow-and-arrow technology in the prehistoric Great Basin. *Am. Antiq.* 64: 231–242.

Biele, G., J. Rieskamp, and R. Gonzalez. 2009. Computational models for the combination of advice and individual learning. *Cognit. Sci.* 33: 206–242.

Bikhchandani, S., D. Hirshleifer, and I. Welch. 1992. A theory of fads, fashion, custom, and cultural change as informational cascades. *J. Pol. Econ.* 100: 992–1026.

———. 1998. Learning from the behavior of others: Conformity, fads, and informational cascades. *J. Econ. Perspect.* 12:151–170.

Biondi, L. M., G. O. Garcia, M. S. Bo, et al. 2010. Social learning in the Caracara Chimango, *Milvago chimango* (*Falconiformes*): An age comparison. *Ethology* 116: 722–735.

Biro, D., N. Inoue-Nakamura, R. Tanooka, G. Yamakoshi, C. Sousa, et al. 2003. Cultural innovation and transmission of tool use in wild chimpanzees: Evidence from field experiments. *Anim. Cognit.* 6: 213–223.

Biro, D., C. Sousa, and T. Matsuzawa. 2006. Ontogeny and cultural propagation of tool use by wild chimpanzees at Bossou, Guinea: Case studies in nut cracking and leaf folding. In *Cognitive Development in Chimpanzees*, edited by T. Matsuzawa, M. Tomonaga, and M. Tanaka, 476–508. Tokyo: Springer.

Blackmore, S. 1999. *The Meme Machine*. Oxford, U.K.: Oxford University Press.

Bloch, M. 2000. A well-disposed social anthropologist's problems with memes. In *Darwinizing Culture: The Status of Memetics as a Science*, edited by R. Aunger, 189–203. Oxford, U.K.: Oxford University Press.

Boakes, R. 1984. *From Darwin to Behaviourism: Psychology and the Minds of Animals*. Cambridge, U.K.: Cambridge University Press.

Boesch, C. 1993. Toward a new image of culture in wild chimpanzees? *Behav. Brain Sci.* 16: 514.

Bolhuis, J., and M. Gahr. 2006. Neural mechanisms of birdsong memory. *Nat. Rev.* 7: 347–357.

Bond, R. 2005. Group size and conformity. *Group Process Interg.* 8: 331–354.

Bond, R, and P. B. Smith. 1996. Culture and conformity: A meta-analysis of studies using Asch's (1952b, 1956) line judgment task. *Psychol. Bull.* 119:111–137.

Bonner, J. T. 1980. *The Evolution of Culture in Animals*. Princeton, NJ: Princeton University Press.

Boogert, N. J., S. M. Reader, and K. N. Laland. 2006. The relationship between social rank, neophobia and individual learning ability in starlings (*Sturnus vulgaris*). *Anim. Behav.* 72: 1229–1239.

Boogert, N., S. M. Reader, W.J.E. Hoppitt, and K. N. Laland. 2008. The origin and spread of innovations in starlings. *Anim. Behav.* 75: 1509–1518.

Borgerhoff Mulder, M., C. L. Nunn, and M. C. Towner. 2006. Macroevolutionary studies of cultural trait transmission. *Evol. Anthro.* 15: 52–64.

Box, H. O. 1984. *Primate Behaviour and Social Ecology*. London: Chapman and Hall.

Boyd, R., and P. J. Richerson. 1985. *Culture and the Evolutionary Process*. Chicago: University of Chicago Press.

———. 1988. An evolutionary model of social learning: The effects of spatial and temporal variation. In *Social Learning: Psychological and Biological Perspectives*, edited by T. R. Zentall and B. G. Galef, Jr., 29–48. Hillsdale, NJ: Lawrence Erlbaum Assoc.

———. 1995. Why does culture increase human adaptability? *Ethol. Sociobiol.* 16:125–143.

Boyd, R., H. Gintis, and S. Bowles. 2010. Coordinated punishment of defectors sustains cooperation and can proliferate when rare. *Science* 328: 617.

Boyd, R., H. Gintis, S. Bowles.,and P. Richerson. 2003. The evolution of altruistic punishment. *Proc. Nat. Acad. Sci USA* 100: 3531–3535.

Boyer, P. 1994. *The Naturalness of Religious Ideas: A Cognitive Theory of Religion*. Berkeley, CA: University of California Press.

Brass, M., and C. M. Heyes. 2005. Imitation: Is cognitive neuroscience solving the correspondence problem? *Trends Cogn. Sci.* 9: 489–495.

Brown, G. E., and E. P. Chivers. 2006. Learning about danger: Chemical alarm cues and the assessment of predation by fishes. In *Fish Cognition and Behavior*, edited by C. Brown, K. Laland, and J. Krause, 49–69. Oxford, U.K.: Blackwell.

Brown, C., and K. N. Laland. 2002. Social learning of a novel avoidance task in the guppy: Conformity and social release. *Anim. Behav.* 64: 41–47.

Buccino, G., S. Vogt, A. Ritzl, G. R. Fink, K. Zilles, et al. 2004. Neural circuits underlying imitation of hand actions: An event related fMRI study. *Neuron* 42: 323–334.

Buchsbaum, D., A. Gopnik, T. L. Griffiths, and P. Shafto. 2011. Children's imitation of causal action sequences is influenced by statistical and pedagogical evidence. *Cognition* 120: 331–340.

Bugnyar, T., and L. Huber. 1997. Push or pull: An experimental study on imitation in marmosets. *Anim. Behav.* 54: 817–831.

Burke, C., P. Tobler, M. Baddeley, and W. Schultz. 2010. Neural mechanisms of observational learning. *Proc. Natl. Acad. Sci. USA* 107: 14431–14436.

Burnham, K. P., and D. R. Anderson. 1998. *Model Selection and Inference.* New York: Springer-Verlag.

———. 2002. *Model Selection and Multimodel Inference: A Practical Information-Theoretical Approach.* 2nd ed. New York: Springer-Verlag.

Buss, D. M. 1989. Sex differences in human mate preferences: Evolutionary hypotheses tested in 37 cultures. *Behav. Brain Sci.* 12: 1–49.

———. 2008. *Evolutionary Psychology: The New Science of the Mind.* 3rd ed. New York: Pearson.

———. 2012. *Evolutionary Psychology: The New Science of the Mind.* 4th ed. New York: Pearson.

Butcher, S. H. 1922. *The Poetics of Aristotle.* 4th ed. London, Macmillan.

Butts C. T. 2010. sna: Tools for Social Network Analysis. R package version 2.2-0. Accessed Oct. 25, 2012, http://CRAN.R-project.org/package=sna/.

Byrne, R. W. 1994. The evolution of intelligence. In *Behaviour and Evolution*, edited by P.J.B. Slater and T. R. Halliday, 223–265. Cambridge, U.K.: Cambridge University Press.

———. 1995. *The Thinking Ape.* Oxford, U.K.: Oxford University Press.

———. 1999. Imitation without intentionality. Using string parsing to copy the organization of behaviour. *Anim. Cognit.* 2: 63–72.

———. 2002. Imitation of novel complex actions: What does the evidence from animals mean? *Adv. Stud. Behav.* 31: 31–77.

———. 2003. Imitation as behaviour parsing. *Phil. Trans. R. Soc.* 358: 529–536.

Byrne, R. W., and A. E. Russon. 1998. Learning by imitation: A hierarchical approach. *Behav. Brain Sci.* 21: 667–721.

Byrne, R. W., and J. E. Tanner. 2006. Gestural imitation by a gorilla: Evidence and nature of the capacity. *Int. J. Psychol. Psychol. Ther.* 6: 215–231.

Byrne, R. W., and M. Tomasello. 1995. Do rats ape? *Anim. Behav.* 50: 1417–1420.

Byrne, R. W., and A. Whiten, eds. 1988. *Machiavellian Intelligence: Social Expertise and the Evolution of Intellect in Monkeys, Apes, and Humans.* New York: Oxford University Press.

Caldwell, M. C., and D. K. Caldwell. 1972. Vocal mimicry in the whistle mode by an Atlantic bottlenose dolphin. *Cetology* 9: 1–8.

Caldwell, C. A., and A. E. Millen. 2008. Experimental models for testing hypotheses about cumulative cultural evolution. *Evol. Hum. Behav.* 29: 165–171.

———. 2009. Social learning mechanisms and cumulative cultural evolution: Is imitation necessary? *Psychol. Sci.* 12: 1478–1453.

Caldwell, C. A., and A. Whiten. 2003. Scrounging facilitates social learning in common marmosets, *Callithrix jacchus*. *Anim. Behav*. 65: 1085–1092.

Call, J. 2001. Body imitation in an encultured orangutan (*Pongo pygmaeus*). *Cybern. Syst*. 32: 97–119.

Call, J., and M. Carpenter. 2002. Three sources of information in social learning. In *Imitation in Animals and Artifacts*, edited by K. Dautenhahn and C. Nehaniv, 211–228. Cambridge, MA: MIT Press.

Call, J., M. Carpenter, and M. Tomasello. 2005. Copying results and copying actions in the process of social learning: Chimpanzees (*Pan troglodytes*) and human children (*Homo sapiens*). *Anim. Cognit*. 8: 151–163.

Campbell, D. T. 1960. Blind variation and selective retention in creative thought as in other knowledge processes. *Psychol. Rev*. 67: 380–400.

Campbell, F. M., and C. M. Heyes. 2002. Rats smell: Odour-mediated local enhancement, in a vertical movement two-action test. *Anim. Behav*. 63: 1055–1063.

Campbell-Meiklejohn, D. K., D. R. Bach, A. Roepstorff, R. J. Dolan, and C. D. Frith. 2010. How others influence our value of objects. *Curr. Biol*. 20: 1165–1170.

Campobello, D., and S. Sealy. 2011. Nest defence against avian brood parasites is promoted by egg-removal events in a cowbird-host system. *Anim. Behav*. 82: 885–891.

Caneiro, R. L. 2003. *Evolutionism in Cultural Anthropology*. Boulder, CO: Westview Press.

Cao, H. H., B. Han, and D. Hirshleifer. 2011. Taking the road less traveled by: Does conversation eradicate pernicious cascades? *J. Econ. Theor*. 146: 1418–1436.

Caro, T. M. 1980a. Predatory behaviour in domestic cat mothers. *Behaviour* 74: 128–147.

———. 1980b. Effects of the mother, object play and adult experience on predation in cats. *Behav. Neural Biol*. 29: 29–51.

Caro, T. M., and M. D. Hauser. 1992. Is there teaching in non-human animals? *Quart. Rev. Biol*. 67: 151–174.

Carpenter, A. 1887. Monkeys opening oysters. *Nature* 36: 53.

Carpenter, M. 2006. Instrumental, social, and shared goals and intentions in imitation. In *Imitation and the Social Mind: Autism and Typical Development*, edited by S. J. Rogers and J.H.G. Williams, 48–70. New York: Guilford Press.

Casanova, C. 2008. Innovative social behavior in chimpanzees (*Pan troglodytes*). *Am. J. Primatol*. 70: 54–61.

Catchpole, C. K., and P.J.B. Slater. 1995. *Bird Song: Biological Themes and Variations*. Cambridge, UK: Cambridge University Press.

———. 2008. *Bird Song*, 2nd ed. Cambridge, U.K.: Cambridge University Press.

Catmur, C., V. Walsh, and C. M. Heyes. 2009. Associative sequence learning: The role of experience in the development of imitation and the mirror system. *Phil. Trans. R. Soc. Lond. B* 364: 2369–2380.

Cavalli-Sforza, L. L., and M. W. Feldman. 1973. Models for cultural inheritance. I. Group mean and within group variation. *Theo. Pop. Biol*. 4: 42–55.

———. 1981. *Cultural Transmission and Evolution*. Princeton, NJ: University of Princeton Press.

Cavalli-Sforza, L. L., and W.S.-Y. Wang. 1986. Spatial distance and lexical replacement. *Language* 62: 38–55.

Cavalli-Sforza, L. L., E. Minch, and J. L. Mountain. 1992. Coevolution of genes and languages revisited. *Proc. Natl. Acad. Sci. USA* 89: 6020–6024.

Chivers, D. P., and R.J.F. Smith. 1995. Chemical recognition of risky habitats is culturally transmitted among fathead minnows, *Pimephales promelas* (Osteichthyes, Cyprinidae). *Ethology* 99: 286–296.

Choloris, E., A. Clipperton-Allen, A. Phan, and M. Kavaliers. 2009. Neuroendocrinology of social information processing in rats and mice. *Front. Neuroendocrinol.* 30: 442–459.

Choleris, E., A. Clipperton-Allen, D. Gray, S. Diaz-Gonzalez, and R. Welsman. 2011. Differential effects of dopamine receptor D1-type and D2-type antagonists and phase of the estrous cycle on social learning of food preferences, feeding and social interactions in mice. *Neuropsychopharmacology* 36: 1689–1702.

Chou, L. S., and P. J. Richerson. 1992. Multiple models in social transmission of food selection by Norway rats, (*Rattus norvegicus*). *Anim. Behav.* 44: 337–343.

Church, R. M. 1968. Applications of behavior theory to social psychology. In *Social Facilitation and Imitative Behavior*, edited by E. C. Simmel, R. A. Hoppe, and G. A. Milton. Boston: Allyn and Bacon.

Clarke, J. A. 2010. White-tailed ptarmigan food calls enhance chick diet choice: Learning nutritional wisdom? *Anim. Behav.* 79: 25–30.

Clipperton, A. E., J. M. Spinato, C. Chernets, D. W. Pfaff, and E. Choleris. 2008. Differential effects of estrogen receptor alpha and beta specific agonists on social learning of food preferences in female mice. *Neuropsychopharmacology* 33: 2362–2375.

Clutton-Brock, T., and P. Harvey. 1980. Primates, brains and ecology. *J. Zool.* 190: 309–323.

Cole, L. W. 1907. Concerning the intelligence of raccoons. *J. Comp. Neurol. Psychol.* 17: 211–261.

Cook, M., S. Mineka, B. Wolkenstein, and K. Laitsch. 1985. Observational conditioning of snake fear in unrelated rhesus monkeys. *J. Abnorm. Psychol.* 93: 355–372.

Coolen, I., R. I. Day, and K. N. Laland. 2003. Species difference in adaptive use of public information in sticklebacks. *Proc. R. Soc. Lond. B* 270: 2413–2419.

Coolen, I., A.J.W. Ward, P.J.B. Hart, and K. N. Laland. 2005. Foraging nine-spined sticklebacks prefer to rely on public information over simpler social cues. *Behav. Ecol.* 16: 865–870.

Coop, G., J. K. Pickrell, J. Novembre, S. Kudaravalli, J. Li, et al. 2009. The role of geography in human adaptation. *PLoS Genet.* 5: e1000500, doi:10.1371/journal .pgen.1000500.

Cornwallis, C., S. West, K. Davis, and A. Griffin. 2010. Promiscuity and the evolutionary transition to complex societies. *Nature* 466: 969–972.

Cosmides, L., and J. Tooby. 1987. From evolution to behaviour: Evolutionary psychology as the missing link In *The Latest on the Best: Essays on Evolution and Optimality*, edited by J. Dupre, 277–306. Cambridge, MA: MIT Press.

Coultas, J. 2004. When in Rome . . . an evolutionary perspective on conformity. *Group Processes and Intergroup Relations* 7: 317–331.

Coussi-Korbel, S., and D. M. Fragaszy. 1995. On the relation between social dynamics and social learning. *Anim. Behav.* 50: 1441–1453.

Couzin, I. D. 2009. Collective cognition in animal groups. *Trends Cogn. Sci.* 13: 36–43.

Cox, D. R., and D. Oakes. 1984. *Analysis of Survival Data*. London: Chapman and Hall.

Croft, D. P., R. James, and J. Krause. 2008. *Exploring Animal Social Networks*. Princeton, NJ: Princeton University Press.

Curio, E. 1988. Cultural transmission of enemy recognition by birds. In *Social Learning: Psychological and Biological Perspectives*, edited by B. G. Galef and T. R. Zentall, 75–97. Hillsdale, NJ: Erlbaum.

Curio, E., U. Ernst, W. and Vieth. 1978. The adaptive significance of avian mobbing, II. Cultural transmission of enemy recognition in blackbirds: Effectiveness and some constraints. *Z. Tierpsychol.* 48: 184–202 .

Currie, T. E., S. J. Greenhill, R. D. Gray, T. Hasegawa, and R. Mace. 2010. Rise and fall of political complexity in island South-East Asia and the Pacific. *Nature* 467: 801–804.

Custance, D. M., A. Whiten, and K. A. Bard. 1995. Can young chimpanzees (*Pan trologydtes*) imitate arbitrary actions? Hayes and Hayes revisited. *Behaviour* 132: 837–859.

Custance, D. M., A. Whiten, and T. Fredman. 1999. Social learning of an artificial fruit task in capuchin monkeys (*Cebus apella*). *J. Comp. Psychol.* 113: 13–23.

Darwin, C. 1871. *The Descent of Man, and Selection in Relation to Sex*. London: John Murray. Reprint (1981) of the first edition, with introduction by J. T. Bonner and R. M. May. Princeton, NJ: Princeton University Press.

Dautenhahn, K., and C. L. Nehaniv, eds. 2002. *Imitation in Animals and Artifacts*. Cambridge, MA: MIT Press.

Davies, N. B., and J. A. Welbergen. 2009. Social transmission of a host defense against cuckoo parasitism. *Science* 324: 1318–1320.

Dawkins, R. 1976. *The Selfish Gene*. Oxford, U.K.: Oxford University Press.

Dawson, B. V., and B. M. Foss. 1965. Observational learning in budgerigars. *Anim. Behav.* 13: 470–474.

Day, R., T. MacDonald, C. Brown, K. N. Laland, and S. M. Reader. 2001. Interactions between shoal size and conformity in guppy social foraging. *Anim. Behav.* 62: 917–925.

Day R. L., R. L. Coe, J. R. Kendal, and K. N. Laland. 2003. Neophilia, innovation and social learning: A study of intergeneric differences in Callitrichid monkeys. *Anim. Behav.* 65: 559–571.

Deaner, R. O., C. L. Nunn, and C. P. van Shaik. 2000. Comparative tests of primate cognition: Different scaling methods produce different results. *Brain Behav. Evol.* 55: 44–52.

Dekker, D., D. Krackhardt, and T. Snijders. 2007. Sensitivity of MRQAP tests to collinearity and autocorrelation conditions. *Psychometrika* 72: 563–581.

Dennett, D. 1991. *Consciousness Explained*. London: Penguin Books.

———. 1995. *Darwin's Dangerous Idea: Evolution and the Meanings of Life*. London: Penguin Books.

Denny, M. R., R. C. Bell, and C. F. Clos. 1983. Two-choice, observational learning and reversal in the rat: S-S versus S-R effects. *Anim. Learn. Behav.* 11: 223–228.

Denny, M. R., C. F. Clos, and R. C. Bell. 1988. Learning in the rat of a choice response by observation of S-S contingencies. In *Social Learning: Psychological and Biological Perspectives*, edited by B. G. Galef and T. R. Zentall, 207–224. Hillsdale, NJ: Erlbaum.

Denny, A. J., J. Wright, and B. Grief. 2001. Foraging efficiency in the wood ant (*Formica rufa*): Is time of the essence in trail following? *Anim. Behav.* 61: 139–146.

de Nooy, W. 2010. Networks of action and events over time. A multilevel discrete-time event history model for longitudinal network data. *Soc. Networks* 33: 31–40.

Deutsch, M., and H. B. Gerard, H. B. 1955. A study of normative and informational social influences upon individual social judgement. *J. Abnorm. Soc. Psychol.* 51: 629–636.

de Waal, F. 2001. *The Ape and the Sushi Master.* New York: Basic Books.

Dietz, E. J. 1983. Permutation tests for association between two distance matrices. *Syst. Zool.* 32: 21–26.

Dindo, M., A. Whiten, and F.B.M. de Waal. 2009. In-group conformity sustains different foraging traditions in capuchin monkeys (*Cebus apella*). *PloS ONE* 4: Artn E7858, doi:10.1371/journal.pone.0018283.

Doligez, B., É. Danchin, and J. Clobert. 2002. Public information and breeding habitat selection in a wild bird population. *Science* 297: 1168–1170.

Doligez B., C. Cadet, E. Danchin, and T. Boulinier. 2003. When to use public information for breeding habitat selection: The role of environmental predictability and density dependence. *Anim. Behav.* 66: 973–988.

Dorrance, B. R., and T. R. Zentall. 2002. Imitation of conditional discriminations in pigeons (*Columba livia*). *J. Comp. Psychol.* 116: 277–285.

Dowsett-Lemaire, F. 1979. The imitative range of the song of the marsh warbler (*Acrocephalus palustris*), with special reference to imitations of African birds. *Ibis* 121: 453–468.

Drea, C. M., and K. Wallen. 1999. Low status monkeys 'play dumb' when learning in mixed social groups. *Proc. Natl. Acad. Sci. USA* 96: 12965–12969.

Duffy, G. A., T. W. Pike, and K. N. Laland. 2009. Size-dependent directed social learning in nine-spined sticklebacks. *Anim. Behav.* 78: 371–375.

Dugatkin, L. A. 1996. Copying and mate choice. In *Social Learning in Animals: The Roots of Culture*, edited by C. M. Heyes and B. G. Galef, 85–105. New York: Academic Press.

Dugatkin, L. A., and J.G.J. Godin. 1993. Female mate copying in the guppy (*Poecilia reticulata*)—age-dependent effects. *Behav. Ecol.* 4: 289–292.

Dunbar, R.I.M. 1995. Neocortex size and group size in primates: A test of the hypothesis. *J. Hum. Evol.* 28: 287–296.

———. 1998. The social brain hypothesis. *Evol. Anthro.* 6: 178–190.

———. 2003. The social brain: Mind, language and society in evolutionary perspective. *Ann. Rev. Anthrop.* 32: 163–181.

Durham, W. H. 1991. *Coevolution: Genes, Culture, and Human Diversity.* Stanford, CA: Stanford University Press.

Efferson C., R. Lalive, P. J. Richerson, R. McElreath, and M. Lubell. 2007. Conformists and mavericks: The empirics of frequency-dependent cultural transmission. *Evol. Hum. Behav.* 29: 56–64.

Efferson, C., R. Lalive, and E. Fehr. 2008. The coevolution of cultural groups and ingroup favoritism. *Science* 321: 1844–1849.

Enquist, M., K. Eriksson, and S. Ghirlanda. 2007. Critical social learning: A solution to Roger's paradox of nonadaptive culture. *Am. Anthropol.* 109: 727–734.

Eriksson K., M. Enquist, and S. Ghirlanda. 2007. Critical points in current theory of conformist social learning. *J. Evol. Psychol.* 5: 67–87.

Eriksson, K., and J. C. Coultas. 2009. Are people really conformist-biased? An empirical test and a new mathematical model. *J. Evol. Psychol.* 7: 5–21.

Fadiga, L., L. Fogassi, G. Pavesi, and G. Rizzolatti. 1995. Motor facilitation during action observation—a magnetic stimulation study. *J. Neurophysiol.* 73: 2608–2611.

Faraway, J. J. 2006. *Extending the Linear Model with R: Generalized Linear, Mixed Effects and Nonparametric Regression Models*. Boca Raton, FL: CRC.

Fawcett, T. W., A.M.J. Skinner, and A. R. Goldsmith. 2002. A test of imitative learning in starlings using a two-action method with an enhanced ghost control. *Anim. Behav.* 64: 547–556.

Fawcett, T.W., S, Hamblin, and L. A. Giraldeau. 2012. Exposing the behavioral gambit: The evolution of learning and decision rules. *Behav. Ecol.*, doi:10.1093/beheco/ars085.

Fay, N., S. Garrod, and L. Roberts. 2008. The fitness and functionality of culturally evolved communication systems. *Phil. Trans. R. Soc. B* 363: 3553–3561.

Fehr, E., and U. Fischbacher. 2003. The nature of human altruism. *Nature* 425: 785–91.

Feldman, M. W., K. Aoki, and J. Kumm. 1996. Individual versus social learning: Evolutionary analyses in a fluctuating environment. *Anthropol. Sci.* 104: 209–231.

Feldman, M. W., and L. L. Cavalli-Sforza. 1976. Cultural and biological evolutionary processes, selection for a trait under complex transmission. *Theor. Popul. Biol.* 9: 238–259.

———. 1989. On the theory of evolution under genetic and cultural transmission with application to the lactose absorption problem. In *Mathematical Evolutionary Theory*, edited by M. W. Feldman, 145–173. Princeton, NJ: Princeton University Press.

Feldman, M. W., and K. N. Laland. 1996. Gene-culture coevolutionary theory. *Trends Ecol. Evol.* 11: 453–457.

Feldman, M. W., and S. P. Otto. 1997. Twin studies, heritability, and intelligence. *Science* 278:1383–1384.

Ferrari, P., S. Rozzi, and L. Fogassi. 2005. Mirror neurons responding to observation of actions made with tools in monkey ventral premotor cortex. *J. Cogn. Neurosci.* 17: 212–226.

Ferrari, P., E. Visalberghi, A. Paukner, L. Fogassi, A. Ruggiero, and S. Suomi. 2006. Neonatal imitation in Rhesus macaques. *PLoS Biology* 4: e302, doi:10.1371/journal.pbio.0040302.

Fisher, J., and R. A. Hinde. 1949. The opening of milk bottles by birds. *Br. Birds* 42: 347–357.

Flemming, A., C. Kuchera, A. Lee, and G. Winocur. 1994. Olfactory-based social learning varies as a function of parity in female rats. *Psychobiology* 22: 37–43.

Fogarty, L, P. Strimling, and K. N. Laland. 2011. The evolution of teaching. *Evolution* 65: 2760–2770.

Forkman, B. 1991. Some problems with current patch-choice theory: A study on the Mongolian gerbil. *Behaviour* 117: 243–254.

Forsman, J., and J. Seppanen. 2011. Learning what (not) to do: Testing rejection and copying of simulated hetrospecific behavioural traits. *Anim. Behav.* 81: 879–883.

Fowler, J. H., and N. A. Christakis. 2008. Dynamic spread of happiness in a large social network: Longitudinal analysis over 20 years in the Framingham Heart Study. *Br. Med. J.* 337: 1–9.

Fragaszy, D. 2012a. Community resources for learning: How capuchin monkeys construct technical traditions. *Biol. Theory* 6: 231–240, doi: 10.1007/s13752-012-0032-8.

———. 2012b. Community resources for learning: How capuchin monkeys construct technical traditions. In "Cultural Niche Construction," edited by M. O'Brien and K. Laland, special issue, *Biological Theory*.

Fragaszy, D. M., and S. Perry. 2003. *The Biology of Traditions: Models and Evidence.* Cambridge, U.K.: Cambridge University Press.

Fragaszy, D. M., and E. Visalberghi. 1990. Social processes affecting the appearance of innovative behaviours in capuchin monkeys. *Folia Primatol.* 54: 155–165.

Franks, N. R., and T. Richardson. 2006. Teaching in tandem-running ants. *Nature* 439: 153.

Franz, M., and L. J. Matthews. 2010. Social enhancement can create adaptive, arbitrary and maladaptive cultural traditions *Proc. R. Soc. B* 277: 3363–3372.

Franz, M., and C. L. Nunn. 2009. Network-based diffusion analysis: A new method for detecting social learning. *Proc. R. Soc. B* 276: 1829–1836.

———. 2010. Investigating the impact of observation errors on the statistical performance of network-based diffusion analysis. *Learn. Behav.* 38: 235–242.

Freeberg, T. M. 1996. Assortative mating in captive cowbirds is predicted by social experience. *Anim. Behav.* 52: 1129–1142.

———. 1998. The cultural transmission of courtship patterns in cowbirds (*Molothrus ater*). *Anim. Behav.* 56: 1063–1073.

Fritz, J., and K. Kotrschal. 1999. Social learning in common ravens (*Corvus corax*). *Anim. Behav.* 57: 785–793.

Fritz, J., A. Bisenberger, and K. Kotrschal. 2000. Stimulus enhancement in greylag geese: Socially mediated learning of an operant task. *Anim. Behav.* 59: 1119–1125.

Gagneux, P., M. K. Gonder, T. L. Goldberg, and P. A. Morin. 2001. Gene flow in wild chimpanzee populations: What genetic data tell us about chimpanzee movement over space and time. *Phil. Trans. R. Soc. Lond. B* 356: 889–897.

Galef, B. G., Jr. 1976. Social transmission of acquired behavior: A discussion of tradition and social learning in vertebrates. *Adv. Stud. Behav.* 6: 77–100.

———. 1989. Enduring social enhancement of rats' preferences for the palatableand the piquant. *Appetite* 13: 81–92.

———. 1988. Imitation in animals: History, definition and interpretation of the data from the psychological laboratory. In *Social Learning: Psychological and Biological Perspectives*, edited by B. G. Galef, Jr. and T. R. Zentall, 3–28. Hillsdale, NJ: Erlbaum.

———. 1992. The question of animal culture. *Hum. Nat.* 3: 157–178.

———. 1995. Why behaviour patterns that animals learn socially are locally adaptive. *Anim. Behav.* 49: 1325–1334.

———. 1996. Social enhancement of food preferences in Norway rats: A brief review. In *Social Learning and Imitation: The Roots of Culture*, edited by C. M. Heyes and B. G. Galef, Jr., 49–64. New York: Academic Press.

———. 2003. Traditional behaviors of brown and black rats (*R. norvegicus* and *R. rattus*). In *The Biology of Traditions: Models and Evidence*, edited by S. Perry and D. Fragaszy, 159–186. Chicago: University of Chicago Press.

———. 2009. Culture in animals? In *The Question of Animal Culture*, edited by K. N. Laland and B. G. Galef, Jr., 222–246. Cambridge, MA: Harvard University Press.

———. 2009. Strategies for social learning: testing predictions from formal theory. *Adv. Study Behav.* 39: 117–151.

Galef, B. G., Jr., and C. Allen. 1995. A new model system for studying animal tradition. *Anim. Behav.* 50: 705–717.

Galef, B. G., Jr., and M. Beck. 1985. Aversive and attractive marking of toxic and safe foods by Norway rats. *Behav. Neural Biol.* 43: 298–310.

Galef B. G., Jr., K. E. Dudley, and E. E. Whiskin. 2008. Social learning of food preferences in 'dissatisfied' and 'uncertain' Norway rats. *Anim. Behav.* 75: 631–637.

Galef, B. G., Jr., and P. J. Durlach. 1993. Absence of blocking, overshadowing, and latent inhibition in social enhancement of food preferences. *Anim. Learn. Behav.* 21: 214–220.

Galef, B. G., Jr., and L.-A. Giraldeau. 2001. Social influences on foraging in vertebrates: Behavioural mechanisms and adaptive functions. *Anim. Behav.* 61: 3–15.

Galef, B. G., Jr., D. J. Kennett, and S. W. Wigmore. 1984. Transfer of information concerning distant foods in rats: A robust phenomenon. *Anim. Learn. Behav.* 12: 292–296.

Galef, B. G. Jr., D. J. Kennett, and M. Stein. 1985. Demonstrator influence on observer diet preference: Effects of simple exposure and the presence of a demonstrator. *Anim. Learn. Behav.* 13: 25–30.

Galef, B. G., Jr., J. R. Mason, G. Preti, and N. J. Bean. 1988. Carbon disulfide: A semiochemical mediating socially-induced diet choice in rats. *Physiol. Behav.* 42: 119–124.

Galef, B. G., Jr., and E. E. Whiskin. 2006. Increased reliance on socially acquired information while foraging in risky situations. *Anim. Behav.* 72: 1169–1176.

———. 2008a. 'Conformity' in Norway rats? *Anim. Behav.* 75: 2035–2039.

———. 2008b. Use of social information by sodium- and protein-deficient rats: A test of a prediction (Boyd and Richerson 1988). *Anim. Behav.* 75: 627–630.

Galef, B. G., Jr., E. E. Whiskin, and E. Bielavska. 1997. Interaction with demonstrator rats changes observer rats' affective responses to flavors. *J. Comp. Psychol.* 111: 393–398.

Galef, B. G., Jr., and D. J. White. 1998. Mate-choice copying in Japanese quail (*Coturnix coturnix japonica*). *Anim. Behav.* 55: 545–552.

Galef, B. G., Jr., and S. W. Wigmore. 1983. Transfer of information concerning distant foods: A laboratory investigation of the 'information-centre' hypothesis. *Anim. Behav.* 31: 748–758.

Galef, B. G., Jr., and N. Yarkovsky. 2009. Further studies of reliance on socially acquired information when foraging in potentially risky situations. *Anim. Behav.* 77: 1329–1335.

Gallese, V., L. Fadiga, L. Fogassi, and G. Rizzolatti. 1996. Action recognition in the premotor cortex. *Brain* 119: 593–609.

Gallese, V., and A. Goldman. 1998. Mirror neurons and the simulation theory of mind-reading. *Trends Cogn. Sci.* 12: 493–501.

Gelman, A., J. Carlin, H. Stern, and H. Rubin. 2004. *Bayesian Data Analysis*. New York: Chapman and Hall.

Gerbault, P., A. Liebert, Y. Itan, A. Powell, M. Currat, J. Burger, D. M. Swallow, and M. G. Thomas. 2011. Evolution of lactase persistence: An example of human niche construction. *Phil. Trans. R. Soc. B* 366: 863–877.

Gerard, R. W., C. Kluckhohn, and A. Rapoport. 1956. Biological and cultural evolution. *Behav. Sci.* 1: 6–34.

Gergely, G., H. Bekkering, and I. Király. 2002. Rational imitation in preverbal infants. *Nature* 415: 755.

Ghirlanda S., and M. Enquist. 2007. Cumulative culture and explosive demographic transitions. *Qual. Quant.* 41: 581–600.

Gintis, H. 2000. *Game Theory Evolving*. Princeton, NJ: Princeton University Press.

———. 2004. The genetic side of gene-culture coevolution: Internalization of norms and prosocial emotions. *J. Econ. Behav. Org.* 53: 57–67.

Giraldeau, L.-A., and G. Beauchamp. 1999. Food exploitation: Searching for the optimal joining policy. *Trends Ecol. Evol.* 14: 102–106.

Giraldeau, L.-A., and T. Caraco. 2000. *Social Foraging Theory*. Princeton, NJ: Princeton University Press.

Giraldeau, L. A., T. Caraco, and T. J. Valone. 1994. Social foraging: Individual learning and cultural transmission of innovations. *Behav. Ecol.* 5: 35–43.

Giraldeau, L.-A., and L. Lefebvre. 1986. Exchangeable producer and scrounger roles in a captive flock of feral pigeons: A case for the skill pool effect. *Anim. Behav.* 34: 777–783.

———. 1987. Scrounging prevents cultural transmission of food-finding behavior in pigeons. *Anim. Behav.* 35: 387–394.

Giraldeau, L.-A., and Templeton, J. J. 1991. Food scrounging and diffusion of foraging skills in pigeons (*Columba livia*): The importance of tutor and observer rewards. *Ethology* 89: 63–72.

Giraldeau, L. A., T. J. Valone, and J. J. Templeton. 2002. Potential disadvantages of using socially acquired information. *Phil. Trans. R. Soc. Lond. B* 357: 1559–1566.

Goeree, J. K., T. R. Palfrey, B. W. Rogers, R. D. McKelvey. 2007. Self-correcting information cascades. *Rev. Econ. Stud.* 74: 733–762.

Goodall, J. 1986. *The Chimpanzees of Gombe: Patterns of Behavior*. Cambridge, MA: Harvard University Press.

Gopnik, A., and J. Tenenbaum. 2007. Bayesian networks, Bayesian learning and cognitive development. *Dev. Sci.* 10: 281–287.

Goslee, S. C., and D. L. Urban. 2007. The ecodist package for dissimilarity-based analysis of ecological data. *J. Stat. Software* 22: 1–19.

Gould, S. J. 1991. *Bully for Brontosaurus: Reflections in Natural History*. New York: Norton.

Gower, J. C. 1971. A general coefficient of similarity and some of its properties. *Biometrics* 27: 857–871.

Grafen, A. 1984. Natural selection, kin selection and group selection. In *Behavioural Ecology: An Evolutionary Approach*, edited by J. R. Krebs and N. B. Davies, 62–84. Oxford, U.K.: Blackwell.

Grafton, S. T., M. A. Arbib, L. Fadiga, and G. Rizzolatti. 1996. Localization of grasp representations in humans by PET: 2. Observation compared with imagination. *Exp. Brain Res.* 112: 103–111.

Gray, R. D., and Q. D. Atkinson. 2003. Language-tree divergence times support the Anatolian theory of Indo-European origin. *Nature* 426: 435–439.

Gray, R. D., and F. M. Jordan. 2000. Language trees support the express-train sequence of Austronesian expansion. *Nature* 405: 1052–1055.

Greenberg, R., and C. Mettke-Hofman. 2001. Ecological aspects of neophobia and neophilia in birds. *Curr. Ornithol.* 16: 119–178.

Griffin, A., and K. Haythorpe. 2011. Learning from watching alarmed demonstrators: Does the cause of alarm matter? *Anim. Behav.* 81: 1163–1169.

Griffiths, S. W. 2003. Learned recognition of conspecifics by fishes. *Fish Fish.* 4: 256–268.

Griffiths, T. L., B. R. Christian, and M. L. Kalish. 2008. Using category structure to test iterated learning as a method for identifying inductive biases. *Cogn. Sci.* 32: 68–107.

Grimes, B. F. 2002. *Ethnologue: Languages of the World*, 14th ed. Dallas, TX: Summer Institute of Linguistics.

Grüter, C., T. Czaczkes, and F.L.W. Ratnieks. 2011. Decision making in ant foragers (*Lasius niger*) facing conflicting private and social information. *Behav. Ecol. Sociobiol.* 65: 141–148.

Grüter, C., E. Leadbeater, and F.L.W. Ratnieks. 2010. Social learning: The importance of copying others. *Cur. Biol.* 20: R683–R685.

Grüter, C., and F.L.W. Ratnieks. 2011. Honeybee foragers increase the use of waggle dance information when private information becomes unrewarding. *Anim. Behav.* 81: 949–954.

Guarino, A., and M. Cipriani. 2008. *Herd Behavior in Financial Markets: An Experiment with Financial Market Professionals*. SSRN eLibrary. J. Euro. Econ. Assoc. 7: 206–233.

Haggerty, M. E. 1909. Imitation in monkeys. *J. Comp. Neurol. Psychol.* 19: 337–455.

Hahn, M. W., and R. A. Bentley. 2003. Drift as a mechanism for cultural change: An example from baby names. *Biol. Let.* 270: S120–S123.

Haldane, J.B.S. 1964. A defense of beanbag genetics. *Pers. Biol. Med.* 7: 343–359.

Hall, K.R.L. 1951. The effect of names and titles upon the serial reproduction of pictorial and verbal material. *Br. J. Psychol.* 41: 109–121.

Hamilton, W. 1964. The genetical evolution of social behaviour: I. *J. Theor. Biol.* 7: 1–16.

Harris, P. L. 2007. Trust. *Trust. Dev. Sci.* 10: 135–138.

Hartig, F., J. M. Calabrese, B. Reineking, T. Wiegand, and A. Huth. 2011. Statistical inference for stochastic simulation models—theory and application. *Ecol. Litt.* 14: 816–827.

Haslinger, B., P. Erhard, E. Altenmuller, U. Schroeder, H. Boecker, et al. 2005 Transmodal sensorimotor networks during action observation in professional pianists. *J. Cogn. Neurosci.* 17: 282–293.

Haun, D., Y. Rekers, and M. Tomasello. 2012. Majority-biased transmission in chipanzees and human children, but not orangutans. *Curr. Biol.* 22: 727–731, doi: 10.1016/j.cub.2010.03.006.

Hauser, M. D. 1988. Invention and social transmission: New data from wild vervet monkies. In *Machiavellian Intelligence: Social Expertise and the Evolution of Intellect in Monkeys, Apes, and Humans*, edited by R. W. Byrne and A. Whiten, 327–343. New York: Oxford University Press.

Hawks, J., E. T. Wang, G. M. Cochran, H. C. Harpending, and R. K. Moyzis. 2007. Recent acceleration of human adaptive evolution. *Proc. Natl. Acad. Sci. USA* 104: 20753–20758.

Hayes, K. J., and C. Hayes. 1952. Imitation in a home-raised chimpanzee. *J. Comp. Physiol. Psychol.* 45: 450–459.

Heath, C., C. Bell, and E. Sternberg. 2001. Emotional selection in memes: The case of urban legends. *J. Person. Soc. Psychol.* 81: 1028–1041.

Helfman, G. S., and E. T. Schultz. 1984. Social transmission of behavioural traditions in a coral reef fish. *Anim. Behav.* 32: 379–384.

Hennig, C. 2010. Flexible procedures for cluster analysis (fps). R package version 2.0-3 Accessed Oct. 25, 2012, http://CRAN.r-project.org/package=fpc/.

Henrich, J. 2001. Cultural transmission and the diffusion of innovations: Adoption dynamics indicate that biased cultural transmission is the predominate force in behavioral change. *Am. Anthropol.* 103: 992–1013.

Henrich, J., and R. Boyd. 1998. The evolution of conformist transmission and between-group differences. *Evol. Hum. Behav.* 19: 215–242.

———. 2001. Why people punish defectors: Conformist transmission stabilizes costly enforcement of norms in cooperative dilemmas. *J. Theor. Biol.* 208: 79–89.

Henrich, J., and J. Broesch, J. 2011. On the nature of cultural transmission networks: Evidence from Fijian villages for adaptive learning biases. *Phil. Trans. R. Soc. B* 366: 1139–1148.

Henrich, J., and F. J. Gil-White. 2001. The evolution of prestige: Freely conferred deference as a mechanism for enhancing the benefits of cultural transmission. *Evol. Hum. Behav.* 22: 165–196.

Henrich, J., and N. Henrich. 2010 The evolution of cultural adaptations: Fijian food taboos protect against dangerous marine toxins. *Proc. R. Soc. B* 277: 3715–3724.

Henrich, J., and R. McElreath. 2003. The evolution of cultural evolution. *Evol. Anthropol.* 12: 123–135.

———. 2007 Dual inheritance theory: The evolution of human cultural capacities and cultural evolution. In *Oxford Handbook of Evolutionary Psychology*, edited by R. Dunbar and L. Barrett, 555–570. Oxford, U.K.: Oxford University Press.

Henrich, N., and J. Henrich. 2007. *Why Humans Cooperate: A Cultural and Evolutionary Explanation*. Oxford, U.K.: Oxford University Press.

Herman, L. M. 2002. Vocal, social and self-imitation by bottlenosed dolphins. In *Imitation in Animals and Artifacts*, edited by K. Dautenhahn and C. L. Nehaniv, 63–108. Cambridge, MA: MIT Press.

Herrmann, E., J. Call, M. V. Hernandez-Lloreda, B. Hare, and M. Tomasello. 2007. Humans have evolved specialized skills of social cognition: The cultural intelligence hypothesis. *Science* 317: 1360–1366.

Herzog, H. A., R. A. Bentley, and M. W. Hahn. 2004. Random drift and large shifts in popularity of dog breeds. *Proc. R. Soc. B* 271: S353–S356.

Heyes, C. M. 1994. Social learning in animals: Categories and mechanisms. *Biol. Rev.* 69: 207–231.

———. 2002. Transformational and associative theories of imitation. In *Imitation in Animals and Artifacts*, edited by K. Dautenhahn and C. L. Nehaniv, 501–524. Cambridge, MA: MIT Press.

———. 2009. Evolution, development and intentional control of imitation. *Phil. Trans. R. Soc. B* 364: 2293–2298.

———. 2011. What's social about social learning? *J. Comp. Psychol.* 126: 193–202.

———. 2012. Grist and mills: On the cultural origins of cultural learning. *Phil. Trans. R. Soc.* 367: 2181–2191.

Heyes, C. M., and G. R. Dawson. 1990. A demonstration of observational learning using a bidirectional control. *Quart. J. Exp. Psychol.* 42: 59–71.

Heyes, C. M., G. R. Dawson, and T. Noakes. 1992. Imitation in rats: Initial responding and transfer evidence. *Quart. J. Exp. Psychol.* 45: 81–92.

Heyes, C. M., L. Huber, G. Gergely, and M. Brass, eds. 2009. "Evolution, Development and Intentional Control of Imitation," special issue, *Phil. Trans. R. Soc. B* 364.

Heyes, C. M., E. Jaldow, and G. R. Dawson. 1994. Imitation in rats: Conditions of occurrence. *Learn. Motiv.* 25: 276–287.

Heyes, C. M., and E. D. Ray. 2000. What is the significance of imitation in animals? *Adv. Stud. Behav.* 29: 215–245.

Heyes, C. M., and A. Saggerson. 2002. Testing for imitative and non-imitative social learning in the budgerigar using a two-object/two-action test. *Anim. Behav.* 64: 851–859.

Hobhouse, L. T. 1901. *Mind in Evolution*. London: Macmillan.

Holden, C., and R. Mace. 1997. Phylogenetic analysis of the evolution of lactose digestion in adults. *Hum. Biol.* 69: 605–628.

———. 2003. Spread of cattle led to the loss of matrilineal descent in Africa: A co-evolutionary analysis. *Proc. R. Soc. Lond. B* 270: 2425–2433.

Holsinger, K. E., and B. Weir. 2009. Genetics in geographically structured populations: Defining, estimating and interpreting FST. *Natl. Rev. Gen.* 10: 639–650.

Holt, E. B. 1931. *Animal Drive*. London: Williams and Norgate.

Hoogland, R., D. Morris, and N. Tinbergen. 1957. The spines of sticklebacks (*Gasterosteus and Pygosteus*) as means of defence against predators (*Pera and Esox*). *Behaviour* 10: 205–237.

Hopper, L. M. 2010. 'Ghost' experiments and the dissection of social learning in humans and animals. *Biol. Rev.* 85: 685–701.

Hoppitt, W.J.E. 2005. Social Processes Influencing Learning: Combining Theoretical and Empirical Approaches. PhD diss., University of Cambridge.

Hoppitt, W.J.E., L. Blackburn, and K. N. Laland. 2007. Response facilitation in the domestic fowl. *Anim. Behav.* 73: 229–238.

Hoppitt, W.J.E., N. J. Boogert, and K. N. Laland. 2010. Detecting social transmission in networks. *J. Theor. Biol.* 263: 544–555.

Hoppitt, W.J.E., G. Brown, R. L. Kendal, L. Rendell, A. Thornton, et al. 2008. Lessons from animal teaching. *Trends Ecol. Evol.* 23: 486–493.

Hoppitt, W.J.E., A. Kandler, J. R. Kendal, and K. N. Laland. 2010. The effect of task structure on diffusion dynamics: Implications for diffusion curve and network-based analyses. *Learn. Behav.* 38: 243–251.

Hoppitt, W.J.E., and K. N. Laland. 2008a. Social processes influencing learning in animals: A review of the evidence. *Adv. Stud. Behav.* 38: 105–165.

———. 2008b. Social processes affecting feeding and drinking in the domestic fowl. *Anim. Behav.* 76: 1529–1543.

———. 2011. Detecting social learning using networks: A user's guide. *Am. J. Primatol.* 73: 834–844.

Hoppitt, W.J.E., J. Samson, K. N. Laland, and A. Thornton, A. 2012. Identification of learning mechanisms in a wild meerkat population. *PLoS ONE* 7: e42044, doi:10.1371/journal.pone.0042044.

Horner, V., D. Proctor, K. E. Bonnie, A. Whiten, and F.B.M. de Waal. 2010. Prestige affects cultural learning in chimpanzees. *PloS ONE* 5:e10625, doi:10.1371/journal.pone.0010625.

Huber, L., F. Range, B. Voelkl, A. Szucsich, Z. Virányi, et al. 2009. The evolution of imitation: What do the capacities of non-human animals tell us about the mechanisms of imitation? *Phil. Trans. R. Soc. B* 364: 2299–2309.

Humle, T., and T. Matsuzawa. 2002. Ant dipping among the chimpanzees of Bossou, Guinea, and comparisons with other sites. *Am. J. Primatol.* 58: 133–148.

Humle, T., C. T. Snowdon, and T. Matsuzawa. 2009. Social influences on ant-dipping acquisition in the wild chimpanzees (*Pan troglodytes verus*) of Bossou, Guinea, West Africa. *Anim. Cogn.* 12: 37–48.

Humphrey, N. K. 1976. The social function of intellect. In *Growing Points in Ethology*, edited by P.P.G. Bateson and R. A. Hinde, 303–317. Cambridge, U.K.: Cambridge University Press.

Hurley, S., and N. Chater, eds. 2005. *Perspectives on Imitation*. Boston, MA: MIT Press.

Iacoboni, M., R. P. Woods, M. Brass, H. Bekkering, J. C. Mazziotta, et al. 1999. Cortical mechanisms of human imitation. *Science* 286: 2526–2528.

Iacoboni, M., L. M. Koski, M. Brass, H. Bekkering, R. P. Woods, et al. 2001. Reafferent copies of imitated actions in the right superior temporal cortex. *Proc. Natl. Acad. Sci. USA* 98:13995–13999.

Ihara, Y. 2008. Spread of costly prestige-seeking behavior by social learning. *Theo. Pop. Biol.* 73: 148–157.

Imanishi, K. 1952. Evolution of humanity. [In Japanese.] In *Man*, edited by K. Imanishi, 36–94. Tokyo: Mainichi-Shinbun-sha.

Ingold, T. 2007. The trouble with 'evolutionary biology.' *Anthropol. Today* 23: 13–17.

Insko, C. A., R. Gilmore, S. Drenan, A. Lipsitz, D. Moehle, et al. 1983. Trade versus expropriation in open groups. *J. Pers. Soc. Psychol.* 44: 977–999.

Insko, C. A., R. Gilmore, D. Moehle, A. Lipsitz, S. Drenan, et al. 1982. Seniority in the generational transition of laboratory groups: The effects of social familiarity and task experience. *J. Exp. Soc. Psychol.* 18: 577–580.

Insko, C. A., J. W. Thibaut, D. Moehle, M. Wilson, W. D. Diamond, et al. 1980. Social evolution and the emergence of leadership. *J. Pers. Soc. Psychol.* 39: 431–448.

Jacobs, R., and D. Campbell. 1961. The perpetuation of an arbitrary norm tradition through several generations of laboratory microculture. *J. Abnorm. Soc. Psychol.* 62: 649–658.

Janik, V. M., and P.J.B. Slater. 1997. Vocal learning in mammals. *Adv. Stud. Behav.* 26: 59–99.

———. 2000. The different roles of social learning in vocalcommunication. *Anim. Behav.* 60: 1–11.

Jaynes, E. T. 1998. *Probability Theory: The Logic of Science*. Cambridge, MA: Cambridge University Press.

Jones, B. C., L. M. DeBruine, A. C. Little, R. P. Burriss, and D. R. Feinberg. 2007. Social transmission of face preferences among humans. *Proc. R. Soc. B* 274: 899–903.

Jordan, P., and S. Shennan. 2003 Cultural transmission, language, and basketry traditions amongst the Californian Indians. *J. Anthropol. Archaeol.* 22: 42–74.

Kaiser, D. H., T. R. Zentall, and B. G. Galef, Jr. 1997. Can imitation in pigeons be explained by local enhancement together with trial-and-error learning? *Psychol. Sci.* 8: 459–460.

Kalish, M. L., T. L. Griffiths, and S. Lewandowsky. 2007. Iterated learning: Intergenerational knowledge transmission reveals inductive biases. *Psychonomic Bull. Rev.* 14: 288–294.

Kameda, T., and D. Nakanishi. 2002. Cost-benefit analysis of social/cultural learning in a nonstationary uncertain environment: An evolutionary simulation and an experiment with human subjects. *Evol. Hum. Behav.* 23: 373–393.

———. 2003. Does social/cultural learning increase human adaptability? Roger's question revisited. *Evol. Hum. Behav.* 24: 242–260.

Kandel, A., Martins, A. and Pacheco, R. 1995. Discussion: On the very real distinction between fuzzy and statistical methods. *Technometrics* 37: 276–281.

Kandler, A., and K. N. Laland. 2009. An investigation of the relationship between innovation and cultural diversity. *Theor. Pop. Biol.* 76: 59–67.

Kandler, A., and J. Steele. 2008. Ecological models of language competition. *Biol. Theor.*
3: 164–173.

Kashima, Y. 2000. Maintaining cultural stereotypes in the serial reproduction of narratives. *Pers. Soc. Psychol. Bull.* 26: 594–604.

Katsnelson, E., U. Motro, M. W. Feldman, et al. 2008. Early experience affects producer-
scrounger foraging tendencies in the house sparrow. *Anim. Behav.* 75:
1465–1472.

Kaufman, L., and P. Rousseeuw. 2005. *Finding Groups in Data: An Introduction to Cluster
Analysis.* New York: Wiley.

Kavaliers, M., E. Choleris, A. Agmo, W. Braun, D. Colwell, et al. 2006. Inadvertent social
information and the avoidance of parasitized male mice: A role for oxytocin.
Proc. Natl. Acad. Sci. USA 103: 4293–4298.

Kavaliers, M., D. Colwell, and E. Choleris. 2001a. NMDA-mediated social learning of
fear-induced conditioned analgesia to biting flies. *NeuroReport* 12: 663–667.

———. 2001b. Learning from others to cope with biting flies: Asocial learning of fear-
induced conditioned analgesia and active avoidance. *Behav. Neurosci.* 115:
661–674.

———. 2003. Learning to fear and cope with a natural stressor: Individually and socially
acquired corticosterone and avoidance responses to biting flies. *Horm. Behav.*
43: 99–107.

Kavaliers, M., N. Devidze, E. Choleris, M. Fudge, J. A. Gustafsson, et al. 2008. Estrogen
receptors alpha and beta mediate different aspects of the facilitatory effects of
female cues on male risk taking. *Psychoneuroendocrinology* 33: 634–642.

Kawai, M. 1965. Newly-acquired pre-cultral behavior of the natural troop of Japanese
monkeys on Koshima islet. *Primates* 6: 1–30.

Keesing, R. M. 1974. Theories of culture. *An. Rev. Anthropol.* 3: 73–97.

Kelley, J. L., J. P. Evans, I. W. Ramnarine, and A. E. Magurran. 2003. Back to school: Can
antipredator behaviour in guppies be enhanced through social learning? *Anim.
Behav.* 65: 655–662.

Kendal, J. R., L. A. Giraldeau, and K. N. Laland. 2009. The evolution of social learning
rules: Payoff-biased and frequency-dependent biased transmission. *J. Theo.
Biol.* 260: 210–219.

Kendal, J, R., R. L. Kendal, and K. N. Laland. 2007. Quantifying and modelling social learning processes in monkey populations. *Int. J. Psychol. Psychol. Ther.* 7: 123–138.

Kendal, R. L., R. L. Coe, and K. N. Laland. 2005. Age differences in neophilia, exploration and innovation in family groups of Callitrichid monkeys. *Am. J. Primatol.*
66: 167–188.

Kendal, R. L., I. Coolen, and K. N. Laland. 2004. The role of conformity in foraging
when personal and social information conflict. *Behav. Ecol.* 15: 269–277.

Kendal, R. L., I. Coolen, Y. van Bergen, and K. N. Laland. 2005. Tradeoffs in the adaptive
use of social and asocial learning. *Adv. Stud. Behav.* 35: 333–379.

Kendal, R. L., D. M. Custance, J. R. Kendal, G. Vales, T. S. Stoinski, N. L. Rakotomalala,
H. Rasamimanana. 2010. Evidence for social learning in wild lemurs (*Lemur
catta*). *Learn. Behav.* 38: 220–234.

Kendal, R. L., B. G. Galef, Jr., and C. van Schaik. 2010. Social learning research outside
the laboratory: How and why? *Learn. Behav.* 38: 187–194.

Kendal, R. L., J. R. Kendal, W. Hoppitt, and K. N. Laland. 2009. Identifying social learning in animal populations: A new 'option-bias' method. *PloS ONE* 4: e6541, doi:10.1371/journal.pone.0006541.

Kenward, B., A. Weir, C. Rutz, and A. Kacelnik. 2005. Tool manufacture by naïve juvenile crows. *Nature* 433: 121.

Kerr, B., C. Neuhauser, B.J.M. Bohannan, and A. M. Dean. 2006. Local migration promotes competitive restraint in a host-pathogen 'tragedy of the commons'. *Nature* 442: 75–78.

Keysers, C., and D. I. Perrett. 2004. Demystifying social cognition: A Hebbian perspective. *Trends Cogn. Sci.* 8: 501–507.

King, A. J., and G. Cowlishaw. 2007. When to use social information: The advantage of large group size in individual decision making. *Biol. Let.* 3: 137–139.

Kirby, S., H. Cornish, and K. Smith. 2008. Cumulative cultural evolution in the laboratory: An experimental approach to the origins of structure in human language. *Proc. Nat. Acad. Sci. USA* 105: 10681–10686.

Kirkpatrick, M., and L. A. Dugatkin. 1994. Sexual selection and the evolutionary effects of copying mate choice. Behav. *Ecol. Sociobiol.* 34: 443–449.

Kishino, H., and M. Hasegawa, M. 1989. Evaluation of the maximum likelihood estimate of the evolutionary tree topologies from DNA sequence data, and the branching order in Hominoidea. *J. Mol. Evol.* 29: 170–179.

Kitowski, I. 2009. Social learning of hunting skills in juvenile marsh harriers (*Circus aeruginosus*). *J. Ethol.* 27: 327–332.

Klein, E. D., and T. R. Zentall. 2003. Imitation and affordance learning by pigeons (*Columba livia*). *J. Comp. Psychol.* 117: 414–419.

Klucharev, V., K. Hytonen, M. Rijpkema, A. Smidts, and G. Fernandez. 2009. Reinforcement learning signal predicts social conformity. *Neuron* 61: 140–151.

Kokko, H. 2007. *Modelling for Field Biologists and Other Interesting People.* Cambridge, U.K.: Cambridge University Press.

Koops, M. A. 2004. Reliability and the value of information. *Anim. Behav.* 67: 103–111.

Krause, J. 1993. Transmission of fright reaction between different species of fish. *Behaviour* 127: 37–48.

Krebs, C. R. 1989. *Ecological Methodology.* New York: Harper and Row.

Krebs, J. R., and N. B. Davies, eds. 1991. *Behavioural Ecology: An Evolutionary Approach,* 3rd ed. Oxford, U.K.: Blackwell Scientific Publications.

Krebs, J. R., and N. B. Davies. 1993. *An Introduction to Behavioural Ecology,* 3rd ed. Oxford, U.K.: Blackwell Scientific Publications.

Krebs, J. R., M. H. MacRoberts, and J. M. Cullen. 1972. Flocking and feeding in the great tit (*Parus major*): An experimental study. *Ibis* 114: 507–530.

Kroeber, A. L., and C. Kluckholm. 1952. Culture: A critical review of the concepts and definitions. *Papers of the Peabody Museum of American Archaeology and Ethnology* 47: 1–223.

Krützen, M., J. Mann, M. R. Heithaus, R. C. Connor, L. Bejder, et al. 2005. Cultural transmission of tool use in bottlenose dolphins. *Proc. Natl. Acad. Sci. USA* 102: 8939–8943.

Krützen, M., C. van Schaik, and A. Whiten. 2007. The animal cultures debate: Response to Laland and Janik. *Trends Ecol. Evol.* 22: 6.

Kübler, D., and G. Weizsäcker. 2004. Limited depth of reasoning and failure of cascade formation in the laboratory. *Rev. Econ. Stud.* 71: 425–441.

Kuhl, P. K., and A. N. Meltzoff. 1996. Infant vocalizations in response to speech: Vocal imitation and developmental change. *J. Acoust. Soc. Am.* 100: 2425–2438.

Kumm, J., K. N. Laland, and M. W. Feldman. 1994. Gene-culture coevolution and sex ratios: The effects of infanticide, sex-selective abortion, and sex-biased parental investent on the evolution of sex ratios. *Theor. Pop. Biol.* 46: 249–278.

Kummer, H., and J. Goodall. 1985. Conditions of innovative behaviour in primates. *Phil. Trans. R. Soc. B* 308: 203–214.

Kuper, A. 2000. If memes are the answer, what is the question? In *Darwinizing Culture: The Status of Memetics as a Science*, edited by R. Aunger, 175–188. Oxford, U.K.: Oxford University Press.

Lachlan, R., L. Crooks, and K. N. Laland. 1998. Who follows whom? Shoaling preferences and social learning of foraging information in guppies. *Anim. Behav.* 56: 181–190.

Lachlan, R. F., and M. W. Feldman. 2003. Evolution of cultural communication systems: The coevolution of cultural signals and genes encoding learning preferences. *J. Evol. Biol.* 16: 1084–1095.

Lachlan, R. F., and P.J.B. Slater. 1999. The maintenance of vocal learning by gene-culture interaction: The cultural trap hypothesis. *Proc. R. Soc. Lond. B* 266: 701–706.

Laland, K. N. 1994. Sexual selection with a culturally transmitted mating preference. *Theo. Pop. Biol.* 45: 1–15.

Laland, K. N. 2004. Social learning strategies. *Learn. Behav.* 32: 4–14.

Laland, K. N. 2008. Exploring gene-culture interactions: Insights from handedness, sexual selection and niche construction case studies. In "Cultural Transmission and the Evolution of Human Behaviour," edited by K. Smith special issue, *Phil. Trans. R. Soc.* 363: 3577–3589.

Laland, K. N., and P.P.J. Bateson. 2001. The mechanisms of imitation. *Cybern. Syst.* 32: 195–224.

Laland, K. N., and G. R. Brown. 2011. *Sense and Nonsense: Evolutionary Perspectives on Human Behaviour*, 2nd ed. Oxford, U.K.: Oxford University Press.

Laland, K. N., and B. G. Galef, Jr., eds. 2009. *The Question of Animal Culture*. Cambridge, MA: Harvard University Press.

Laland, K. N., and W.J.E. Hoppitt. 2003. Do animals have culture? *Evol. Anthropol.* 12: 150–159.

Laland, K. N., and V. Janik. 2006. The animal cultures debate. *Trends Ecol. Evol.* 21: 542–547.

Laland, K. N., and V. Janik. 2007. Response to Krützen et al.: Further problems with the 'method of exclusion.' *Trends Ecol. Evol.* 22: 7.

Laland, K. N., and J. R. Kendal. 2003. What the models say about social learning. In *The Biology of Traditions: Models and Evidence*, edited by D. Fragaszy and S. Perry, 33–55. Cambridge, U.K.: Cambridge University Press.

Laland, K. N., J. R. Kendal, and R. L. Kendal. 2009. Animal culture: Problems and solutions. In *The Question of Animal Culture*, edited by K. N. Laland and B. G. Galef, Jr. 174–197. Cambridge, MA: Harvard University Press.

Laland, K. N., J. Kumm, J. D. Van Horn, and M. W. Feldman. 1995. A gene-culture model of handedness. *Behav. Genet.* 25: 433–445.

Laland, K. N., J. Kumm, and M. W. Feldman. 1995. Gene-culture coevolutionary theory: A test case. *Curr. Anthropol.* 36: 131–56.

Laland, K. N., F. J. Odling-Smee, and M. W. Feldman. 2000. Niche construction, biological evolution and cultural change. *Behav. Brain Sci.* 23: 131–146.

———. 2001. Cultural niche construction and human evolution. *J. Evol. Biol.* 14: 22–33.

Laland, K. N., F. J. Odling-Smee, and S. Myles. 2010. How culture has shaped the human genome: Bringing genetics and the human sciences together. *Natl. Rev. Genet.* 11: 137–148.

Laland, K. N., and H. C. Plotkin. 1990. Social learning and social transmission of foraging information in Norway rats (*Rattus norvegicus*). *Anim. Learn. Behav.* 18: 246–251.

———. 1991. Excretory deposits surrounding food sites facilitate social learning of food preferences in Norway rats. *Anim. Behav.* 41: 997–1005.

———. 1992. Further experimental analysis of the social learning and transmission of foraging information amongst Norway rats. *Behav. Processes* 27: 53–64.

———. 1993. Social transmission of food preferences amongst Norway rats by marking of food sites, and by gustatory contact. *Anim. Learn. Behav.* 21: 35–41.

Laland, K. N., and S. M. Reader. 1999. Foraging innovation in the guppy. *Anim. Behav.* 52: 331–340.

———. 2009. Comparative perspectives on human innovation. In *Innovation in Cultural Systems*, edited by M. O'Brien and S. Shennan, 37–51. Cambridge, U.K.: MIT Press.

Laland, K. N., P. J. Richerson, and R. Boyd. 1993. Animal social learning: Towards a new theoretical approach. In *Behavior and Evolution*, edited by P.P.G. Bateson, P. H. Klopfer, and N. S. Thompson. Vol. 10, *Perspectives in Ethology*. New York: Plenum Press.

Laland, K. N., and K. Williams. 1997. Shoaling generates social learning of foraging information in Guppies. *Anim. Behav.* 53: 1161–1169.

———. 1998. Social transmission of maladaptive information in the guppy. *Behav. Ecol.* 9:493–499.

Langergraber, K. E., C. Boesch, E. Inoue, M. Inoue-Marayama, J. C. Mitani, et al. 2010. Genetic and 'cultural' similiarity in chimpanzees. *Proc. R. Soc. B* 278: 408–416.

Latané, B. 1981. The psychology of social impact. *Am. Psychol.* 36: 34–356.

Latané, B. 1996. Dynamic social impact: The creation of culture by communication. *J. Comm.* 4: 13–25.

Latané B., and S. Wolf. 1981. The social impact of majorities and minorities. *Psychol. Rev.* 88: 438–453.

Leadbeater, E., and L. Chittka. 2007. The dynamics of social learning in an insect model, the bumblebee (*Bombus terrestris*). *Behav. Ecol. Sociobiol.* 61: 1789–1796.

Leadbeater, E., N. E. Raine, and L. Chittka. 2006. Social learning: Ants and the meaning of teaching. *Curr. Biol.* 16: R323e–R325.

Leenders, R. 2002. Modeling social influence through network autocorrelation: Constructing the weight matrix. *Soc. Net.* 24: 21–47.

Lefebvre, L. 1986. Cultural diffusion of a novel food-finding behaviour in urban pigeons: An experimental field test. *Ethology* 71: 295–304.

———. 1995. The opening of milk bottles by birds: Evidence for accelerating learning rates, but against the wave-of-advance model of cultural transmission. *Behav. Proc.* 34: 43–53.

Lefebvre, L., and L. Giraldeau. 1994. Cultural transmission in pigeons is affected by the number of tutors and bystanders present. *Anim. Behav.* 47: 331–337.

———. 1996. Is social learning an adaptive specialisation? In *Social Learning: The Roots of Culture*, edited by C. M. Heyes and B. G. Galef, Jr., 107–128. London: Academic Press.

Lefebvre, L., and B. Palameta. 1988. Mechanisms, ecology and population diffusion of socially learned, food-finding behaviour in feral pigeons. In *Social Learning: Psychological and Biological Perspectives*, edited by B. G. Galef, Jr. and T. R. Zentall, 141–164. Hillsdale, NJ: Erlbaum.

Lefebvre, L., S. M. Reader, and D. Sol. 2004. Brains, innovations and evolution in birds and primates. *Brain Behav. Evol.* 63: 233–246.

Lefebvre, L., J. Templeton, W. Brown, and M. Koelle. 1997. Carib grackles imitate conspecific and Zenaida dove tutors. *Behaviour* 134: 1003–1017.

Legendre, P., and L. Legendre. 1998. *Numerical Ecology*, 2nd English ed. Amsterdam: Elsevier.

Leggio, M., M. Molinari, P. Neri, A. Graziano, L. Mandolesi, et al. 2000. Representation of actions in rats: The role of cerebellum in learning spatial performances by observation. *Proc. Natl. Acad. Sci. USA* 97: 2320–2325.

Lindeyer, C. M., and S. M. Reader. 2010. Social learning of escape routes in zebrafish and the stability of behavioural traditions. *Anim. Behav.* 79: 827–834.

Little, A. C., R. P. Burriss, B. C. Jones, L. M. DeBruine, and C. C. Caldwell. 2008. Social influence in human face preference: men and women are influenced more for longterm than short-term attractiveness decisions. *Evol. Hum. Behav.* 29: 140–146.

Lonsdorf, E. V. 2006. What is the role of mothers in the acquisition of termite-fishing behaviors in wild chimpanzees (*Pan troglodytes schweinfurthii*)? *Anim. Cogn.* 9: 36–46.

Lonsdorf, E. V., and K. E. Bonnie. 2010. Opportunities and constraints when studying social learning: Developmental approaches and social factors. *Learn. Behav.* 38: 195–205.

Lonsdorf, E. V., E. A. Pusey, and L. Eberly. 2004. Sex differences in learning in chimpanzees. *Nature* 428: 715–716.

Lumsden, C. J., and E. O. Wilson. 1981. *Genes, Mind, and Culture: The Coevolutionary Process*. Cambridge, MA: Harvard University Press.

Lusseau, D., H. Whitehead, and S. Gero. 2008. Incorporating uncertainty into the study of animal social networks. *Anim. Behav.* 75: 1809–1815.

Lycett, S. J. 2010. The importance of history in definitions of 'culture': Implications from phylogenetic approaches to the study of social learning in chimpanzees. *Learn. Behav.* 38: 252–264.

Lycett, S. J., M. Collard, and W. C. McGrew. 2007. Phylogenetic analyses of behavior support existence of culture among wild chimpanzees. *Proc. Natl. Acad. Sci. USA* 104: 17588–17592.

———. 2009. Cladistic analyses of behavioral variation in wild Pan troglodytes: Exploring the chimpanzee culture hypothesis. *J. Hum. Evol.* 57: 337–349.

———. 2011. Correlations between genetic and behavioural dissimilarities in wild chimpanzees (*Pan troglodytes*) do not undermine the case for culture. *Proc. R. Soc. B* 278: 2091–2093.

Lyons, D., A. Young and F. Keil. 2007. The hidden structure of overimitation. *Proc. Natl. Acad. Sci. USA* 104, 19751–19756.

Mace, R., and C. J. Holden. 2005. A phylogenetic approach to cultural evolution. *Trends Ecol. Evol.* 20: 116–121.

Madden, J. R. 2008. Do bowerbirds exhibit cultures? *Anim. Cogn.* 11: 1–12.

Madden, J. R., T. J. Lowe, H. V. Fuller, K. K. Dasmahapatra, and R. L. Coe. 2004. Local traditions of bower decoration by spotted bowerbirds in a single population. *Anim. Behav.* 68: 759–765.

Maechler, M., P. Rousseeuw, A. Struyf, M. Hubert, and K. Hornik. 2012. cluster: Cluster Analysis Basics and Extensions. Software package in 'R.' Version 1.14.2, http://CRAN.R-project.org/package=cluster/.

Maloney, R. F., and I. G. MacLean. 1995. Historical and experimental learned predator recognition in free-living New Zealand robins. *Anim. Behav.* 50: 1193–1201.

Manly , B. 2008. *Randomization, Bootstrap and Monte Carlo Methods in Biology*, 3rd ed. London: Chapman and Hall.

Marler, P., and S. S. Peters. 1989. Species differences in auditory responsiveness in early vocal learning. In *The Comparative Psychology of Audition: Perceiving Complex Sounds*, edited by S. Hulse and R. Dooling, 243–273. Hillsdale, NJ: Lawrence Erlbaum.

Marler, P., and M. Tamura. 1964. Culturally transmitted patterns of vocal behaviour in sparrows. *Science* 146: 1483–1486.

Mason, J. R. 1988. Direct and observational learning by red-winged blackbirds (*Agelaius Phoeniceus*): The importance of complex stimuli. In *Social Learning: Psychological and Biological Perspectives*, edited by B. G. Galef, Jr. and T. R. Zentall, 99–117. Hillsdale, NJ : Lawrence Erlbaum.

Matsuzawa, T. 1994. Field experiments on use of stone tools by chimpanzees in the wild. In *Chimpanzee Cultures*, edited by R. Wrangham, W. McGrew, F.B.M. de Waal, P. Heltne and J. Goodall, 351–370. Cambridge, MA: Harvard University Press.

Matthews, L. J. 2009. Intragroup behavioral variation in whitefronted capuchin monkeys (*Cebus albifrons*): Mixed evidence for social learning inferred from new and established analytical methods. *Behaviour* 146: 295–324.

———. 2012. Variations in sexual behavior among capuchin monkeys function for conspecific mate recognition: A phylogenetic analysis and a new hypothesis for female proceptivity in tufted capuchins. *Am. J. Primatol.* 74: 287–298.

Maxwell, R. S. 1936. Remembering in different social groups. *Br. J. Psychol.* 27: 30–40.

Maynard Smith, J. 1982. *Evolution and the Theory of Games*. Cambridge, U.K.: Cambridge University Press.

McDougall, W. 1936. *An Outline of Psychology*, 7th ed. London: Methuen.

McDougall, W. 1945. *An Introduction to Social Psychology*, 26th ed. London: Methuen.

McElreath, R., M. Lubell, P. J. Richerson, T. Waring, W. Baum, et al. 2005. Applying evolutionary models to the laboratory study of social learning. *Evol. Hum. Behav.* 26: 483–508.

McElreath, R., A. Bell, C. Efferson, et al. 2008. Beyond existence and aiming outside the laboratory: Estimating frequency-dependent and pay-off-biased social learning strategies. *Phil. Trans. R. Soc. B* 363: 3515–3528.

McElreath R.M. & Boyd R. 2007. Mathematical models of social evolution: a guide for the perplexed. University of Chicago Press.

McGregor, P. K., T. M. Peake, and H. M. Lampe. 2001. Fighting fish Betta splendens extract relative information from apparent interactions: What happens when what you see is not what you get. *Anim. Behav.* 62: 1059–1065.

McGregor, A., A. Saggerson, J. Pearce, and C. Heyes. 2006. Blind imitation in pigeons (*Columba livia*). *Anim. Behav.* 72: 287–296.

McGrew, W. C. 1992. *Chimpanzee Material Culture: Implications for Human Evolution*. Cambridge, U.K.: Cambridge University Press.

McGrew, W. C. 2004. *The Cultured Chimpanzee: Reflections on Cultural Primatology*. Cambridge, U.K.: Cambridge University Press.

McGrew, W. C. 2009. Ten dispatches from the chimpanzee culture wars, plus postscript (revisiting the battlefronts). In *The Question of Animal Culture*, edited by K. N. Laland and B. G. Galef, Jr., 41–69. Cambridge, MA: Harvard University Press.

McGrew, W.C., and C.E.G. Tutin. 1978. Evidence for a social custom in wild chimpanzees? *Man* 13: 234–251.

McPhail, E. M. 1982. *Brain and Intelligence in Vertebrates*. Oxford, U.K.: Oxford Science Publications.

McQuoid, L. M., and B. G. Galef, Jr. 1993. Social stimuli influencing feeding behaviour of Burmese fowl: A video analysis. *Anim. Behav.* 46: 13–22.

Meltzoff, A. N., and M. K. Moore. 1977. Imitation of facial and manual gestures by human neonates. *Science* 198: 75–78.

Meltzoff, A. N., and W. Prinz, eds. 2002. *The Imitative Mind*. Cambridge, U.K.: University of Cambridge Press.

Menzel, E. W., Jr. 1973a. Leadership and communication in young chimpanzees. In *Precultural Primate Behavior*, edited by E. W. Menzel, Jr., 192–225. Basel: Karger.

———. 1973b. *Precultural Primate Behavior*. Vol. 1, *Precultural Primate Behavior*. Berlin: Karger.

———. 1974. A group of young chimpanzees in a one acre field. In *Behavior of Nonhuman Primates: Modern Research Trends*, edited by A. M. Schrier and F. Stollnitz, 83–153. Vol. 5, *Precultural Primate Behavior*. New York: Academic Press.

Menzel, E. W., R. K. Davenport, and C. M. Rogers. 1972. Protocultural aspects of chimpanzees responsiveness to novel objects. *Folia Primatol.* 17: 161–170.

Mesoudi, A. 2008. An experimental simulation of the 'copy-successful-individuals' cultural learning strategy: Adaptive landscapes, producer-scrounger dynamics, and informational access costs. *Evol. Hum. Behav.* 29: 350–363.

———. 2011. *Cultural Evolution: How Darwinian Theory can Explain Human Culture and Synthesize the Social Sciences*. Chicago: University of Chicago Press.

Mesoudi, A., and K. N. Laland. 2007. Extending the behavioral sciences framework: Clarifying methods, predictions and concepts. *Behav. Brain Sci.* 30: 1–61.

Mesoudi, A., and S. Lycett. 2009. Random copying, frequency-dependent copying and cultural change. *Evol. Hum. Behav.* 30: 41–48.

Mesoudi, A., and M. J. O'Brien. 2008. The learning and transmission of hierarchical cultural recipes. *Biol. Theory* 3: 63–72.

Mesoudi, A., and A. Whiten. 2004. The hierarchical transformation of event knowledge in human cultural transmission. *J. Cogn. Cult.* 4: 1–24.

Mesoudi, A., A. Whiten, and K. N. Laland. 2004. Is human cultural evolution Darwinian? Evidence reviewed from the perspective of The Origin of Species. *Evolution* 58: 1–11.

———. 2006. Towards a unified science of cultural evolution. *Behav. Brain Sci.* 29: 329–383.

Miller, N. E., and J. Dollard. 1941. *Social Learning and Imitation*. New Haven, CT: Yale University Press.

Milligan, G. W., and M. C. Cooper. 1985. An examination of procedures for determining the number of clusters in a data set. *Psychometrika* 50: 159–179.

Mineka, S., and M. Cook. 1988. Social learning and the acquisition of snake fear in monkeys. In *Social Learning: Psychological and Biological Perspectives*, edited by B. G. Galef, Jr. and T. R. Zentall, 51–73. Hillsdale, NJ: Lawrence Erlbaum.

Mitchell, C. J., C. M. Heyes, M. R. Gardner, and G. R. Dawson. 1999. Limitations of a bidirectional control procedure for the investigation of imitation in rats: Odour cues on the manipulandum. *Q. Rev. Exp. Psychol.* 52: 193–202.

Moore, B. R. 1992. Avian movement imitation and a new form of mimicry: Tracing the evolution of a complex form of learning. *Behaviour* 122: 231–263.

———. 1996. The evolution of imitative learning. In *Social Learning in Animals: The Roots of Culture*, edited by C. M. Heyes and B. G. Galef, Jr., 245–266. San Diego: Academic Press.

Morand-Ferron, J., and L.-A. Giraldeau. 2010. Learning behaviorally stable solutions to producer-scrounger games. *Behav. Ecol.* 21: 343–348.

Morand-Ferron, J., and J. L. Quinn. 2011. Larger groups of passerines are more efficient problem-solvers in the wild. *Proc. Natl. Acad. Sci. USA* 108:15898–15903.

Morand-Ferron, J., E. Cole, J. Rawles, et al. 2011. Who are the innovators? A field experiment with 2 passerine species. *Behav. Ecol.* 22: 1241–1248.

Morgan, B.J.T. 2010. *Applied Stochastic Modelling*, 2nd ed. Boca Raton, FL: CRC.

Morgan, C. L. 1896a. On modification and variation. *Science* 4: 733–740.

———. 1896b. *Habit and Instinct*. London: Edward Arnold.

———. 1900. *Animal Behaviour*. London: Edward Arnold.

Morgan, L. H. 1877. *Ancient Society, or Researches in the Lines of Human Progress from Savagery through Barbarism to Civilization*. New York: Holt.

Morgan, T.J.H., L. E. Rendell, M. Ehn, W. Hoppitt, and K. N. Laland. 2012. The evolutionary basis of human social learning. *Proc. R. Soc. B* 279: 653–662, doi: 10.1098/rspb.2011.1172.

Morrell, L. J., D. P. Croft,, J.R.G. Dyer., B. B. Chapman, J. L. Kelley, et al. 2008. Association patterns and foraging behaviour in natural and artificial guppy shoals. *Anim. Behav.* 76: 855–864.

Moscarini, G., M. Ottaviani, and L. Smith. 1998. Social learning in a changing world. *Econ. Theory* 11: 657–665.

Mueller, C., and M. Cant. 2010. Imitation and traditions in wild banded mongooses. *Curr. Biol.* 20: 1171–1175.

Mullen, B. 1985. Strength and immediacy of sources: A meta-analytic evaluation of the forgotten elements of social impact theory. *J. Pers. Soc. Psychol.* 48: 1458–1466.

Mundinger, P. C. 1980. Animal cultures and a general theory of cultural evolution. *Ethol. Sociobiol.* 1: 183–223.

Munger, S., T. Leinders-Zufall, L. McDougall, R. Cockerham, A. Schmid, et al. 2010. An olfactory subsystem that detects carbon disulfide and mediates food-related social learning. *Curr. Biol.* 20: 1438–1444.

Nagell, K., R. Olguin, and M. Tomasello. 1993. Processes of social learning in the tool use of chimpanzees (*Pan troglodytes*) and human children (*Homo sapiens*). *J. Comp. Psychol.* 107:174–186.

Nakahashi, W. 2007. The evolution of conformist transmission in social learning when the environment changes periodically. *Theor. Popul. Biol.* 72: 52–66.

Nehaniv, C. L., and K. Dautenhahn. 2002. The correspondence problem. In *Imitation in Animals and Artifacts*, edited by K. Dautenhahn and C. Nehaniv, 41–61. Cambridge, MA: MIT Press.

Nesse, R. M. 1990. Evolutionary explanations of emotions. *Hum. Nat.* 1: 261–289.

Newman, M.E.J. 2010. *Networks: An Introduction*. Oxford, U.K.: Oxford University Press.

Nicol, C. J., and S. J. Pope. 1994. Social learning in small flocks of laying hens. *Anim. Behav.* 47: 1289–1296.

———. 1996. The maternal display of domestic hens is sensitive to perceived chick error. *Anim. Behav.* 52: 767–774.

Nishida, T. 1987 Local traditions and cultural transmission. In *Primate Societies*, edited by B. Smuts, D. Cheney, R. Seyfarth , R. Wrangham, and T. Struhsaker, 462–474. Chicago: University of Chicago Press.

Noad, M. J., D. H. Cato, M. M. Bryden, M. N. Jenner, and K. C. Jenner. 2000. Cultural revolution in whale songs. *Nature* 408: 537.

Norenzayan. A., S. Atran, J. Faulkner, and M. Schaller. 2006. Memory and mystery: The cultural selection of minimally counterintuitive narratives. *Cogn. Sci.* 30: 531–553.

Northway, M. L. 1936. The influence of age and social group on children's remembering. *Br. J. Psychol.* 27: 11–29.

Nottebohm, F. 1981. A brain for all seasons: Cyclical anatomical changes in song control nuclei of the canary brain. *Science* 214: 1368–1370.

Nowak, M. A., and R. M. May. 1992. Evolutionary games and spatial chaos. *Nature* 359: 826–829.

Nowak, A., J. Szamrej, and B. Latane. 1990. From private attitude to public-opinion: A dynamic theory of social impact. *Psychol. Rev.* 97: 362–376.

Nunn, C. L., C. Arnold, L. Matthews, and M. Borgerhoff Mulder. 2010. Simulating trait evolution for cross-cultural comparison. *Phil. Trans. R. Soc. B* 365: 3807–3819.

O'Brien, M. J., and R. L. Lyman. 2000. *Applying Evolutionary Archaeology*. New York: Kluwer Academic.

———. 2003. *Cladistics and Archaeology*. Salt Lake City, UT: University of Utah Press.

O'Brien, M. J., J. Darwent, and R. L. Lyman. 2001. Cladistics is useful for reconstructing archaeological phylogenies: Palaeoindian points from the southeastern United States. *J. Archaeol. Sci.* 28: 1115–1136.

Odling-Smee, F. J., K. N. Laland, and M. W. Feldman. 2003. *Niche Construction. The Neglected Process in Evolution*. Monographs in Population Biology 37. Princeton, NJ: Princeton University Press.

Olsson, A., and E. Phelps. 2007. Social learning of fear. *Natl. Neurosci.* 10: 1095–1102.

Olsson, A., K. I. Nearing, E. A. Phelps. 2007. Learning fears by observing others: The neural systems of social fear transmission. *Soc. Cogn. Affect*. Neurosci. 2: 3–11.

O'Malley, A. J., and N. A. Christakis. 2011. Longitudinal analysis of large social networks: Estimating the effect of health traits on changes in friendship ties. *Stats. Med*, 30: 950–964, doi: 10.1002/sim.4190.

Osborn, H. F. 1896. Ontogenic and phylogenic variation. *Science* 4: 786–789.

Otto, S. P., and T. Day. 2007. *A Biologist's Guide to Mathematical Modeling in Ecology and Evolution*. Princeton, NJ: Princeton University Press.

Otto, S. P., F. B. Christiansen, and M. W. Feldman. 1995. *Genetic and Cultural Inheritance of Continuous Traits*. Morrison Institute for Population and Resource Studies Working Paper no. 64. Stanford CA: Stanford University Press.

Overington, S., J. Morand-Ferron, N. Boogert, et al. 2009. Technical innovations drive the relationship between innovativeness and residual brain size in birds. *Anim. Behav.* 78: 1001–1010.

Pagel, M. 2012. *Wired for Culture: The Natural History of Human Cooperation*. London: Allen Lang.

Pagel, M., Q. D. Atkinson, and A. Meade. 2007. Frequency of word-use predicts rates of lexical evolution throughout Indo-European history. *Nature* 449: 717–720.

Palameta, B., and L. Lefebvre. 1985. The social transmission of a food-finding technique in pigeons: What is learned? *Anim. Behav.* 33: 892–896.

Pasqualone, A. A., and J. M. Davis. 2011. The use of conspecific phenotypic states as information during reproductive decisions. *Anim. Behav.* 82: 281–284, doi:10.1016/j.anbehav.2011.05.002

Payne, R. B. 1985. Behavioral continuity and change in local song populations of villiage indigobirds (*Vidua chalybeate*). *Z. Tierpsychol.* 70: 1–44.

Payne, K., and R. Payne. 1985. Large scale changes over 19 years in songs of humpback whales in Bermuda. *Z. Tierpsychol.* 68: 89–114.

Pearl, J. 1988. *Probabilistic Reasoning in Intelligent Systems: Networks of Plausible Inference.* San Mateo, CA: Morgan Kauffmann.

———. *Causality: Models Reasoning and Inference.* 2nd ed. Cambridge, U.K.: Cambridge University Press.

Pearson, A. T. 1989. *The Teacher: Theory and Practice in Teacher Education.* New York: Routledge.

Perry, S. 2009. Social influence and the development of food processing techniques in wild white-faced capuchin monkeys (*Cebus capucinus)* at Lomas Barbudal, Coasta Rica. *Amer. Soc. Primatol.* 71: 99.

Perry, S., M. Baker, L. Fedigan, J. Gros-Louis, K. Jack, et al. 2003. Social conventions in wild white-faced capuchins: Evidence for traditions in a Neotropical primate. *Curr. Anthropol.* 44: 241–268.

Petrosini, L., A. Graziano, L. Mandolesi, P. Neri, M. Molinari, et al. 2003. Watch how to do it! New advances in learning by observation. *Brain Res. Rev.* 42: 252–264.

Piaget, J. 1962. *Play, Dreams, and Imitation in Childhood.* New York: Norton.

Pike, T., and K. N. Laland. 2010. Conformist learning in nine-spined sticklebacks' foraging decisions. *Biol. Let.* 6: 466–468.

Pike, T. W., J. R. Kendal, L. E. Rendell, and K. N. Laland. 2010. Learning by proportional observation in a species of fish. *Behav. Ecol.* 21: 570–575.

Pinasco, J. P., and L. Romanelli. 2006. Coexistence of language is possible. *Physica A* 361: 355–360.

Pinheiro, J. C., and D. M. Bates. 2000. *Mixed-Effects Models in S and S-PLUS.* New York: Springer.

Pinker, S. 1997. *How the Mind Works.* London: Allen Lane.

Plotkin, H. 1994. *Darwin Machines and the Nature of Knowledge.* New York: Penguin.

Pongrácz, P., Á. Miklósi, E. Kubinyi, J. Topál, and V. Csányi. 2003. Interaction between individual experience and social learning in dogs. *Anim. Behav.* 65: 595–603.

Popik, P., and J. M. van Ree. 1993. Social transmission of flavored tea preferences: Facilitation by a vasopressin analog and oxytocin. *Behav. Neur. Biol.* 59: 63–68.

Powell, A., S. Shennan, and M. Thomas. 2009. Late Pleistocene demography and the appearance of modern human behavior. *Science* 324: 1298–1301.

Prather, J. F., S. Peters, S. Nowicki, and R. Mooney. 2008. Precise auditory-vocal mirroring in neurons for learned vocal communication. *Nature* 451: 305–310.

Preyer, W. T. 1889. *The Mind of the Child,* 2nd ed. New York: D Appleton.

R Core Development Team. 2008. *R: A Language and Environment for Statistical Computing.* Vienna: R Foundation for Statistical Computing.

R Core Development Team. 2011. *R: A Language and Environment for Statistical Computing,* 2nd ed. Vienna: R Foundation for Statistical Computing.

Raafat, R. M., N. Chater, and C. Frith. 2009. Herding in humans. *Trends Cogn. Sci.* 13: 420–428.

Raihani, N. J., and A. R. Ridley. 2008. Experimental evidence for teaching in wild pied babblers. *Anim. Behav.* 75: 3–11.

Ramsey, G., M. Bastian, and C. van Schaik. 2007. Animal innovation defined and operationalized. *Behav. Brain Sci.* 30: 393–437.

Range, F., Z. Viranyi, and L. Huber. 2007. Selective imitation in domestic dogs. *Curr. Biol.* 17: 1–5.

Rapaport, L. M., and G. R. Brown. 2008. Social influences on foraging behaviour in young non-human primates: Learning what, where, and how to eat. *Evol. Anthropol.* 17: 189–201.

Ray, E. D. 1997. Social and Associative Learning. PhD diss., University of London. London, England.

Reader, S. M. 2000. Social Learning and Innovation: Individual Differences, Diffusion Dynamics and Evolutionary Issues. University of Cambridge, Cambridge, U.K.

Reader, S. M. 2004. Distinguishing social and asocial learning using diffusion dynamics. *Learn. Behav.* 32: 90–104.

Reader, S. M., and D. Biro. 2010. Experimental identification of social learning in wild animals. *Learn. Behav.* 38: 265–283.

Reader, S. M., M. Bruce, and S. Rebers. 2008. Social learning of novel route preferences in adult humans. *Biol. Let.* 4: 37–40.

Reader, S. M., Y. Hager, and K. N. Laland. 2011. The evolution of primate general and cultural intelligence. *Phil. Trans. R. Soc. Lond. B* 366: 1017–1027.

Reader, S. M., and K. N. Laland. 2000. Diffusion of foraging innovation in the Guppy. *Anim. Behav.* 60: 175–180.

———. 2001. Primate innovation: Sex, age and social rank differences. *Intl. J. Primatol.* 22: 787–805.

———. 2002. Social intelligence, innovation and enhanced brain size in primates. *Proc. Natl. Acad. Sci. USA* 99: 4436–4441.

———, eds. 2003a. *Animal Innovation.* Oxford, U.K.: Oxford University Press.

———. 2003b. Animal innovation: An introduction. In *Animal Innovation*, edited by S. M. Reader and K. N. Laland, 3–38. Oxford, U.K.: Oxford University Press.

Rendell, L., R. Boyd, D. Cownden, M. Enquist, K. Eriksson, et al. 2010. Why copy others? Insights from the social learning strategies tournament. *Science* 327: 208213.

Rendell, L., L. Fogarty, W.J.E. Hoppitt, T.J.H. Morgan, M. M. Webster, et al. 2011. Cognitive culture: Theoretical and empirical insights into social learning strategies. *Trends Cogn. Sci.* 15: 68–76.

Rendell, L., L. Fogarty, and K. N. Laland. 2009. Roger's paradox recast and resolved: Population structure and the evolution of social learning strategies. *Evolution* 63: 534–548.

Rendell, L., S. L. Mesnick, M. L. Dalebout, J. Burtenshaw, and H. Whitehead. 2011. Can genetic differences explain vocal dialect variation in Sperm whales (*Physeter macrocephalus*)? *Behav. Genet.* 42: 332–343, doi: 10.1007/s10519-011-9513-y

Rendell, L., and H. Whitehead. 2001. Culture in whales and dolphins. *Behav. Brain Sci.* 24: 309–324.

Reyes-García, V., T. McDade, V. Vadez, T. Huanca, W. R. Leonard, S. Tanner, et al., 2008. Nonmarket returns to traditional human capital: Nutritional status and traditional knowledge in a native Amazonian society. *J. Dev. Stud.* 44: 217–232.

Richerson, P., and R. Boyd. 2005. *Not by Genes Alone: How Culture Transformed Human Evolution.* Chicago: University of Chicago Press.

Richerson, P., R. Boyd, and J. Henrich. 2010. Gene-culture coevolution in the age of genomics. *Proc. Natl. Acad. Sci. USA* 107: 8985–8992.

Rizzolatti, G., and M. A. Arbib. 1998. Language within our grasp. *Trends Neurosci.* 21: 188–194.

Rizolatti, G., and L. Craighero. 2004. The mirror-neuron system. *Annu. Rev. Neurosci.* 27: 169–192.

Rizzolatti, G., L. Fadiga, M. Matelli, V. Bettinardi, E. Pauleus, et al. 1996. Localization of grasp representation in humans by PET: 1. Observation versus execution. *Exp. Brain Res.* 111: 246–252.

Rogers, A. R. 1988. Does biology constrain culture? *Am. Anthropol.* 90: 819–31.

Rogers, E. 1995. *Diffusion of Innovations.* New York: The Free Press.

Roitblatt, H. L. 1998. Mechanisms of imitation: The relabelled story. *Behav. Brain. Sci.* 21: 701–702.

Romanes, G. J. 1882. *Animal Intelligence.* London: Kegan, Paul, Trench.

Romanes, G. J. 1884. *Mental Evolution in Animals.* New York: Appleton.

Roper, T. J. 1986. Cultural evolution of feeding behaviour in animals. *Sci. Prog.* 70: 571–583.

Rousseeuw, P. J. 1995, Fuzzy clustering at the intersection. *Technometrics* 37: 283–286.

Ryan, B., and N. C. Gross. 1943. The diffusion of hybrid seed corn in two Iowa communities. *Rural Sociol.* 8: 15–24.

Saggerson, A. L., D. N. George, and R. C. Honey. 2005. Imitative learning of stimulus-response and response-outcome associations in pigeons. *J. Exp. Psychol. Anim. Behav. Process.* 31: 289–300.

Sarin, S., and R. Dukas. 2009. Social learning about egg laying substrates in fruit flies. *Proc. R. Soc. B* 276: 4323–4328.

Sargeant, B. L., and J. Mann. 2009. Developmental evidence for foraging traditions in wild bottlenose dolphins. *Anim. Behav.* 78: 715–721.

Sargeant, B. L., A. J. Wirsing, M. R. Heithaus, and J. Mann. 2007. Can environmental heterogeneity explain individual foraging variation in wild bottlenose dolphins (*Tursiops sp.*)? *Behav. Ecol. Sociobiol.* 61: 679–688.

Scheines, R., P. Spirtes, C. Glymour, C. Meek, and T. Richardson. 1998. The TETRAD Project: Constraint Based Aids to Causal Model Specification, *Multivariate Behav. Res.* 33: 65–117.

Schlag K. H. 1998. Why imitate and if so, how? A boundedly rational approach to multi-armed bandits. *J. Econ. Theo.* 78: 130–156.

———. 1999. Which one should I imitate? *J. Math. Econ.* 31: 493–522.

Schmidt, R. A. 1991. *Motor Behavior and Control.* Leeds: Human Kinetics Europe Ltd.

Seppanen, J., J. Forsman, M. Monkkonen, et al. 2011. New behavioural trail adopted or rejected by observing heterospecific tutor fitness. *Proc. R. Soc. B* 278:1736–1741.

Shennan, S. J., and J. R. Wilkinson. 2001. Ceramic style change and neutral evolution: A case study from Neolithic Europe. *Am. Antiq.* 66: 577–594.

Sherif, M. 1936. *The Psychology of Social Norms.* Oxford, UK: Harper.

Sherry, D. F., B. G. Galef, Jr. 1984. Cultural transmission without imitation: Milk bottle opening by birds. *Anim. Behav.* 32: 937–938.

Shettleworth, S. J. 1998. *Cognition, Evolution and Behaviour.* Oxford, U.K.: Oxford University Press.

Shettleworth, S. J. 2001. Animal cognition and animal behaviour. *Anim. Behav.* 61: 277–286.

Shettleworth, S. J. 2010. *Cognition, Evolution and Behaviour*, 2nd ed. Oxford, U.K.: Oxford University Press.

Shipley, B., 1999. Testing causal explanations in organismal biology: Causation, correlation and structural equation modelling. *Oikos* 86: 374–382.

Shipley, B., 2000. *Cause and Correlation in Biology: A User's Guide to Path Analysis, Structural Equations and Causal Inference.* Cambridge, U.K.: Cambridge University Press.

Silver, M., and E. Di Paolo. 2006 Spatial effects favour the evolution of niche construction. *Theor. Popul. Biol.* 20: 387–400.

Simonton, D. K. 1999. *Origins of Genius: Darwinian Perspectives on Creativity.* Oxford, U.K.: Oxford University Press.

Simkin, M. V., and V. P. Roychowdhury. 2003. Read before you cite! *Complex Sys.* 14: 269.

Sirot, E. 2001. Mate-choice copying by females: The advantages of a prudent strategy. *J. Evol. Biol.* 14: 418–423.

Skinner, B. F. 1953. *Science and Human Behavior.* New York: Macmillan.

Slagsvold, T., and K. L. Wiebe. 2007. Learning the ecological niche. *Proc. R. Soc. B* 274: 19–23.

Slater, P.J.B. 1985. *Introduction to Ethology.* Cambridge, U.K.: Cambridge University Press.

———. 1986. The cultural transmission of bird song. *Trends Ecol. Evol.* 1: 94–97.

Slater, P.J.B., S. A. Ince, and P. W. Colgan. 1980. Chaffinch song types: Their frequencies in the population and distribution between repertoires of different individuals. *Behaviour* 75: 207–218.

Smouse, P. E., J. C. Long, and R. R. Sokal. 1986. Multiple regression and correlation extensions of the Mantel test of matrix correspondence. *Syst. Zool.* 35: 627–632.

Sol, D. 2003. Behavioral innovation: A neglected issue in the ecological and evolutionary literature? In *Animal Innovation*, edited by S. M. Reader and K. N. Laland, 63–82. Oxford, U.K.: Oxford University Press.

Sol, D., and L. Lefebvre. 2000. Behavioural flexibility predicts invasion success in birds introduced to New Zealand. *Oikos* 90: 599–605.

Sol, D., R. Duncan, T. Blackburn, P. Cassey, and L. Lefebvre. 2005. Big brains, enhanced cognition and response of birds to novel environments. *Proc. Natl. Acad. Sci. USA* 102: 5460–5465.

Sousa, C., D. Biro, and T. Matsuzawa. 2009. Leaf-tool use for drinking water by chimpanzees (*Pan troglodytes*): Acquisition patterns and handedness. *Anim. Cogn.* 12: S115–S125.

Spalding, D. A. 1873. Instinct, with original observations on young animals. *Macmillans Mag.* 27: 282–93.

Spence, K. W. 1937. Experimental studies of learning and higher mental processes in infrahuman primates. *Psychol. Bull.* 34: 806–850.

Spencer, H. 1857. Progress: its law and cause. *Westminster Review* 67: 445–485.

Sperber, D. 1996. *Explaining Culture: a Naturalistic Approach.* Oxford, U.K.: Blackwell.

Spiegelhalter, D., A. Thomas, N. Best, and D. Lunn. 2003. WinBUGS User Manual Version 1.4. MRC Biostatistics Unit. http://www.mrc-bsu.cam.ac.uk/bugs/winbugs/manual14.pdf.

Spiess, E. B. 1968. Low frequency advantage in mating of Drosophila psuedoobscura karyotypes. *Am. Natl.* 102: 363–79.

Spirtes, P. 1995. Directed cyclic graphical representation of feedback models. In *Proceedings of the 11th Conference on Uncertainty in Artificial Intelligence*, edited by Philippe Besnard and Steve Hanks, 491–498. San Mateo, CA: Morgan Kauffmann.

Spirtes, P., C. Glymour, and R. Scheines. 2001. *Causation, Prediction, and Search*, 2nd ed. Cambridge, MA: MIT Press.

Stanley, E. L., R. L. Kendal, J. R. Kendal, S. Grounds, and K. N. Laland. 2008. The effects of group size, rate of turnover and disruption to demonstration on the stability of foraging traditions in fish. *Anim. Behav.* 75: 565–572.

Stephens, D. 1991. Change, regularity and value in the evolution of learning. *Behav. Ecol.* 2: 77–89.

Stephens, D. W., and J. R. Krebs. 1986. *Foraging Theory*. Princeton, NJ: Princeton University Press.

Sterelny, K. 2009. Peacekeeping in the culture wars. In *The Question of Animal Culture*, edited by K. N. Laland and B. G. Galef, Jr., 288–304. Cambridge, MA: University of Harvard Press.

———. 2012. *The Evolved Apprentice*. Cambridge, MA: MIT Press

Sternberg, R., ed. 1999. *Handbook of Creativity*. Cambridge, U.K.: Cambridge University Press.

Strum, S. C. 1975. Primate predation: Interim report on the development of a tradition in a troop of olive baboons. *Science* 187: 755–757.

Strupp, B. J., M. Bunsey, B. Bertsche, D. A. Levitsky, M. Kesler. 1990. Enhancement and impairment of memory retrieval by a vasopressin metabolite: An interaction with the accessibility of the memory. *Behav. Neurosci.* 104: 268–276.

Subiaul, F., J. F. Cantlon, R. L. Holloway, and H. S. Terrace. 2004. Cognitive imitation in rhesus macaques. *Science* 305: 407–410.

Suboski, M. D. 1990. Releaser-induced recognition learning. *Psychol. Rev.* 9: 271–284.

Suboski, M. D., and J. J. Templeton. 1989, Life skills training for hatchery fish: Social learning and survival. *Fish. Res.* 7: 343–352.

Sugita, Y. 1980. Imitative choice behavior in guppies. *Jap. Psychol. Res.* 22: 7–12.

Swaney, W., J. R. Kendal, H. Capon, C. Brown, and K. N. Laland. 2001. Familiarity facilitates social learning of foraging behaviour in the guppy. *Anim. Behav.* 62: 591–598.

Symons, D. 1979. *The Evolution of Human Sexuality*. New York: Oxford University Press.

Takahata, Y., M. Hiraiwa-Hasegawa, H. Takasaki, and R. Nyundo. 1986. Newly-acquired feeding habits among the chimpanzees of the Mahale Mountains National Park, Tanzania. *Hum. Evol.* 1: 277–284.

Takasaki, H. 1983. Mahale chimpazees taste mangoes—toward acquisition of a new food item? *Primates* 24: 273–275.

Tanaka, M. M., J. R. Kendal, and K. N. Laland. 2009. From traditional medicine to witchcraft: Why medical treatments are not always efficacious. *PloS ONE* 4: e5192, doi:10.1371/journal.pone.0005192.

Tanford, S, and S. Penrod. 1984. Social influence model: A formal integration of research on majority and minority influence processes. *Psychol. Bull.* 95: 189–225.

Tarde, G. 1903. *The Laws of Imitation*. New York, Holt.

Tehrani, J., and M. Collard, M. 2002. Investigating cultural evolution through biological phylogenetic analyses of Turkmen textiles. *J. Anthropol. Archaeol.* 21: 443–463.

Telle, H. J. 1966. Beitrag zur Kenntnis der Verhaltensweise von Ratten, vergleichand dargestellt bei *Rattus norvegicus* and *Rattus rattus. Z. Angewandte Zool.* 53: 179–196.

Templeton, J. J., and L.-A. Giraldeau. 1996. Vicarious sampling: The use of personal and public information by starlings foraging in a simple patchy environment. *Behav. Ecol. Sociobiol.* 38: 105–114.

Terkel, J. 1995. Cultural transmission in the black rat: Pine-cone feeding. *Adv. Stud. Behav.* 24: 119–154.

———. 1996. Cultural transmission of feeding behaviour in the black rat (*Rattus rattus*). In *Social Learning in Animals*, edited by C. Heyes and B. G. Galef, Jr., 17–47. San Diego, CA: Academic Press.

Therneau, T. M., and P. Grambsch. 2000. *Modeling Survival Data: Extending the Cox Model.* New York: Springer-Verlag.

Thorndike, E. L. 1898. Animal intelligence: An experimental study of the associative processes in animals. *Psychological Review Monographs* Supplements 2 (whole no. 8), June.

Thorndike, E. L. 1911. *Animal Intelligence.* New York: Macmillan.

Thorndike, E. L. 1913. *Educational Psychology. The Original Nature of Man.* Vol. 1, *Educational Psychology.* New York: Columbia University Press.

Thornton, A., and A. Malapert. 2009. The rise and fall of an arbitrary tradition: An experiment with wild meerkats. *Proc. R. Soc. B* 276: 1269–1276.

Thornton, A., and K. McAuliffe. 2006. Teaching in wild meerkats. *Science* 313: 227–229.

Thornton, A., and N. J. Raihani. 2008. The evolution of teaching. *Anim. Behav.* 75: 1823–1836.

———. Identifying teaching in wild animals. *Learn. Behav.* 38: 297–309.

Thornton, A., N. J. Raihani, and A. N. Radford. 2007. Teachers in the wild: Some clarification. *Trends Cog. Sci.* 11: 272–273.

Thorpe, W. H. 1956. *Learning and Instinct in Animals.* London: Methuen.

———. 1963. *Learning and instinct in animals*, 2nd ed. London: Methuen.

Tilman, D., 1982. *Resource Competition and Community Structure.* Princeton, NJ: Princeton University Press.

Toelch, U., M. J. Bruce, M.T.H. Meeus, and S. M. Reader. 2010. Humans copy rapidly increasing choices in a multiarmed bandit problem. *Evol. Hum. Behav.* 31: 326–333.

Tomasello, M. 1990. Cultural transmission in the tool use and communicatory signalling ofchimpanzees? In *'Language' and Intelligence in Monkeys and Apes: Comparative Developmental Perspectives*, edited by S. T. Parker and K. R. Gibson, 274–311. Cambridge: Cambridge University Press.

———. 1994. The question of chimpanzee culture. In *Chimpanzee Cultures*, edited by R. Wrangham, W. McGrew, F.B.M. de Waal, and P. Heltne, J. Goodall, 301–317. Cambridge, MA: Harvard University Press.

———. 1996. Do Apes Ape?. In *Social Learning in Animals: The Roots of Culture*, edited by C. M. Heyes and B. G. Galef, Jr., 319–346. New York: Academic Press.

———. 1999. *The Cultural Origins of Human Cognition.* Cambridge, MA: Harvard University Press.

Tomasello, M., and J. Call. 1997. *Primate Cognition.* New York: Oxford University Press.

Tooby, J., and L. Cosmides. 1989. Evolutionary psychology and the generation of culture, part I. Theoretical considerations. *Ethol. Sociobiol.* 10: 29–49.

———. 1992. The psychological foundations of culture. In *The Adapted Mind: Evolutionary Psychology and the Generation of Culture*, edited by J. H. Barkow, L. Cosmides, and J. Tooby, 19–136. New York: Oxford University Press.

Tramontin, A. D., and E. A. Brenowitz. 1999. A field study of seasonal neuronal incorporation into the song control system of a songbird that lacks adult song learning. *J. Neurobiol.* 40: 316–326.

———. 2000. Seasonal plasticity in the adult brain. *Trends Neurosci.* 23: 251–258.

Trivers, R. L. 1971. The evolution of reciprocal altruism. *Q. Rev. Biol.* 46: 35–57.

Tylor, E. B. 1865. *Researches into the Early History of Mankind and the Development of Civilization.* London: John Murray.

———. 1871. *Primitive Culture: Researches into the Development of Mythology, Philosophy, Religion, Art, and Custom.* 2 Vols. London: John Murray.

Valente, T. 2005. Network models and methods for studying the diffusions of innovations. In *Models and Methods in Social Network Analysis*, edited by P. Carrington, J. Scott, and S. Wasserman, 98–116. Cambridge, U.K.: Cambridge University Press.

Valone, T. J. 1989. Group foraging, public information, and patch estimation. *Oikos* 56: 357–363.

van Baaren, R., L. Janssen, T. L. Chartrand, and A. Dijksterhuis. 2009. Where is the love? The social aspects of mimicry. *Phil. Trans. R. Soc. B* 364: 2381–2389.

van Bergen, Y., J. Coolen, and K. N. Laland. 2004. Nine-spined sticklebacks exploit the most reliable source when public and private information conflict. *Proc. R. Soc. B* 271: 957–962.

van de Casteele, T., P. Galbusera, and E. Matthysen. 2001. A comparison of microsatellite-based pairwise relatedness estimators. *Mol. Ecol.* 10: 1539–1549.

van der Post, D. J., and P. Hogeweg. 2006. Resource distributions and diet development by trial-and-error learning. *Behav. Ecol. Sociobiol.* 61: 65–80.

———. 2008. Diet traditions and cumulative cultural processes as side-effects of grouping. *Anim. Behav.* 75: 133–144 .

———. 2009. Cultural inheritance and diversification of diet in variable environments. *Anim. Behav.* 78: 155–166.

Van De Waal, E., N. Renevey, C. M. Favre, and R. Bshary. 2010. Selective attention to philopatric models causes directed social learning in wild vervet monkeys. *Proc. R. Soc. B* 277: 2105–2111.

van Schaik, C. P. 2009. Geographic variation in the behavior of wild great apes: Is it really Cultural? In *The Question of Animal Culture*, edited by K. N. Laland and B. G. Galef, Jr., 70–98. Cambridge, U.K.: Cambridge University Press.

van Schaik, C. P., and J. M. Burkart. 2011. Social learning and evolution: The cultural intelligence hypothesis. *Phil. Trans. R. Soc. B* 366: 1008–1016.

van Schaik, C. P., M. Ancrenaz, G. Borgen, B. Galdikas, C. D. Knott, et al. 2003. Orangutan cultures and the evolution of material culture. *Science* 299: 102–105.

van Schaik, C. P., M. A. van Noordwijk, and S. Wich. 2006. Innovation in wild Bornean orangutans (*Pongo pygmaeus wurmbii*). *Behaviour* 143: 839–876.

van Schaik, C. P., K. Isler, and J. M. Burkart. 2012. Explaining brain size variation: From social to cultural brain. *Trends Cog. Sci.* 16: 277–284.

Verma, T., and J. Pearl. 1991. Equivalence and synthesis of causal models. In *Proceedings of the Sixth Conference on Uncertainty in Artificial Intelligence*, edited by P. Bonissone, M. Henrion, L. Kanal, and J. Lemmer, 255–268. New York: Elsevier.

Visalberghi, E., and E, Adessi. 2000. Seeing group members eating a familiar food enhances the acceptance of novel foods in capuchin monkeys. *Anim. Behav.* 60: 69–76.

Visalberghi, E., and D. Fragaszy. 1995. The behaviour of capuchin monkeys (*Cebus paella*) with food: The role of social context. *Anim. Behav.* 49:1089–1095.

———. 2002. 'Do monkeys ape?' Ten years after. In *Imitation in Animals and Artefacts*, edited by K. Dautenhahn and C. Nehaniv, 471–499. Cambridge, MA: MIT Press.

Visalberghi, E., D. Fragaszy, E. Ottoni, P. Izar, M. G. de Oliveira, et al. 2007. Characteristics of hammer stones and anvils used by wild bearded capuchin monkeys (*Cebus libidinosus*) to crack open palm nuts. *Am. J. Phys. Anthropol.* 132: 426–444.

Voelkl, B., and L. Huber. 2007. Imitation as faithful copying of a novel technique in marmoset monkeys. *PloS ONE* 2: e611, doi: 10.1371/journal.pone.0000611.

Waite, R. K. 1981. Local enhancement for food finding by rooks (*Corvus frugeligus*) foraging on grassland. *Z. Tierpsychol.* 57: 15–36.

Wakano, J. Y., and K. Aoki. 2007. Do social learning and conformist bias coevolve? Henrich and Boyd revisited. *Theor. Popul. Biol.* 72: 504–512.

Wallace , A. R. 1864. The origin of human races and the antiquity of man as deduced from the theory of 'natural selection.' *Anthropol. Rev.* 2: clviii–clxx.

———. 1870. The limits of natural selection as applied to man. In *Contributions to the Theory of Natural Selection. A Series of Essays*, 333–371. MacMillan and Company, London.

Want, S. C., and P. L. Harris. 1998. Indices of program-level comprehension. *Behav. Brain. Sci.* 21: 706–707.

Ward, A., D. Sumpter, I. Couzin, P. Hart, and J. Krause. 2008. Quorum decision-making facilitates information transfer in fish shoals. *Proc. Natl. Acad. Sci. USA* 105: 6948-6953, doi: 10.1073/pnas.0710344105.

Ward, T.H.G. 1949. An experiment on serial reproduction with special reference to the changes in the design of early coin types. *Bri. J. Psychol.* 39: 142–147.

Warden, C. J., and T. A. Jackson. 1935. Imitative behaviour in the rhesus monkey. *J. Genet. Psychol.* 46: 103–125.

Warner, R. R. 1988. Traditionality of mating-site preferences in a coral reef fish. *Nature* 335: 719–721.

———. 1990. Male versus female influences on mating-site determination in a coral-reef fish. *Anim. Behav.* 39: 540–548.

Wasserman, S., and K. Faust. 1994. *Social Network Analysis: Methods and Applications.* Cambridge, U.K.: Cambridge University Press.

———. 1998. *Social Network Analysis: Methods and Applications.* 2nd ed. Cambridge, U.K.: Cambridge University Press.

Watanabe, K. 1989. Fish: A new addition to the diet of the Japanese macaques on Koshima Island. *Folia Primatol.* 52: 124:131.

Watson, J. B. 1908. Imitation in monkeys. *Psychol. Bull.* 5: 169–179.

———. 1919. Psychology from the Standpoint of a Behaviorist. Philadelphia: Lippincott.

Watts, D. 2002. A simple model of global cascades on random networks. *Proc. Natl. Acad. Sci. USA* 99: 5766–5771.

Webster, M. M., and P.J.B. Hart. 2006. Subhabitat selection by foraging threespine stickleback (*Gasterosteus aculeatus*): Previous experience and social conformity. *Behav. Ecol. Sociobiol.* 60: 77–86.

Webster, M. M., and K. N. Laland. 2008. Social learning strategies and predation risk: Minnows copy only when using private information would be costly. *Proc. R. Soc. B* 275: 2869–2876.

———. 2011. Reproductive state affects reliance on public information in sticklebacks. *Proc. R. Soc. B* 278: 619–627.

Weisberg, S. 2005. *Applied Linear Regression*, 3rd ed. New York: Wiley-Blackwell.

West, M. J., and A. P. King. 1988. Female visual displays affect the development of male song in the cowbird. *Nature* 334: 244–246.

White, D. J. 2004. Influences of social learning on mate-choice decisions. *Learn. Behav.* 32: 105–113.

White, D. J., and B. G. Galef, Jr. 1999. Mate-choice copying and conspecific cueing in Japanese quail, *Coturnix coturnix japonica. Anim. Behav.* 57: 465–473.

———. 2000. 'Culture' in quail: Social influences on mate choices of female *Coturnix japonica. Anim. Behav.* 59: 975–979.

Whitehead, H. 2008. *Analyzing Animal Societies.* Chicago: University of Chicago Press.

———. 2009. How might we study Culture? A perspective from the ocean. In *The Question of Animal Culture*, edited by K. N. Laland and B. G. Galef, Jr., 125– 151. Cambridge, MA: Harvard University Press

Whiten, A. 1998. Imitation of the sequential structure of actions by chimpanzees (*Pan Trologdytes*). *J. Comp. Psychol.* 112: 270–281.

———. 2005. The second inheritance system of chimpanzees and humans. *Nature* 437: 52–55.

———. 2009. The identification of culture in chimpanzees and other animals: From natural history to diffusion experiments. In: *The Question of Animal Culture*, edited by K. N. Laland and B. G. Galef, Jr, 99–124. Cambridge, MA: Harvard University Press.

Whiten, A., and D. Custance. 1996. Studies of imitation in chimpanzees and children. In *Social Learning in Animals: The Roots of Culture*, edited by C. M. Heyes and B. G. Galef, Jr., 291–318. San Diego: Academic Press.

Whiten, A., D. M. Custance, J.-C. Gomez, P. Texidor, and K. A. Bard. 1996. Imitative learning of artificial fruit processing in children (*Homo sapiens*) and chimpanzees (*Pan troglodytes*). *J. Comp. Psychol.* 110: 3–14.

Whiten, A., and E. Flynn. 2010. The transmission and evolution of experimental microcultures in groups of young children. *Dev. Psychol.* 46: 1694–1709.

Whiten, A., J. Goodall, W. C. McGrew, T. Nishida, V. Reynolds, et al. 1999. Cultures in chimpanzees. *Nature* 399: 682–685.

Whiten, A., J. Goodall, W. C. McGrew, T. Nishida, V. Reynolds, Y. Sugiyama, et al. 2001. Charting cultural variation in chimpanzees. *Behaviour* 138: 1481–1516.

Whiten, A., and R. Ham. 1992. On the nature and evolution of imitation in the animal kingdom: Reappraisal of a century of research. *Adv. Stud. Behav.* 21: 239–283.

Whiten, A., R. Hinde, K. N. Laland, and C. Stringer. 2011. Introduction. *Phil. Trans. R. Soc. B* 366: 938–948.

Whiten, A., and A. Mesoudi. 2008. Establishing an experimental science of culture: Animal social diffusion experiments. *Phil. Trans. R. Soc. B* 363: 3477–3488.

Whiten, A., A. Spiteri, V. Horner, K. E. Bonnie, S. P. Lambeth, S. J. Schapiro, and F.B.M. de Waal. 2007. Transmission of multiple traditions within and between chimpanzee groups. *Curr. Biol.* 17:1038–1043.

Whiten, A., and C. P. van Schaik. 2007. The evolution of animal 'cultures' and social intelligence. *Phil. Trans. R. Soc. B* 363: 603–620.

Wilkinson, G. 1992. Information transfer at evening bat colonies. *Anim. Behav.* 44: 501–518.

Williams, J., A. Whiten, T. Suddendorf, and D. Perrett. 2001. Imitation, mirror neurons and autism. *Neurosci. Behav. Rev.* 25: 287–295.

Williams, J., G. Waiter, A. Gilchrist, D. I. Perrett, A. Murray, and D. A. Whiten. 2006. Neural mechanisms of imitation and 'Mirror Neuron' functioning in autistic spectrum disorder. *Neuropsychologia* 44: 610–621.

Williams, J.H.G., A. Whiten, G. D. Waiter, S. Pechey. and D. Perrett. 2007. Cortical and sub-cortical mechanisms at the core of imitation. *Soc. Neurosci.* 2: 66–78.

Wilson, A. C. 1985. The molecular basis of evolution. *Sci. Am.* 253: 148–157.

———. 1991. From molecular evolution to body and brain evolution. In Perspectives on Cellular Regulation: From Bacteria to Cancer, edited by J. Campisi and A. B. Pardee, 331–340. New York: J. Wiley/A. R. Liss.

Wood, L. A., R. L. Kendal, and E. G. Flynn. 2012. Context-dependent model-based biases in cultural transmission: Children's imitation is affected by model age over model knowledge state. *Evol. Hum. Behav.* 33(4): 387–394.

Wrangham, R., W. McGrew, F. de Waal, P. Heltne, and J. Goodall. 1994. *Chimpanzee Cultures.* Cambridge, MA: Harvard University Press.

Wright, S. 1921. Correlation and causation. *J. Agric. Res.* 20: 162–177.

Wyles, J. S., J. G. Kunkel, and A. C. Wilson. 1983. *Proc. Natl. Acad. Sci. USA* 80: 4394–4397.

Zajonc, R. B. 1965. Social facilitation. *Science* 149: 269–274.

Zentall, T. R. 1996. 'An analysis of imitative learning in animals.' In *Social Learning in Animals: The Roots of Culture*, edited by C. M. Heyes and B. G Galef, Jr., 221–234. London: Academic Press.

———. 2001. Imitation in animals: Evidence, function and mechanism. *Cybern. Syst.* 32: 53–96.

Zentall, T. R., Sutton, J. E., and Sherburne, L. M. 1996. True imitative learning in pigeons. *Psychol. Sci.* 7: 343–346.

Ziegler, H. P., and P. Marler, eds. 2008. *Neuroscience of Birdsong.* Cambridge, UK: Cambridge University Press.

Zucker, L. G. 1977. The role of institutionalization in cultural persistence. *Am. Sociol. Rev.* 42: 726–743.

Index